EVOLUTION FROM ROCKS

ENCYCLOPAEDIA OF EVOLUTIONARY BIOLOGY SERIES

EVOLUTION FROM ROCKS

By

Richa Arora

ANMOL PUBLICATIONS PVT. LTD.

NEW DELHI - 110 002 (INDIA)

ANMOL PUBLICATIONS PVT. LTD.
4374/4B, Ansari Road, Daryaganj
New Delhi - 110 002
Phones: 23261597, 23278000
Visit us at: www.anmolpublications.com

Evolution from Rocks

First Edition, 2004

ISBN 81-261-1525-4

PRINTED IN INDIA

Published by J.L. Kumar for Anmol Publications Pvt. Ltd., New Delhi and Printed at Mehra Offset Press, Delhi.

Contents

Preface

All living organisms share a master plan of structural and functional organization because they arose from a common ancestor long ago by the process of evolution. The tremendous work conducted by scientists in this field has increased our knowledge of evolution many folds. In the limited space, it is not possible to go into the details. Hence, a brief and comprehensive account has been produced in the light of recent work and latest information available from various sources.

The present title is an attempt to present an account of what is known or what is postulated about the life, and the subsequent evolution of the main animal groups. Certain basic scientific facts are included, it is true, but the majority of the contents consist of hypothesis which are never likely to be experimentally verified; and the reader should treat the book as a source of facts, ideas and theories on which thought and discussion can be based.

The present title claims to be a complete, thorough and authentic exposition of the topics dealt with. Emphasis has been laid on introducing basic as well as modern concepts. The language of the book is kept simple in lucid style. The subject matter has been interwoven for keeping up interest and curiosity of readers.

Though care has been taken to verify all passages quoted, but it is hardly likely that, in so large a mass of material, all errors shall have been avoided. The author and the publishers would welcome as a favour any suggestions or corrections submitted by interested readers.

Thanks are due to all friends and colleagues whose continuous inspiration have initiated me to bring out the present work.

There can be no claim to originality except in the manners of treatment and much of the information has been obtained from the books and scientific journals available in the different libraries.

—Author

Chapter—1

Igneous Rocks and Igneous Activity

Igneous rocks make up about 80% of the mass of the earth's crust. It is not surprising then, that the subject of igneous rocks holds a very prominent place in physical geology.

Igneous rock is formed from *magma,* a high temperature mixture of the chemical ingredients of silicate minerals. Magma normally included substances in liquid, solid, and gaseous states. A large proportion of most magmas consists of a hot liquid, or *melt,* present because the temperature of the magma is above the melting points of certain magma ingredients. In a melt, the metallic ions move about more or less freely, without being organized into crystal lattice structures. However, the silicon-oxygen tetrahedra probably remain largely intact and may be linked to one another in complex but irregular chains and networks. Suspended in the melt of most magmas are crystals of minerals formed during early stages of magma cooling. If the proportion of suspended crystals to liquid is high, it gives the magma some of the physical properties of a solid. It may help to visualize this mixture as resembling slush, produced by rapid melting of snow, in which ice crystals are mixed with liquid water, Slush has enough strength to hold its form on a flat pavement but contains a large proportion of free water. In addition to both liquids and solids, magma also contains various gases, which are dissolved in the melt.

Magma that appears at the earth's surface coming into direct contact with air or water, is called *lava,* and its emergence from the solid earth provides us with sure proof that magma exists. Because lava is rapidly cooling and losing its dissolved gases, however, it can not provide us with an accurate picture of the magma that lies far beneath the earth's surface. But it is important to note that magma formed deep within the crust undergoes various chemical and physical changes as it moves upward through the crust. Some of the chemical components originally present may be separated out, while new chemical components may be added from surrounding rock with which the magma comes in contact. The study of magma inferences based on indirect scientific evidence. The study of plate tectonics has shed a great deal of light on major sources of magmas and why the composition of emerging magma often varies from place to place.

There is general agreement among geologists that magma can be formed by the melting of solid rock in certain depth zones in the asthenosphere. You will recall from previous chapter that the asthenosphere is a soft layer of the upper mantle within which rock temperature is close to its melting point. Evidence from earthquake waves, indicates that the mantle is closest to its melting point in the depth of 100 to 200 km. It seems most probable that only a small fraction of the mantle rock at this depth range actually melts, while the larger proportion of the mass remain crystalline. This phenomenon is called *partial melting*. The melted fraction, which is a liquid less dense than the crystalline fraction, tends to rise through the sponge like crystalline mass and to collect in magma pockets at shallower depths. It is supposed that the enormous volume of mafic magma continually rising along the spreading plate boundaries collects in magma chambers near the base of the oceanic crust at a depth of 4 to 6 km below the ocean floor. Magma that rises from a location above a descending lithospheric plate at a subduction (converging) boundary is produced under quite different conditions and depths.

TEMPERATURES AND PRESSURES IN MAGMA

Temperature increases rapidly with depth into the earth. This increase in rock temperature, as measured directly with thermometers in deep mines, averages about 1°C per 30 m. Downward through the crust and into the upper mantle temperatures can only be estimated. The graph showing the estimated increase in temperature with depth. Two curves are show, one typical of conditions under the continents, the other for conditions under the ocean basins. Both curves show that while the temperature increases, the rate at which it does so falls off rapidly after about 300 km. Temperatures within the asthenosphere are largely in the range from 1000°C to 1500°C. For comparison, the melting point of iron at the earth's surface is about 1500°C, while molten iron in a blast furnace reaches a temperature of 2000°C. It might seem like a simple procedure to measure the melting point of a specimen of igneous rock and using the temperature-depth curves, find the depth at which melting is to be expected and magma to be formed.

Unfortunately, the actual relationships are not that simple. The temperature at which silicate minerals will melt depends upon several factors. Each mineral has its won melting point for a given environment. For example, in the surface environment, under the atmospheric pressure that prevails, pure olivine melts in the temperature range of about 1600°C to 1800°C, whereas plagioclase feldspar melts in the range from 1200°C to 1400°C.

Downward into the earth, the confining pressure to which all rock is subjected increases steadily under the load of the overlying rock layer. The unit to pressure used by geologists is the *kilobar,* which is equal to one thousand bars, a *bar* being approximately equal to the pressure of the earth's atmosphere at sea level-about 1 kg per sq cm. The graph to show the approximate increase of confining pressure with depth. An increase in pressure causes the temperature of the melting point of a given

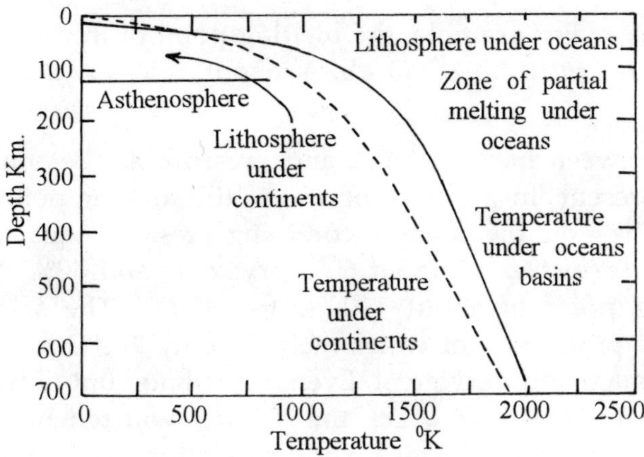

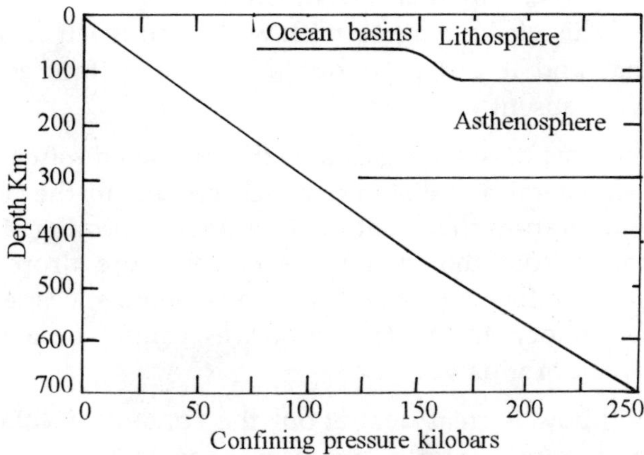

Fig. 1.1: A. Estimated increase in temperature with depth unde continents and under ocean basins. The dashed lines shows the melting point of dy peidotite. B. Estimated increase of confining pressure with depth.

Table 1.1
Composition of Gases from Internal Earth Sources Compared with Total Earth Volatiles.

	Volatiles of Earth's hydrosphere and atmosphere	*Gases in hot springs, fumaroles, and geysers*	*Volcanic gases from Basaltic lava of Mauna Loa and Kilauea*
Water, H_2O	9.28	99.4	57.8
Total carbon, as CO_2	5.1	0.33	23.5
Sulfur, S_2	0.13	0.03	12.6
Nitrogen, N_2	0.24	0.05	5.7
Argon, Ar	trace	trace	0.3
Chlorine, Cl_2	1.7	0.12	0.1
Fluorine, F_2	trace	0.03	—
Hydrogen, H_2	0.07	0.05	0.04

mineral or rock to increase. For example, the melting point of iron at a depth of about 100km, under a confining pressure of 40 kb, is about 1650°, which is 150°C higher than at the surface.

This relationship between melting point and pressure applies only to conditions in which no water is present. In presence of water, the melting-point temperature is considerably lowered. For example, using a confining pressure equal to a depth of 20 km, a dry rock sample consisting of about 60% pryoxene and 40% calcic plagioclase feldspar has a melting point of about 1300°C to 1400°C. The same rock sample containing a substantial proportion of water melts at only 700°C to 1000°C. Thus the importance of water in magma is very great. Even a small amount of water in a magma can greatly lower the temperature at which the silicates will remain molten. A rising magma, experiencing a progressive lowering of temperature as it moves upward into cooler zones of surrounding rock, can remain molten at considerably lower temperatures with water present than if the magma consisted of the same silicates without water. Because of this effect, magma with water can reach closer to the earth's surface before solidifying, and it can pour out upon the surface as lava in greater volumes than if water were absent.

Water in magma occurs as dissolved gas, and the dissolved gases are also present. Termed *Volatiles,* these are chemical substance which remain in the liquid or gaseous state at a much lower temperature than that of the mineral-forming silicates. Thus the volatiles become separated from the magma as temperature drop and the silicate minerals crystallize. We can understand from this why, when a specimens of an igneous rock is melted in the laboratory, the melt is produces cannot be a true copy of the magma from which the rock originated.

Although we do not know a great deal about the content of volatiles in magmas at depth in the crust and upper mantle, much as been learned by sampling gases emitted from volcanoes along with emerging lava.

The emission of volatiles with magma at the earth's surface is a geological process *outgassing;* it has acted throughout all geologic time and continues today. As water is the dominant constituent of the volatiles contained in lava, we infer that it is by volume about 90% of the gas contained in an average magma. Estimates of the proportion of water in magmas show a range from 0.5 to 8%, while fresh igneous rocks commonly contain about 1% water, entrapped in small cavities within the minerals during their crystallization. Besides water, the list of Volcanic volatiles includes carbon (as carbondioxide gas), sulfur, nitrogen, argon, chlorine, fluorine, and hydrogen.

Outgassing is responsible for the formation of the earth's atmosphere and hydrosphere. The *atmosphere* is the gaseous envelop that surrounds the earth, while the *hydrosphere* is the total free water of the globe, including not only the oceans but all fresh water on the land and beneath the land surfaces. Nitrogen, oxygen, argon, and carbon dioxide together make up over 99% of the atmosphere, and it is clear from Table that these ingredients reached the earth's surface through outgassing. Molecular oxygen (O_2) was formed from water and released to the atmosphere through the process of photosynthesis in green plants.

Outgassing of water was also responsible for the growth of the earth world ocean. The process was probably very rapid in the first aeon of geologic time. Salts accumulated in seawater through the decomposition of silicate minerals rich in ions of sodium, magnesium, calcium and potassium. It is thought that as early as 3 billion years ago the salinity (salt concentration) of seawater had reached a value comparable to what it is today. Chlorine, which today makes up about 55% of the total weight of all matter dissolved in sea water, and sulfure, found in much smaller proportion, were contributed as volatiles through outgassing.

CRYSTALLIZATION OF MAGMAS

During the crystallization of magmas, not all minerals from simultaneously. If you example, the magma originally has a basaltic composition, olivine, pyroxene, and calcium plagioclase will be among the earliest minerals to crystallize. Often these first-formed crystals are larger and more perfect than those formed later because, when they formed, there was ample space and an abundance of elements required for growth. Those minerals that crystallize later must fit into the remaining spaces and thus tend to be smaller and less perfect. Also minerals enclosed in other minerals must have formed before the enclosing mineral. By using these observations of size and spatial relationships of minerals as viewed in thin sections of the rock, it is sometimes possible to make a reliable estimate of the order of crystallization of minerals in a magma.

During the first of the present century, an eminent petrologist named *N. L. Bowen* studied the crystallization of artificial silicate melts in the laboratory and provided further evidence of the approximate order of crystallization of in gradually-cooling magmas. Bowen used melts of basaltic composition for his experiments and was able to ascertain the specific temperature range over which particular silicate minerals would being to form. Olivine, for example, would crystallize first and at the highest temperature. Pyroxenes and the more calcium-rich feldspars would form next, followed by hornblende, biotite, and the more sodic feldspars. The order of crystallization, called the Bowen Reaction Series. It is called *reaction series* because early formed crystals react with silica in the remaining liquid and change to the next lower mineral in the series. For example, the silica in the magma would react with olivine crystals by simultaneously dissolving and precipitating the dissolved components so as to form pyroxene (augite), and pyroxene crystals would begin to grow in spaces between earlier formed crystals. At lower temperatures, the silica in the melt would similarly react with pyroxene to form hornblende, and subsequently hornblende would react with the liquid to form biotite. In these transformations, single tetrahedra of olivine utilize silica in the melt to form single chains of pyroxene, which in turn combine to form double chains of hornblende, which then combine to form sheets of biotite. For the complete sequence of changes to occur, however, the original magma must have

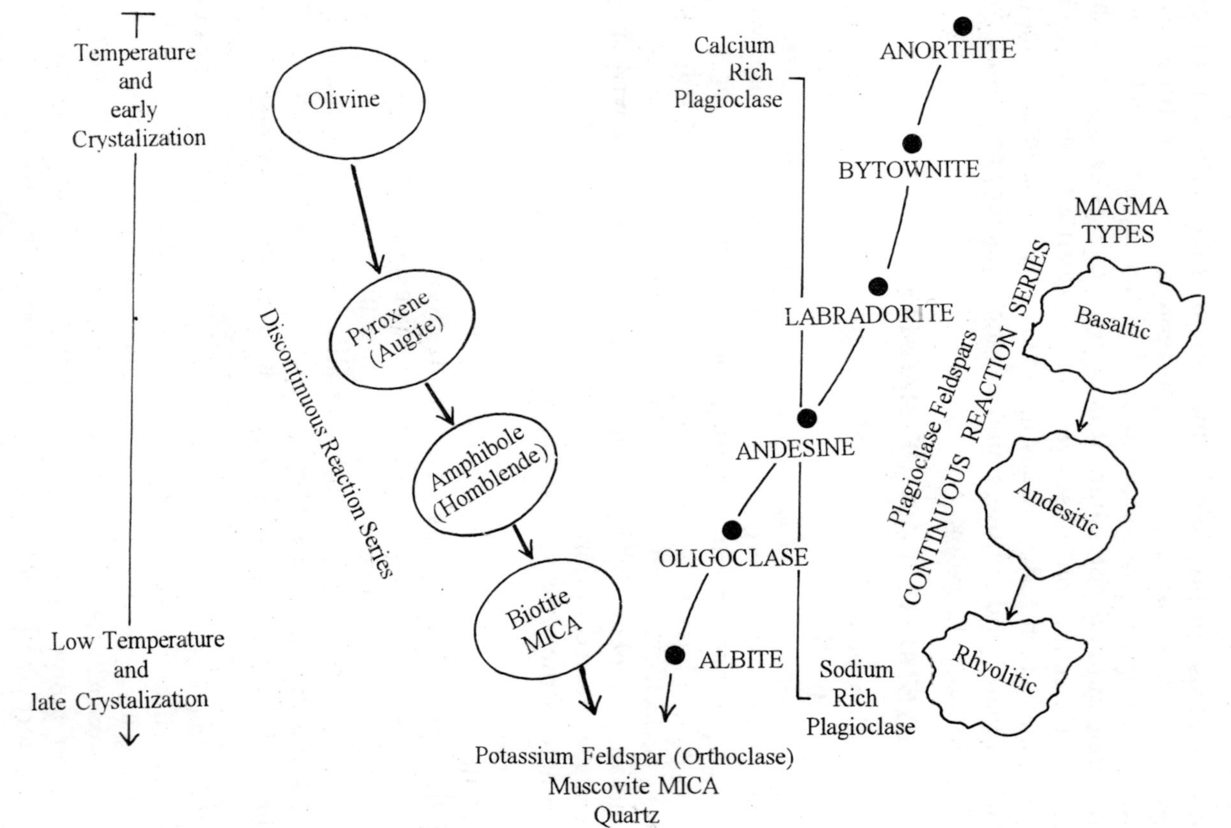

Fig. 1.2: Bowen Reaction Series. Note that the earliest minerals to crystalize from a cooling magma are olivine and calcium-rich plagioclase minerals like bytownite. As crystallization proceeds, olivine reacts with the melt to form pyroxene, pyroxene reacts to form amphibole etc. The plagioclase minerals also react with the remaining liquid, but rather than forming new minerals, change continuously in composition.

sufficient magnesium, iron, and calcium. If the magma is rich in silica, and deficient in these elements, the first-formed crystals might be biotite and sodic feldspars, followed by potassium feldspar, muscovite and quartz. Thus reaction is prevented if the silica in the melt is exhausted in forming the early minerals. It is also prevented if minerals are removed from contact with the magma shortly after they have crystallized.

On the right branch of Bowen's chart, one can trace the changes that occur as calcium-rich plagioclase reacts with the magma to produce varieties of plagioclase that are successively richer in sodium. Because the plagioclase minerals maintain the same basic structure (triclinic) but change continuously in their content of calcium and sodium, the right side of the diagram was named the *continuous series*. The ferromagnesian or left side of the diagram depicts reactions that result in minerals with distinctly different structure. It was therefore termed a *discontinuous series*. The three minerals at the base of the chart do not react with the melt. By the time they crystallize, little liquid remains, and the residue of silicon, aluminum, oxygen, and potassium join to form orthoclase, muscovite, and quartz.

By examining thin sections of igneous rocks with the aid of a microscope, one can often see rather direct evidence of the reactions described above. It is not uncommon, for example, to find plagioclase crystals in which the innermost layers or zones are more calcium-rich and outer zones more sodium-rich. Similarly, reaction rims of pyroxene can be observed around olivine and rims of hornblende around pyroxene.-

During the crystallization of magmas, not all minerals form simultaneously. If, for example, the magma originally has a basaltic composition, olivine, pyroxene, and calcium plagioclase will be among the earliest minerals to crystallize. Often these first-formed crystals are larger and more perfect than those formed later because, when they formed, there was ample space and an abundance of elements required for growth. Those minerals that crystallize later must fit into the remaining spaces and thus tend to be smaller and less perfect. Also, minerals enclosed in other minerals must have formed before the enclosing mineral. By using these observations of size and spatial relationships of minerals as viewed in thin sections of the rock, it is sometimes possible to make a reliable estimate of the order of crystallization of minerals in a magma.

During the first half of the present century, an eminent petrologist named *N. L. Bowen* studied the crystallization of artificial silicate melts in the laboratory and provided further evidence of the approximate order of crystallization of in gradually cooling magmas. Bowen used melts of basaltic composition for his experiments and was able to ascertain the specific temperature range over which particular silicate minerals would being to form. Olivine, for example, would crystallize first and at the highest temperature. Pyroxenes and the more calcium-rich feldspars would form next, followed by hornblende, biotite, and the more sodic feldspars. The order of crystallization, called the Bowen Reaction Series. It is called *reaction series* because early formed crystals react with silica in the remaining liquid and change to the next lower mineral in the series. For example, the silica in the magma would react with olivine crystals by simultaneously dissolving and precipitating the dissolved components so as to form pyroxene (augite), and pyroxene crystals would begin to grow in spaces between earlier formed crystals. At lower temperatures, the silica in the melt would similarly react with pyroxene to form hornblende, and subsequently hornblende would react with the liquid to form biotite. In these transformations, single tetrahedra of olivine utilize silica in the melt to form single chains of pyroxene, which in turn combine to form double chains of hornblende, which then combine to form sheets of biotite. For the complete sequence of changes to occur, however, the original magma must have sufficient magnesium, iron, and calcium. If the magma is rich in silica, and deficient in these elements, the first-formed crystals might be biotite and sodic feldspars, followed by potassium feldspar, muscovite and quartz. Thus reaction is prevented if the silica in the melt is exhausted in forming the early minerals. It is also prevented if minerals are removed from contact with the magma shortly after they crystallized.

On the right branch of Bowen's chart, one can trace the changes that occur as calcium-rich plagioclase reacts with the magma to produce varieties of plagioclase that are successively richer in sodium. Because the plagioclase minerals maintain the same basic structure (triclinic) but change continuously in their content of calcium and sodium, the right side of the diagram was named the *continuous series.* The ferromagnesian or left side of the diagram depicts reactions that result in minerals with distinctly different structure. It was therefore termed a *discontinuous series.* The three minerals at the base of the chart do not react with the melt. By the time they crystallize, little liquid remains, and the residue of silicon, aluminium, oxygen, and potassium join to form orthoclase, muscovite, and quartz.

By examining thin sections of igneous rocks with the aid of a microscipe, one can often see rather direct evidence of the reactions described above. It is not uncommon, for example, to find plagioclase crystals in which the innermost layers or zones are more calcium-rich and outer zones more sodium-rich. Similarly, reaction rims of pyroxene can be observed around olivine and rims of hornblende around pyroxene.

TEXTURES OF IGNEOUS ROCKS

Igneous rocks are classified according to two different classed of information, or categories. The first is according to chemical composition, as apparent from our study of magma and its differentiation into mafic and felsic types. Second is on the basis of the texture of the igneous rock. *Texture,* as the word applies to rocks, relates to the sizes and patterns of the mineral crystals present in the rock. Texture also distinguishes between the crystalline and the noncrystalline (glassy) state of mineral matter.

Geologists use the term texture with somewhat the same meaning as it has in the classification of fabrics. We say that burlap or canvas has a *"coarse texture"*, in contrast to the *"fine texture"* of chiffon. Cloth woven on a loom to produce threads that cross at right angles (warp and woof) contains a texture quite different from that of a knitted fabric. A garment can also be made from a sheet of film plastic that shows no structure. Perhaps this last material can be considered the equivalent of glass in the classification of rock texture.

The size of mineral crystals in a igneous rock depends largely upon the rate of cooling of the magma through the stages crystallization. As a general rule, rapid cooling results in small crystals, slow sooling in large crystals. Extremely sudden cooling will result in formation of a natural glass, which is noncyrtalline

The rate of cooling depends on where the magma is located as it cools. Large bodies of magma trapped deep beneath the surface cool very slowly because the surrounding rock conducts the heat very slowly. Bodies of rock formed this way belong to the class of *intrusive igneous rocks,* whose mineral crystals are usually large enough to be identified with the unaided eye. Rapid cooling occurs in lava, which belongs to the class of *extrusive igneous rocks.* Lava forms thin layers that lose heat rapidly to the atmosphere or to overlying ocean water. Thus lava usually has mineral crystals too small to identify with the unaided eye, or it may solidify as a glass.

Among the crystalline igneous rocks, texture falls into two classes:

(1) *Phaneritic texture* consists of crystals large enough to be seen with the unaided eye or with the help of small hand lens. *Aphanitic texture* consists of crystals too small to be distinguished as individual particles without the aid of a microscope.

The phaneritic igneous rocks have crystal grains ranging in diameter from 0.05 mm to over 10 mm. An accepted grade scale is as follows:

Fine-grained	0.05-1mm
Medium-grained	1-5mm
Coarse-grained	5-10mm
Pegmatite texture over	10 mm

Particules under 0.05 mm fall into the aphanitic texture class. Pegmatite texture consists of crystals that are often several centimeters long and in rare cases may be as long as a few meters.

Though it can be seen with the naked eye, the crystalline fabric of the phaneritic rocks is best studied under a specialized microscope in which very thin slices of the rock are examined under polarized light which makes possible mineral identification based upon distinctive optical properties. For the most part, external crystal forms are lacking or only poorly developed. On the hand, evidences of cleavage show up well as linear markings within the grains.

Where crystals in the rock are all within the same size range, the texture is described as *equigranular.* Where a few large crystals' called *phenocrysts,* are embedded in a matrix, or *groundmass,* of smaller crystals; we have *porphyritic texture;* the rock is designated as a *porphyry* in addition to its proper name. The groundmass may consist of crystals ranging in size from fine-grained to coarse-grained, or it may consist of a glass.

Extrusive igneous rocks have distinctive texture resulting from the rapid expansion of volatile gases are confining pressure is reduced and cooling occurs. The specimen of volcanic *scoria* shown in Figure is full cavities formed by gas bubbles, and is said to have *scoriaceous texture*. Because the cavities go by the name of *vesicles,* an alternative adjective for this texture is *vesicular texture*. An extreme case of scoriaceous texture is seen in *pumice,* formed of magma frothed by expanding gases into a glassy rock of such low density that it easily floats on water. Pieces of pumice are used as abrasive blocks, while powdered pumice serves as an abrasive cleaning powder. In contrast, the volcanic glass also pictured in Figure is dense and free of such cavities. Notice is conchoidal fracture.

IDENTIFICATION OF IGENOUS ROCKS

Texture and mineral composition are the most useful properties for inferring the origin of igneous rocks and are the basis for classification and identification. Although there are hundreds of names used for igneous rocks, most of these represent variations of the relatively few major groups. In a widely use simple scheme, there are ten groups of crystalline igneous rocks: granite, granodiorite, diorite, gabbro, peridotite, rhyolite, dacite, andesite, basalt, and ultramafic extrusives.

In order to identify rocks in each of these groups, it is necessary to recognize texture and a few of the mineral constituents as they appear in the rock itself. To better identify the more common minerals, hold the rock in strong light and turn in back and forth so that you can see reflections from cleavage planes and crystal faces. Quartz can be recognized by its glassy appearance and mostly irregular reflecting surface. Feldspars show cleavage faces, may be white gray, or pink, and often have rectangular outlines. One can often identify the feldspar as plagioclase by thin parallel striations that develop on one cleavage face. Orthoclase feldspar lacks these striations. Ferromagnesian are black or green. Among ferromagnesium, hornblende is more shiny and elongated than augite, and biotite may be distinguished by its smooth gleaming cleavage surfaces.

Rocks of the Granitic Clan

The light-colored igneous rocks are, as we have already noted, derived from high-silica magmas. When such magmas crystallize, they yield a high proportion of orthoclase feldspar and quartz. Granite is a familiar representative of this group and is recognized by its light color, phaneritic texture, and the presence of about 25 per cent quartz grains along with larger quantities of potassium feldspar (orthoclase) and sodium-plagioclase. Muscovite and biotite are present, as are small quantities of hornblende. *Granite,* of course, is well known to most us because it has been used for centuries in the construction of buildings statues, and monuments. It is not, however, the most abundant granitic rock. *Granodiorite,* somewhat less siliceous granitic rock, is more abundant. This sierra Neveda batholith is composed largely of granodiorite. The fine-grained equivalent of granodiorite is *dacite.* The more finely crystalline but compositional equivalent of granite is *rhyolite.* Actually, the basic aphanitic texture of rhyolite may range into the glassy condition. Many rhyolites tend to be porphyritic, with small phenocrysts of orthoclase, sodic plagioclase, quartz, and mica. Some appear to be welded tuffs. In color, rhyolites tend to be white, light gray, pink; or orange.

Rocks of the Dioritic Clan

Igneous rocks of the diorite clan include diorite itself and its aphanitic equivalent known as andesite. Diorite is a coarse-to fine-grained phaneritic rock containing less silica than granite or granodiorite but more than gabbro. Thus, in the gradations of composition across the rock chart, it is an intermediate rock type. The most abundant minerals are plagioclases, and these are compositionally in the mid range and contain both calcium and sodium. Quartz and orthoclase are very rare or absent in diorites. Among the ferromagnesian constituents, hornblende occurs in amounts about equal to plagioclase, and biotite mica is a common accessory mineral. The color of diorite is controlled by its compositional makeup of nearly equal amounts of black ferromagnesians and grayish plagioclase. Thus, it tends to be rather drab gray or greenish gray in color.

The somber color of diorite prevails also in *andesite-* its aphanitic equivalent. Andesites, which were first identified in the summit volcanoes of the Andes Mountains of South American, are mostly found along the more mountainous continental borders and island arcs that surround the Pacific Ocean. In fact, geologists determine the true edge of the Pacific Basin by plotting a line, on the oceanic side of which rocks are principally basalt and on the continental side and site. Geologists refer to this demarcation as the *andesite line.*

A member of the palgioclase family known as andesine is the most abundant mineral constituent of andesite. It is a mineral that contains nearly equal proportions of calcium and sodium. Dark minerals such as biotite, hornblende, and pyroxene are also prevalent in andesites. Andesites are intermediate between basalts and rhyolites in composition and also intermediate in relative abundance on the earth's surface.

Rocks of the Gabbro Clan

In *gabbro,* the chief mineral is labradorite plagioclase, which is darker and more calcium-rich than the plagioclase found in diorite. The typical ferromagnesian constituents are pyroxenes and olivine. Gabbros are typically coarsely crystalline plutonic rocks. One variety called anorthosite is composed mostly of intergrown clacium-rich plagioclase crystals. Anorthosite is characteristic of the lighter-colored crustal areas of the moon. Varieties of gabbro are frequently polished and used as ornamental stone because of their rich dark hue and the iridescent play of colors reflected by the plagioclase crystals.

Basalt is the aphanitic equivalent of gabbro and is the most abundant extrusive igneous rock. Basalts are mostly black or very dark gray and therefore darker than andesites. In thin section, they are found to contain myriads of elongate lathlike crystals of calcic plagioclase interspersed with pyroxene and olivine. Biotite and hornblende, which are common minerals in andesite, are sparse or absent in basalt.

The Ultramafic Clan

Ultramafic rocks are characterized by having a high density (3.3 g/cm^3), low percentage of silica (45 per cent), and a major mineral composition of pyroxenes and olivine. *Peridotite,* in fact, is an ultramafic rock composed of 70 to 90 per cent olivine, while *Pyroxenite* is almost entirely an intergrowth of pyroxene crystals. The ultramafic rocks are not common near or on the surface of the earth. They are occasionally encountered in the lower parts of magmatic bodies where dense ferromagnesian minerals accumulated by fractional crystallization and gravity settling. At several locations around the world, peridotites have pushed their way upward from great depths and penetrated near-surface rocks. Geologists are keenly interested in these peridotites, for they may represent samples of the earth's mantle.

Although peridotites contain a high percentage of olivine, they also contain 10 to 30 per cent pyroxene. When the olivine content in an ultramafic rock exceeds 90 per cent, the rock is called a *dunite.*

KINDS OF INTRUSIVE ROCKS BODIES

Bodies of intrusive igenous rocks that have solidified from magmas located deep within the crust are called *plutons*. The shape, size, and relations of plutons to surrounding rock is quite varied. Some represent injections of once molten rock parallel to preexisting strata. Such plutons are said to be *concordant;* in contrast, a plutton that cuts across layering is spoken of as *dicordant.*

Tabular Plutons: Sills and Dikes

A tabular body of rocks is one shaped like a book or board, that is, having small thickness relative to its length and width. There are several kinds of tabular plutons. The concordant ones, which have been injected along bedding planes of sediment or between older lava flows, include *sills* and *lopoliths.* Sills may be either horizontal or tilted depending on the attitude of the enclosing beds. They range in thickness from only a centimeter to hundreds of meters. The well known Palisades of the Hudson River owe their distinctive appearance to columnar jointing in the exposed edge of the 300 meter-thick palisades sill.

Because both sills and lava flows are sheetlike and parallel to adjacent beds, the distinction between a flow that has been buried by later strata and a sill can be difficult to make. There are, however, a few characteristics that are useful in distinguishing between the two features. Flows bake only the stratum beneath them, may fill the cracks only of the underlying bed, and in their upper portions may contain vesicles (small cavities made by gas bubbles in the lava). In contrast, sills bake the stratum above as well as the one below, penetrate cracks in both underlying and overlying beds, and rarely have vesicular borders.

Lopoliths are tabular pluton in which the roof and floor sag downward so as to give the overall shape of a bowl. Examples of these large-scale structures occur in the Bushveld Igneous complex of South Africa, as well as in the region around Duluth, Minnesota. The Duluth lopolith is estimated to have a diameter of 250 km. and thickness of 15 km. Lopoliths are composed of coarsely crystalline igneous rocks, as is to be expected in deeply buried bodies that have cooled slowly.

Dikes are tabular bodies that are *discordant* in that they cut across preexisting rock layers. Dikes may be many kilometers long and range in thickness from paper-thin to tens of meters. Also, a dike may thin out or altermately pinch and expand along its breadth. Many dikes branch and cross one another so as to form a complex system called a *dike swarm*. They also occur in circular patterns called *ring dikes*. Ring dikes are thought to develop when lava rises into concentric fractures created around an area where overlying rocks sink into a magma chamber below. Although dikes may be composed of nearly any kind of igneous rock, most are basaltic in composition and were derived from magmas of low viscosity.

Batholiths

In many areas of former intrusive igneous activity, great discordant nontabular masses of intrusive rock occur. The largest of such bodies form the cores of great mountain ranges. They are known as batholiths. Among the better known examples are the Idaho Batholith, the Coast Range Batholith of British Columbia, and the Sierra Nevada Batholith of California. Each of these immense bodies of igneous rock actually represents the coalescence of several smaller intrusions of granites and granodiorites. Granitic rocks of this kind form the bulk of most of the world's batholiths.

To be termed a *batholith,* the intrusive body must have an areal extent greater than 100 km^2. In shape, they may be irregular or roughly cylindric. It is difficult to determine the depth to the base of existing batholiths, for at no place on earth is the bottom of batholith exposed for study. However, investigations based on gravity measurements and earthquake data suggest that batholiths begin to form at depth of about 30 km. As intrusive masses, batholiths were once deeply covered by pre-existing rocks. We see batholithic rocks at the surface of the earth today only because great thicknesses of covering rocks have been removed by erosion. Batholiths are usually located in present or former mountainous belts, and such regions are subject to frequent episodes of uplift, which in turn greatly increase the rates of erosion.

Batholiths intrude into *country rock* (pre-existing rock) in a variety of ways. In some places there is evidence that the melt has been forcefully injected. Elsewhere, the heat of the magma appears to have been sufficient to melt surrounding country rock so that the magma melted its way upward. Such a process would, of course, change the composition of the magma. *Stoping,* a process by which magma moves upward as blocks of country rock are wedged loose and fall into the magma chamber, provides another mechanism for batholiths to work their way upward. The upward migration of magma may also be aided by density differences. Magma that is less dense than the enclosing rock to rise bouyantly toward the surface.

As the magma makes it way upward, pieces of the enclosing country rock may spall off and fall into the melt. Some of these fragments may melt and be assimilated by the magma. Included fragments of unmelted country rock are called *xenoliths,* meaning "stranger rocks". Xenoliths also occur in lavas when pieces of rock are torn from the sides of volcanic vents and fissures during eruptions.

Laccoliths are massive concordant plutons that have a mushroom shape. Like sills, they are formed by magma that has been injected into bedding planes. However, unlike sills, they dome-up the overlying rock layers. Some laccoliths represent bulges on the surface of a sill, whereas others are self-contained structures fed by a vent from an underlying magma. Most laccoliths have a granitic composition. Surface exposures of laccoliths can be found in the Henry, LaSal, and Abajo Mountains of Southeastern Utah.

Chapter—2

Metamorphic Rocks and the Continental Crust

Rock metamorphism consists of physical and chemical changes in rock in an environment of high temperature, high confining pressure, or intense shearing action, or some combination of two or three of those factors, but without melting. The definition usually excludes chemical changes caused by hydrothermal solution penetrating rock at shallow depths where pressures are very low. Generally speaking, rock metamorphism occurs under confining pressures of at least 2 kilobars (kb).

Rock metamorphism can result in the formation of new minerals, new rock textures, new rock structures, or a combination of such changes. The formation of new minerals by recrystallization of preexisting minerals is usually the most distinctive and important aspect of metamorphism that affects large masses of rock over wide areas. Crystal lattices are broken down and re-created, using different combinations of the same ions that were present in the earlier minerals. Another aspect of rock metamorphism is the importation of ions and atoms of rock-forming minerals from an outside source, or the export of various substances to an outside region (usually the overlying rock). Geologists refer to this import-export process as *metasomatism*. When it occurs, the rock undergoes corresponding changes in chemical composition. One important example of metasomatism during rock metamorphism is the loss of volatiles-particularly water and carbon dioxide.

THE STANDAD GEOLOGIC TIME SCALE

The early geologists had no way of knowing how many time units would be represented in the completed geologic time scale, nor could they know which fossils would be useful in correlation or which new strata might be discovered at a future time in some distant corner of the globe. Consequently, the time scale grew piecemeal, in an unsystematic manner. Units were named as they were discovered and studied. Sometimes the name for a units was borrowed from local geography, from a mountain range in which rocks of a particular age were well exposed, or from an ancient tribe of Welshmen, sometimes the name was suggested by the kind of rocks that predominated.

DIVISIONS IN THE GEOLOGIC TIME SCALE

The two major divisions in the geologic time scale are termed eons. Approximately seven eighths of all of earth history was expended in the first eon-the Cryptozoic Eon (informally termed "Precambrian"). To a geologist, the biblical phrase "in the beginning" alludes to this long interval of time that began about 4.6 billion years ago.

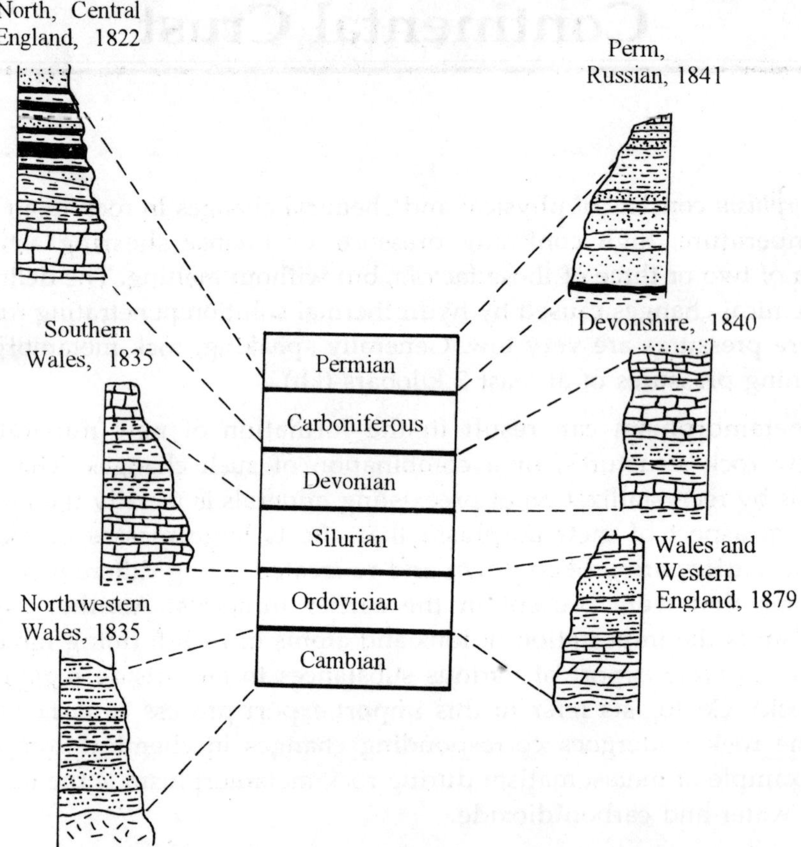

Fig. 2.1: The standard geologic time scale for the Paleogoic and other eras developed without benefit of a grand plan, but rather by the compilation of type section for each of the system.

It was during the Cryptozoic Eon (literally meaning "hidden life") that the earth had completed the process of gathering together most of the rocky substance, possibly from what was part of a much older cloud of turbulent cosmic dust. In addition, it was the interval during which life in earth first appeared. Cryptozoic sequences around

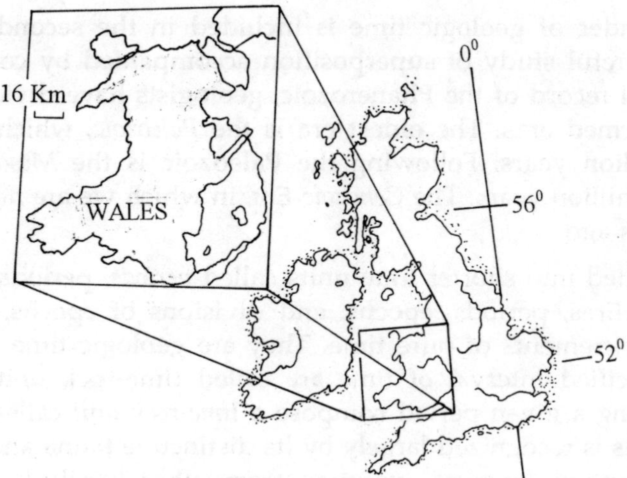

Fig. 2.2: Outcrop of Cambrian strata in northwestern. Wales where these rocks were named.

the worked include great tracts of igneous and metamorphic rocks. The antiquity of these rocks was recognized in the mid- 1700's by Johann Lehman, a professor of mineralogy in Berlin who referred to them as the "primary series." One frequently finds the term in the writing of French and Italian geologists who were contemporaries of Lehman. In 1833, the term appeared again when Sir Charles Lyell used it in his formulation of a surprisingly modern geologic time scale. Lyell and his predecessors recognized these "primary" rocks by their crystalline character and took their uppermost boundary to be an unconformity that separated them from the overlying - and therefore younger-fossiliferous strata.

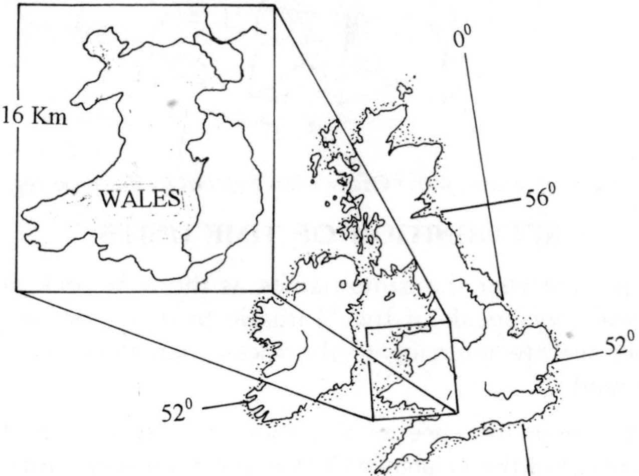

Fig. 2.3: Outcrop of strata of the Silurian System in the classic region of Wales and western England, whee Sir roaderick impey Murchison named them.

All of the remainder of geologic time is included in the second, or *Phanerozoic Eon*. As a result of careful study of superposition accompanied by correlations based on the abundant fossil record of the Phanerozoic, geologists have divided it into three major subdivision, termed eras. The oldest era is the *Paleozoic,* which we now know lasted about 370 million years. Following the Paleozoic is the *Mesozoic Era,* which continued about 170 million years. The *Cenozoic Era,* in which we are now living, began about 60 million years ago.

The eras are divided into shorter time units called *periods;* periods may in turn be divided into epochs. Eras, periods, epochs, and divisions of epochs, called ages, all represent intangible increments of pure time. They are geologic time units. The rocks formed during a specified interval of time are called time-rock units. For example, strata laid down during a given period compose a *time-rock unit* called a *system*. Each of the geologic systems is recognized largely by its distinctivc fauna and flora of fossils. The fossils are different in stage of evolution from other fossils in both older and younger systems. Series is the time-rock term used for rocks deposited during an epoch, whereas stage represents the tangible rock record of an age.

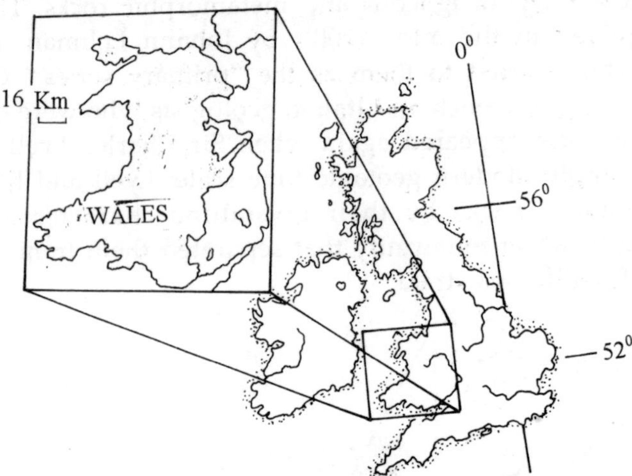

Fig. 2.4: Outcrop pattern of strata of the Ordovician System in the type region of Wales.

RECOGNITION OF TIME UNITS

Units of geologic time bear the same names as the time-rock units to which they correspond. Thus, we may speak of the "Jurassic System" or the "Jurassic Period" according to whether we are referring to the rocks themselves or to the time during which they accumulated.

Time terms have come into use as a matter of convenience. Their definition in necessarily dependent upon the existence of tangible time-rock units. The steps leading to the recognition of time-rock units began with the use of superposition in establishing age relationships. Local sections of strata were used by early geologists to recognize beds of successively different age and, thereby, to record successive evolutionary changes in fauna and flora. (The order and nature of these evolutionary changes could be determined because higher layers are successively younger.) once the faunal and floral succession was deciphered, fossils provided an additional tool for establishing the order of events. They could also be used for correlation, so that strata at one locality could be related to the strata of various other localities. No single place on earth contains a complete sequence of strata from all geologic ages. Hence, correlation to standard sections of many widely distributed local sections was necessary in constructing the geologic time scale. Clearly, the time scale was not conveived as a coherent whole but rather evolved part by part as a result of the individual studies of many earth scientists. Indeed, for some units at the series and stage level, the process continues even today. The fact that the time scale developed in piecemeal fashion is apparent when one reviews its growth and development.

THE CAMBRIAN SYSTEM

The rocks of the Cambrian System take their name from Cambria, the Latin name for Wales. Exposures of strata in Wales provide a standard section with which rocks elsewhere in Europe and on other continents can be correlated. The standard section in Wales is named Cambrian *by definition.* All other sections deposited during the same time as the rocks in Wales are recognized as Cambrian by *comparison* to the standard section.

Adam Sedgwick, a Yorkshire clergyman and professor of geology at Cambridge, named the Cambrian in the 1830's for outcrops of poorly fossiliferous graywackes and dark siltstones sandstones. The area in north Wales that sedgewick studied was noted for its complexity, yet he was able to unravel its geologic history on the basis of spatial relationships and lithology.

THE SILURIAN AND ORDOVICIAN SYSTEMS

At about the same time that Sedgwick was laboring with outcrops that were to become the Cambrian System, another geologist, Sir Rederic Impey Murchison, had begun studies of fossiliferous strata outcropping in the hills of south Wales. Murchison named these rocks the **Silurian,** taking the name from early inhabitants of western England and Wales know as the *Silures.* In 1835, Murchison and Sedgwick jointly presented a paper, *on the Silurian and Cambrian System, Exhibiting the Order in Which the Older Sedimentary Strata Succeed Each Other in England and Wales.* With this publication, the two geologists introduced the basis of the modern time scale. in the years that followed, a controversy arose between the two men that was to sever their friendship. Because Sedgwick had not described fossils distinctive of the Cambrian, the unit could not be recognized in other countries. Murchison argued, therefore, that the Cambrian was not a valid system. During the 1850's, he maintained that all fossiliferous strata above the "primary Series" (the old name for Precambrian) and below the Old Red Sandstone (of Devonian age) belonged within the Silurian System. Sedgwick, of course, disagreed, but his opinion that the Cambrian was a valid system did not receive wide support until fossils were described from the upper part of the sequence. The fossils proved to be similar to faunas in Europe and North America. Hence, the Cambrian did meet the test of recognition outside England. Using these fossils as a basis for reinterpretation, the English geologist Charles Lapworth proposed combining the upper part of Sedgwick's Cambrian and the lower part of Murchison's Silurian into a new system. In 1879, he named the system *Ordovician* after the *Ordovices*, an early Celtic tribe. The first three systems of the Paleozoic were now established.

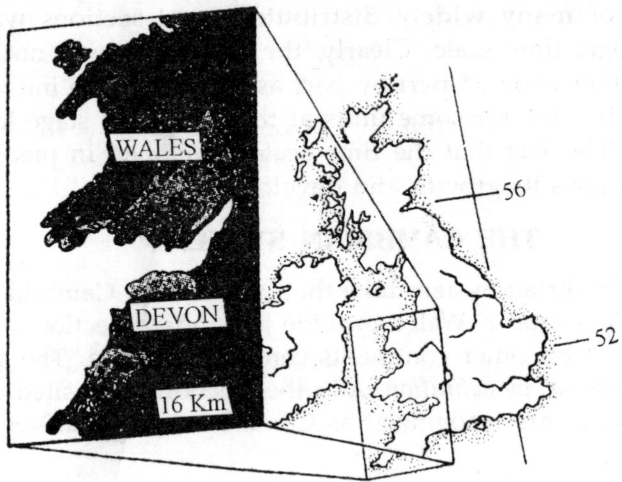

Fig. 2.5: Outcrop of strata of Devonian System in Wales and southwestern England where these rocks were name.

THE DEVONIAN SYSTEM

The *Devonian System* was proposed for outcrops near Devonshire, England, by Sedgwick and Murchison in 1839 (prior to the years of their bitter debate). They based their proposal on the fact that the rocks in question lay beneath the previously recognized Carboniferous System and contained a fauna that was distinctive and different from that of the underlying Silurian and overlying Carboniferous. In their interpretation of the intermediate nature of the fauna, they were aided by the studies of William Lonsdale, a retired army officer who had become a self-taught specialist on fossil corals. Further evidence that the new unit was a valid one came when Murchison and Sedgwick were able to recognize it in the Rhineland region of Europe. The Devonian rocks of Devonshire were also found to be equivalent to the widely known *Old Red Sandstone* of south Wales.

THE CARBONIFEROUS SYSTEM

The term Carboniferous was coined in 1822, by the English geologists William Conybeare and William phillips to designate strata that included beds of coal in north-central England. Subsequently, it became convenient in Europe and Britain to divide the system into a Lower *Carboniferous and Upper Carboniferous*-the latter containing most of the workable coal seams. Two systems in North America, the *Mississippian* and *Pennsylvanian,* are broadly equivalent to these subdivisions. The American geologist Alexander Winchedl formally proposed the name Mississippian in 1870 for the Lower Carboniferous strata that are extensively exposed in the Upper Mississippi River drainage region. In 1891, Henry S. Williams provided the name Pennsylvanian for the Upper carboniferous System. Although both Pennsylvanian and Mississippian Systems are recognized by most United Stated geologists, neither term has been employed outside North America.

THE PERMIAN SYSTEM

The Permian System takes its name from the small Russian town of Perm on the western side of the Ural Mountains. In 1840 and 1841, Murchison, in company with the French paleontologist Edouard de Verneuil and several Russian companions, travelled extensively across western Russia. To his delight, Murchison found he was able to recognize Silurian, Devonian, and Carboniferous rocks by the fossils they contained. As a result he became even more convinced that groups of fossil organisms succeed one another in a definite and determinable order (this finding is now labeled the *principle of biologic succession*). Murchison established the new Permian System for rocks that overlay the Carboniferous System and contained fossils similar to those in German strata (the Zechstein beds), which had the same stratgraphci position as the Magnesian Limestone in England. Field studies had previously shown that the Magnesian Limestone rested upon Carboniferous strata. Thus Murchison was able to include the magnesian Limestone within the Permian by correlation. The fossils of the new system appeared distinctly intermediate between those of the Carboniferous below and the Triassic above. Murchison's establishment of the Permian system provides a fine example of the logic employed by early geologists in putting together the pieces of the standard time scale.

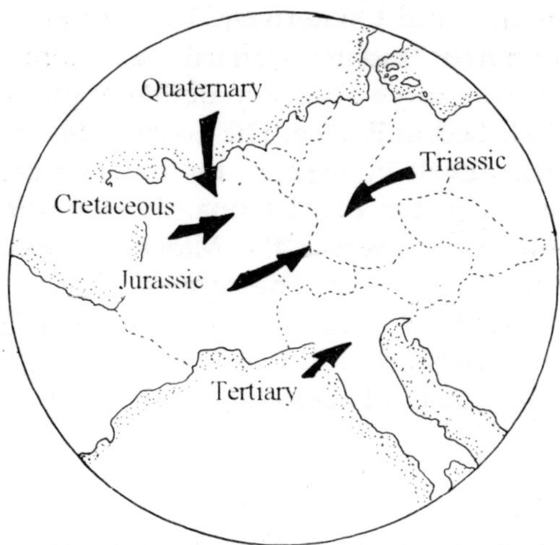

Fig. 2.6: Type areas for the systems of the Mesozoic and Cenozoic.

The Triassic System

The influence of British geologists in providing names for the system of the Paleozoic is by now obvious. However, their presence is not as evident in the development of Mesozoic nomenclature. The Triassic, for example, was applied in 1834 by a German geologist named Frederich von Alberti. The term refers to a threefold division of rocks of this age in Germany. However, because the German strata in the type area are poorly fossiliferous, the standard of reference has been shifted to richly fossiliferous marine strata in the Alps.

The Jurassic System

Another German scientist, Alexander von Humboldt, proposed the term Jurassic for strata of the Jura Mountains between France and Switzerland. However, in 1795 when he used the term, the concept of systems had not been developed. As a result, the Jurassic was redefined as a valid geologic system in 1839 by Leopold von Buch.

The Cretaceous System

During the same year that Conybeare and Phillips were defining the Carboniferous, a Belgian geologist named Omalius d'Halloy proposed the term *Cretaceous* (from the Latin, *Creta,* meaning "chalk") for rock outcrops in France, Belgium, and Holland. Although chalk beds are prevalent in some Cretaceous exposures, the system is actually recognized on the basis of fossils. Indeed, some thick sections of Cretaceous rocks contain no chalks whatsoever.

The Tertiary System

The name Tertiary leads us back to the time when geology was just beginning as a science. Giovanni Arduino suggested a classification with four major divisions: Primary, Secondary, Tertiary, and Quaternary. The Tertiary was derived from his 1759 description of unconsolidated "montes tertiarii" sediments at the food of the Italian Alps. Later, the Tertiary was more precisely defined, and standard sections for series of the Teritary were established in France. The *Eocene, Miocene,* and *Pliocene,* for example, were proposed by Charles Lyell in 1832 on the basis of the proportions of species of living marine invertebrates in the fossil fauna. By definition, 3 per cent of the fossil fauna of the Eocene still live, whereas the Miocene contained 17 per cent, and the Pliocene contained 50 to 67 per cent. The term Oligocene was proposed by August von Beyrich in 1854, and the Paleocene was proposed 20 years later by Wilhelm Schimper. Other system names are also used in place of the Tertiary. Some geologists prefer the terms *Paleogene* (for the Paleocene, Eocene, and Oligocene) and *Neogene* (for the Miocene and Fliocene).

The Quaternary System

In 1829,' the French geologist Jules Desnoyers proposed the term *Quaternary* for certain sediments and volcanics exposed in northern Frence. Although these deposits contained few fossils, Desnoyers was convinced on the basis of field studies that they were younger than Tertiary rocks. In the decade following Desnoyer's establishment of the Quaternary, the unit was divided by Charles Lyell into an older *Pleistocene Series*, composed primarily of deposists formed during the glacial ages, and the younger *Recent Series*.

This belief review describing how geologists drew up a table of geologic time clearly shows a lack of any grand and coherent design. These geologic pioneers were influenced by conspious changes in assemblages of fossils from one sequence of strata to another. In many places in Europe they found that such changes frequently occurred above and below an unconformity. The success of their methods is apparent from the fact that, by and large, the systems have persisted and found wide use even to the present day.

The Agents of Metamorphism

The three principle agents of metamorphism are heat, pressure, and chemically active fluids. Alone or in combination, these agents operate at various intensities to produce metamorphic rocks having distinctive textures and compositions.

Heat

Heat is a major cause of metamorphism. It can reduce the ability of a rock to withstand deformation, and it causes an increase in the rate of most chemical reactions. The heat for metamorphism may be provided by nearby intrusions of magma. It may be associated with the compression of the crust in regions experiencing mountain-building, or it may be induced by increases in pressure resulting from deep burial. The rate of increase in temperature at increasing depths in the earth is called the *geothermal gradient*. Measurements made in deep mines and wells indicate that the *geothermal gradient* for the crust is about 30°C per km of depth. This rate, however, varies from place to place, and is, not unexpectedly, greater near centres of active volcanism. The geothermal gradient is such that at depths of about 35 km temperatures are high enough to melt rock.

Because particular mineral-forming chemical reactions only occur within a specific range of temperature, heata, influences, the ultiamte mineral composition of metamorphic rocks. Also as temperature arises, ions within the atomic lattice of minerals become increasingly agitated. Eventually, they may move to new locations and arrange themselves into structural forms stable under the newer thermal conditions. As an example, consider what might happen to clay minerals in shake strata invaded by a granitic magma. Clay minerals are hydrous aluminum silicates. Heat applied to these minerals from the hot magma would cause a loss of the water contained in the clay and a conversion of clay into a common metamorphic minerals named andalusite. Andalusite is a nonhydrous aluminum silicate having an atomic structure that is stable under the new conditions. The released water in the above example, together with gases such as carbon dioxide, may also participate in the many chemical reactions associated with metamorphism.

As noted in the chapter with igneous geology, if temperatures in a rock mass rise high enough, the melting poins of constituent minerals will be exceeded and magma will begin to form. If the melting becomes pervasive, igneous rather than metamorphic rocks will result.

Pressure

The tremendous pressures that exist several kilometers below the earth's surface (about 1000 kg/cm^2 at a depth of 4 km) or that are associated with collisions of tectonic plates can cause profound changes in deeply buried sedimentary rocks, preexisting metamorphic rocks, or igneous rocks. Mineral grains, for example, may recystallize into new minerals with more tightly packed atomic structure and therefore greater density. Where mineral grains are in contact, the squeezing action may cause melting or solution at the points of contact and precipitation of material along the sides of grains are in contact, the squeezing action may cause melting or solution at the points of contact and precipitation of material along the side of grains that experience less pressure.

Every swimmer knows that as one descends into a body of water, pressure increases with depth because of the progressively greater weight of the column of overlying water. Much of the pressure to which rocks are subjected is similarly caused by the load of overlying rocks. This kind of confining pressure in water is termed *hydrostatic*. When rock rather than water is involved, a more appropriate term is *lithostatic pressure*. Lithostatic pressure is applied nearly equally to all sides of a mass of rock. Like hydrostatic pressure, it increases with depth. In combination with heat from the geothermal gradient, lithostatic pressure causes progressive changes in minerals at progressively increasing depths.

In addition to lithostatic pressure, which affects rocks uniformly, pressure may also be applied to a rock mass along certain preferred directions. Such directional pressure may result in reorientation of grains, the development of tiny shear planes in the rocks, and recrystallization. These processes are largely responsible for the development of the lineated and banded textures characteristic of many metamorphic rocks.

Chemically Active Solutions and Gases

In varying amounts, liquids and gases are always present in regions where metamorphism takes places. They play an important role in metamorphism by increasing the efficiency or recrystallization, serving as solvents, and accelerating the rate of chemical reactions. Laboratory experiments have shown that most minerals react so slowly to increases in temperature that metamorphic reactions would require lengthy spans of time to run their course. By adding only a minute amount of water to a laboratory capsule containing the experimental minerals, however, reaction rates are dramatically increased. One reason for this is that ions in a fluid medium can be brought into close proximity with each other and reach appropriate sites in the atomic structure of minerals more readily than in a dry environment. In some instances, water may enter into the composition of newly forming minerals such as mica, amphibole, and chlorite. An example is provided by the reaction below, in which the attractive green mineral serpentine is formed

$$5Mg_2SiO_4 + 4H_2O \rightarrow 2H_4Mg_3Si_2O_9 + 4MgO + SiO_2$$

(Olivine) (Water) (Serpentine)

(Removed in solution)

The water associated with metamorphic reaction may be derived from several possible sources. Some is water entrapped in parent sedimentary rocks at the time of their deposition. Another source is magma, from which may emanate large quantities of watery liquids and vapours. Smaller amounts of water may be given off by hydrous minerals as they begin to experience the effects of heat and pressure.

In addition to water, the gas carbon dioxide also promotes metamorphism. Carbon dioxide is readily liberated during the heating of limestone. This leaves the remaining oxide of calcium free to combine with silica or other impurities in the limestone. An example of such a reaction involving the liberation of carbon dioxide and the formation of metamorphic mineral known as *wollastonite* is as follows:

$$CaCO_3 + SiO_2 \rightarrow CaSiO_3 + CO_2$$

(limestone) (quartz) (wollastonite) (Carbon dioxide)

Other gases containing fluorine as well as hydrofluoric and hydrochloric acids may be important in particular metamorphic environments.

From Sediments to Sedimentary Rock

The most significant factors involved in the origin of sedimentary rocks are *weathering,* which produces sediment, *transportation* and *deposition* of that sediment, and the *lithification* necessary to convert loose particle of sediment into solid rock. Each of these factors may alter the composition or textural features of sediment and thereby provide a variety of different kinds of sedimentary rocks.

CLASSIFICATION OF THE METAMORPHIC ROCKS

The metamorphic rocks can be broadly grouped into two major classes: cataclastic rocks; recrystallized rocks. The *cataclastic rocks* have experienced mechanical disruption (breaking, crushing) of the original minerals without appreciable chemical change. This process of change can be described as *dynamic metamorphism.* The *recrystallized rocks* have as the name indicates, undergone a recrystallization of the original minerals. Recrystallization is considered to be a chemical change, because it usually produces minerals of chemical formulas and crystal lattice structures different from the parent minerals.

Within the class of recrystallized rocks we must distinguish two subclasses: contact metamorphic rocks; regional metamorphic rocks. The *contact metamorphic rocks* are formed by recrystallization under high temperature in country rock immediately adjacent to an intruding magma that solidifies into a pluton. The rock is not subjected to tectonic forces (bending and breaking) during the process of change, but new mineral substances emanating from the magma can be added to the country rock (i.e., metasomatism can take place). The *regional metamorphic rocks* undergo recrystallization during the process of being deformed by shearing, often under conditions of high pressure or high temperature or both, and often to the accompaniment of loss or gain of mineral components by metasomatism. The adjective "regional" refers to the occurrence of this subclass of rocks over large areas and in great crystal thicknesses. The total process is called *dynamothermal metamorphism.*

A third class of rocks, intermediate between metamorphic rocks, is shown in Table. These are rocks characterized by a banded appearance on in proposed surface caused by layering. The minds represent literature layers of igneous rock and metamorphic rock. It is though that the igneous material penetated or replaced layers of a preexisting rock of sedimentary origin. Various degrees of transition from metamorphic to igneous rock can be included in this class.

Table 2.1

Composition of Gases from Internal Earth Sources Compared with Total Earth Volatiles.

	Volatiles of Earth's hydrosphere and atmosphere	*Gases in hot springs, fumaroles, and geysers*	*Volcanic gases from Basaltic lava of Mauna Loa and Kilauea*
Water, H_2O	92.8	99.4	57.8
Total carbon, as CO_2	5.1	0.33	23.5
Sulfur, S_2	0.13	0.03	12.6
Nitrogen, N_2	0.24	0.05	5.7
Argon, Ar	trace	trace	0.3
Chlorine, Cl_2	1.7	0.12	0.1
Fluorine, F_2	trace	0.03	—
Hydrogen, H_2	0.07	0.05	0.04

Metamorphic Minerals

Recystallization under high temperatures produces a distinctive group of metamorphic minerals. Most of these are different from the minerals we have thus far encountered in our study of igneous and sedimentary rocks. On the other hand, some of the same minerals found in the igneous rocks persist or reappear during recrystallization, such as quartz, biotite mica, pyroxene, amphibole, and feldspar. Calcite and dolomite also persist in recystallized carbonate sedimentary rocks. Metamorphic minerals not encountered in earlier chapters are described in Table. The eleven minerals on this list have been selected because of their importance in the classification and naming of metamorphic rocks and to show a wide range of chemical diversity.

The first three minerals, *kyanite, andalustite,* and *sillimanite,* are of identical composition. All are aluminosilicates with the formula Al_2SiO_5, but each has a different space lattice structure. They are polymorphs of Al_2SO_5. They form by recrystallization of rock with abundant felsic components, such as quartz and feldspar. It shows that each of these minerals forms under a different combination of pressure and temperature. Andalusite forms under conditions of comparatively low pressure and low temperature; kyanite at high pressure and low temperature; sillimanite at high temperature and moderate pressure. For this reason, the polymorphs can serve as indicators of the environment in which recrystallization took place.

Almandite is one of the *garnet group*-aluminosilicates of magnesium, iron, calcium, or manganese that crystallize in the isometric system. Almandite, the iron garnet, is a red mineral familiar as a semiprecious gemstone. Almandite crystals often grow to diameters of several centimetres in metamorphic rock.

Wollastonite, a silicate of calcium, is usually associated with the metamorphism of limestone. Under high temperatures the following reaction occurs:

$$CaCO_3 + SiO_2 \rightarrow CaSiO_3 + CO_2$$

The carbon dioxide is driven off as a volatile gas-an example of one of the forms of metasomatism.

Staurolite, a hydrous aluminosilicate of iron, is formed in a middle range of pressure and temperature. It is a particularly interesting mineral because of is habit of forming as twin crystals penetrating each other at right angles to make a natural cross.

Chlorite, a soft hydrous silicate of iron and magnesium, is similar in some ways to the micas, which sheet structure that forms thin cleavage flakes. Chlorite is associated with metamorphic conditions of comparatively low temperature and is one of the earliest formed of the metamorphic minerals.

Epidote, a hydrous aluminosilicate of calcium and iron, is a green mineral that typically forms in elongate prisms. It is associated with the metamorphism of rocks rich in mafic minerals, such as pyroxene and amphibole, and it is formed in a middle range of temperatures.

Talc is a very soft, scaly mineral, a hydrous silicate of magnesium, and is classed as a clay mineral. Talc occurs in metamorphic rocks that are derived from mafic minerals such as olivine, pyroxene, and amphibole. In massive form as a rock, tale is known as *soapstone.* Because it is immune to action of acids, *soapstone* was once widely used for table tops in chemical laboratories. In powder form, talc has many industrial uses and is perhaps most familiar as talcum powder.

The *serpentine group* consists of two minerals that are hydrous magnesium silicates-antigorite and chrysotile. These minerals are derived from mafic minerals, particularly olivine, pyroxene, and amphibole. Chrysotile is well known as *asbestos,* formed of minute flexible crystals that resemble texrile fibers, widely used in fire-resistant industrial products and as an insulator. It is now established that the minute fibers of asbestos cause irreversible lung diseases (asbestosis and lung cancer) in humans.

Graphite, comparatively minor among the metamorphic minerals, illustrates a substance probably derived from hydrocarbon compounds of organic origin that were present in sedimentary rock prior to its metamorphism. In some instances, seams of coal have been converted partly into graphite during metamorphism.

Cataclastic Rocks

The cataclastic metamorphic rocks result from mechanical deformation without appreciable chemical change and recrystallization. They form under intense shearing stresses which cause grain fragmentation.

Grains that consist of individual mineral crystals or groups of crystals are crushed and pulverized as they are rotated (turned over and over).

Cataclastic rocks are most commonly produced during the progress of faulting. They form a thin layer along the fault plane where strong crushing and grinding action takes place where strong crushing and grinding action takes place. One product of this action is *friction breccia,* a rock consisting of rather large angular fragments in a matrix of small fragments. The largest fragments may measure a meter or more across, while the smallest are a millimeter or smaller in diameter. Where the pulverized rock along a fault plane consists of very small particles-0.01 to 0.1 mm-the rock is called *mylonite.* It is a dense, fine-grained rock often with a streaked or banded appearance, and may outwardly resemble chart. (Chert is a sedimentary form of silica.) Some recystallisation may occur in mylonite, enabling the grains to adapt their shapes so as to fill the entire rock volume.

Contact Metamorphic Rocks

Contact metamorphic rocks form in an environment of high temperature in comparatively shallow crustal locations where confining pressure is not great. The high temperature is provided by intrusive magma, which has entered the country rock to solidify as a pluton, and thus a strong temperature contrast exists between the magma and the country rock. Shearing stress is practically absent under such conditions. The country rock close to the magma body is literally baked, like fired brick or tile, made of mud or clay. The resulting rock is *hornfels,* a word of German origin. The prefix *horn* refers to the hornlike appearance of the rock; the world *fels* means 'rock'. A direct English translation would be close to 'hornstone'. Hornfels has very fine-grained texture because cooling followed rapidly after recrystallization of the parent minerals.

Hornfels forms an *aureole,* which is a metamorphic layer surrounding the magma body. The aureole conforms with the outline of the pluton and may be subdivided into two or more zones of somewhat different mineral content. Where the magma is of granite composition, the aureole contains abundant quartz and potash feldspar. The inner zone may contain wollastonite and andalusite, while the outer zone may have amphibole and mica, which are hydrous minerals.

Texture of Regional Metamorphic Rocks

As regional metamorphism takes place, intense shearing action accompanied by recystallization brings about new textures and structures. These features give the meamorphic rocks their distinctive appearances.

Consider first the structural changes that take place in a thick mass of black shale, a fine-grained sedimentary rock rich in clay minerals such as kaolinite and illite. Shale shows stratification (bedding) resulting from fluctuations in the particle size grades deposited from one layer to the next. Typically, regional metamphrphism of shale results in crumpling of the beds into small, tight folds. This tectonic process results from shear stresses that tend to deform a ductile rock mass. We can visualize shale layers originally in a horizontal attitude as being compressed by stressess acting in a horizontal direction.

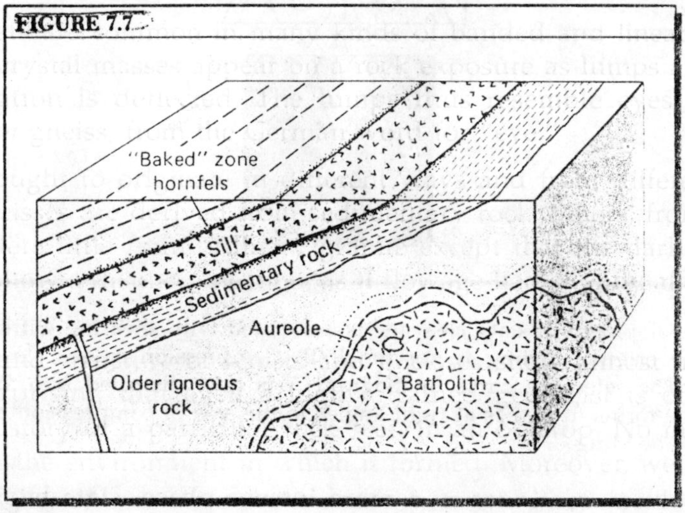

FIGURE 7.7

Fig. 2.7: Schematic block diagram of contact metamorphic aureoles adjacent to plutons. At the upper left is a sill of gabbro intruded between shale layers, forming a narrow zone of hornfels. At the lower right is an aureole adjacent to a granite batholite intruding a mass of older igneous rock.

One of the first structures to develop as a result of tight folding in shales is *slaty cleavage*. As in mineral cleavage, the word 'cleavage' refer to the presence of planes of weakness along which the rock easily split apart. Slaty cleavage consists of innumerable, closely spaced planes weakness, formed in an orientation parallel with the long axis of the reference ellipse in Figure. This is the direction along which the rock has been elongated, or stretched. On a microsopic scale, flakes of clay minerals and mica arrange themselves in parallel layers in the rock, forming planes of weakness in the rock. A fine-grained metamorphic rock derived from shale and exhibiting slaty cleavage is known as slate. Because water has been driven out during deformation, slate is dense and hard. It can be split into thin layers that are strong and rigid, as we know from examining roofing shingles made of slate. Layers of slate a few centimeters thick make ideal flooring slabs for patios and walkways.

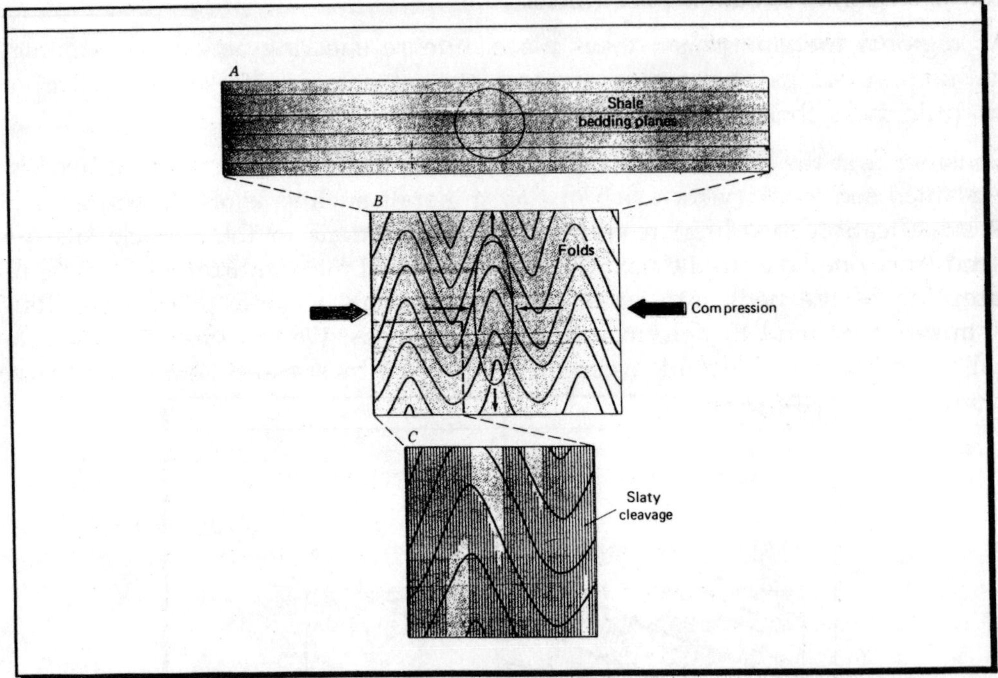

Fig. 2.8: The development of slaty cleavage is associated with shearing within a mass of shale. (A) The original beding is horizonal (B) Compression causes close folding of shale strata. The reference circle in diagram A is now distorted into an ellipse C. Shaty cleavage forms in planes at right angles to the axis of compression and cuts across the original bedding.

As metamorphism continues to a more intense phase, state begins to develop a structure known as *foliátion,* in which a large percentage of the minerals assume a platelike shape and are assembled in parallel orientation in the rock. minerals with a strong tendency to play cleavage—Chlorite, muscovite, biotite, tale-concentrate in layers along which the rock splits easily into parallel sheets, or leaves. (The Latin word for leaves is *folia,* hence our word foliation.) Metamorphic rocks with strongly developed foliation are known as *schists.* The foliation typical of schists is referred to as *schistosity.* A freshly broken piece of schist shows glistening or silky surfaces because of the strong reflection of light from the cleavage places of minerals occupying the foliation planes.

As the foliation structure is developing in schists, crystals of certain minerals such as garnet and staurolite are growing in size. Geologists refer to these large crystals as *porphyroblasts.* An example is the large garnet crystal shown in figure.

Another structure found in metamorphic rocks in *lineation,* the presence of mineral grains drawn out into long, thin, pencillike objects all in parallel alignment. Lineation is associated with massive types of metamorphic rock and does not produce planes of weakness. The elongated mineral grains may have been formed by lengthening as the rock was stretched along one axis, or may have assumed an alignment parallel with the direction of maximum rock stretching.

Another common structure of metamorphic rocks is *banding,* a rough kind of layering in which minerals of different varieties or groups have become segregated into alternate layers. These layers are usually of different shades-light or dark-so that the banding is conspicuous. Metamorphic rock of this description is called gneiss. (The word is also applied to some metamorphic rocks showing lineation, but poor banding.) Within individual bands, the rock of coarsely crystalline and strongly bonded. Orientation of grains parallel with the banding is usually present in the banded gneisses. Weakness may exist between individual bands, so that the rock may tend to break apart in layers and can he said to show coarse foliation.

Porphyroblasts are common in many kinds of banded and lineated gneiss. These large crystals or crystal masses appear on a rock exposure as lumps around which the lineation or foliation is deflected. The lumps thus resemble eyes and the rock is described as *augen* gneiss, from the German word for "eye."

Gneiss is thought to originate in different ways and from different parent rocks. Some banded gneisses are derived from sedimentary rock; others, from igneous rocks, *Granite gneiss* differs little from ordinary granite except that the dark grains of biotite and hornblende show a distinct lineation, as if flowage had slightly affected the granite when it was in a plastic state.

The texture and structure of schists and gneisses show almost infinite variation, and it is not surprising that even a highly trained geologist is often at a loss to reconstruct the history of a particular rock seen in an outcrop. No metamorphic rock can be studied in the environment in which it formed. Moreover, we can see only the final stage in a long series of changes through a changing environment. Detailed laboratory analysis, using sophisticated tools of research, is needed to unravel the mysteries of dynamothermal metamorphic rocks. Chemical analysis of the regional metamorphic rocks sheds a great deal of light on their origin. We therefore turn from metamorphic textures to the assemblages of minerals present in the rock and how they are related to the environment in which metamorphism occurred.

MINERALOGICAL CLASSIFICATION OF THE REGIONAL METAMORPHIC ROCKS

The regional metamorphic rocks involve large-scale dynamothermal processes. Recrystallization takes place along with wholesale shearing of thick masses of crustal rock. The environments in which these processes act span a wide range of temperatures and pressures. The most important concept relating to these rocks is that they are produced in a well-defined sequence according to increasing temperature and conflining pressure. Followed across country, exposures of regional metamorphic rocks show *metamorphic zones,* which reflect the increased temperature that accompanied recrystallization. Geologists found that each zone can be defined by the first appearance of an index mineral, not present in zone of lower temperature. A typical sequence of index minerals is shown in figure; it runs as follows: chloric, biotite, almandite, staurolite, kyanite sillimanite. This sequence would result from metamorphism of a large mass of sedimentary rock made up mostly of shale. The appearance of each mineral can be shown on a map by a line, called an *isograd*.

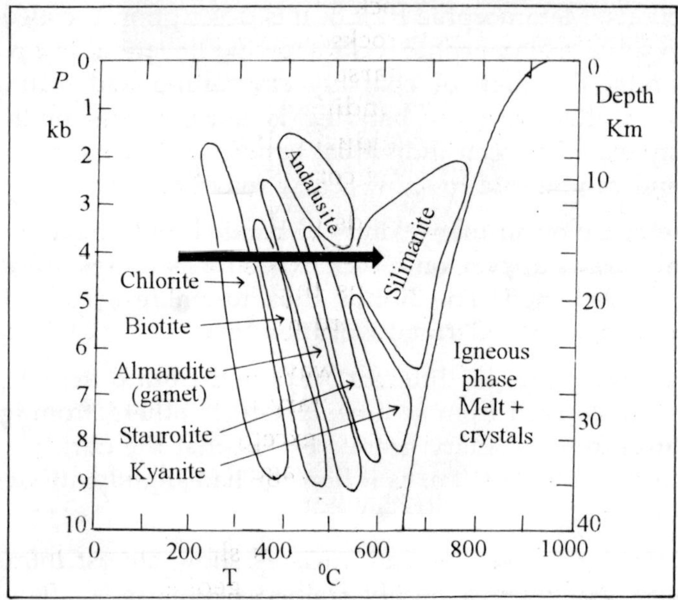

Fig. 2.9: This graph shows how the grade of regional meamorphism are related pressures and temperature. The arrow shows the typical series of changes from lower to higher grades at a given depth.

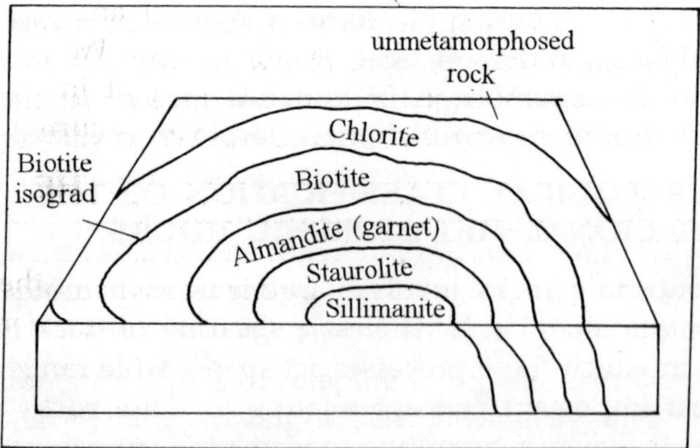

Fig. 2.10: Schematic diagram of isograds of regional metamorphism forming concentric zones around a region of the highest grade of metamorphism. The granite intrusion might represent melting or granitization occuring at a higher temperature.

The metamorphic rock is thus zoned into a succession of *metamorphic grades*, each associated with a particular type of metamorphic rock figure shows typical sequences of metamorphic rock types in terms of pressure and temperature. The broad arrows suggest the sequence of changes that took place as the rock was subjected to increasing temperature. As metamorphism begins at comparatively low temperature. As metamorphism begins at comparatively low temperatures and pressures, a mass of shale develops new minerals of a group called the *zeolites*. We have not referred to the zeolite minerals previously and mention them only briefly at this point. They are hydrous aluminosilicates of calcium and sodium with a rather large water content, and they are soon destroyed as water is driven out of the changing sedimentary clay minerals.

The rock now enters the first major metamorphic grade, *greenschist*, formed under moderate pressure and fairly low temperature (low-grade metamorphism). Minerals dominant in greenschist are chlorite, muscovite mica, biotite, sodic plagioclase feldspar, and quartz. From this point on, the pressure is assumed to hold about constant, while temperature increases.

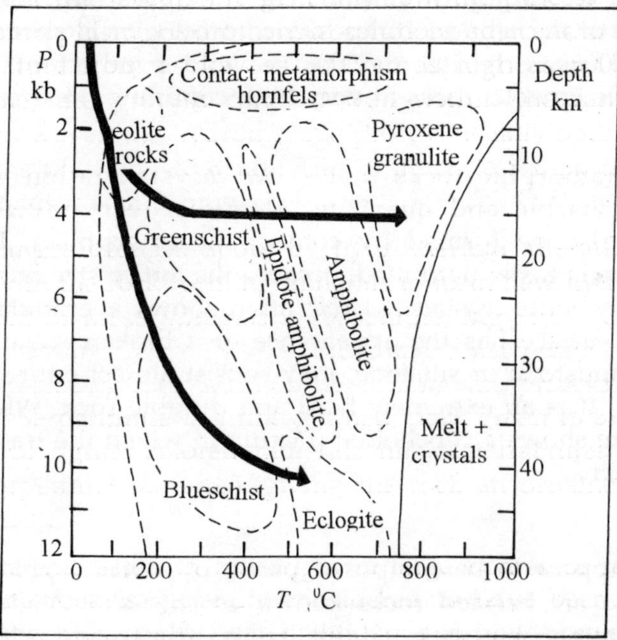

Fig. 2.11: A schematic graph of the major types of metamorphic rocks in relation to pressure (depth) and temperature. The arrows show low possible patterns of evolution in the sequence of regional metamorphic rocks. Contact metamorphism is limited to a shallow zone.

The next higher, or intermediate grade is characterized by a rock called *amphibolite*, named for the presence of hornblende (an amphibole). In some places this grade is preceded by *epidote amphibolite*, characterized by the presence of epidote. Amphibolite may also contain quartz, plagioclase feldspar, almandite garnet, and biotite. The rock composition varies according to the original rock composition, whether felsic or mafic.

At still higher temperature, above about 600 °C, a high-grade metamorphic rock, *pyroxene granulite,* is produced. *Granulite* is a term describing the rock texture, which consists of small mineral grains of more or less equal size, developed through shearing action. Grains of quartz and feldspar show a flattening into platelike shapes. Pyroxene is an important new mineral in the rock, while sillimanite appears as the index mineral. With an increase in temperature above the granulite grade, the melting point of the rock would be reached, and new magma could be produced, yielding an igneous rock.

It is possible for the confining pressure to increase to high levels while temperature remains low. The result is formation of *blueschist,* named for the presence of a blue variety of amphibole (glaucophane).

With an increase in temperature to a middle range (400 °C to 600 °C) and with extremely high pressure (1 to 10 kb or higher), blueschist may become *eclogite.*

Dominant minerals in eclogite are an unusual green pyroxene rich in sodium and calcium and a red-brown variety of garnet rich in magnesium. The rock grains are coarse in texture. The total chemical composition of eclogite is approximately that of gabbro and basalt. Thus, eclogite may be the end product of dynamothermal metamorphism of mafic igneous rocks at very high confining pressures corresponding to great depths.

Two regional metamorphic rocks that show recystallization with little or no chemical change are marble and quartzite. *Marble* is recrystallized limestone or dolostone, and may also be formed by contact metamorphism. Under moderate shearing stresses, calcite is easily deformed, because the lattice structure permits gliding motions to occur easily with crystals. *Marble* often shows a granular structure such that a freshly broken surface has the appearance of a broken sugar cube. *Quartzite,* formed from quartz sandstone or siltstone, may consist almost entirely of quartz with only minor impurities. It is an extremely hard and durable rock. When struck a hard hammer blow, quartzite shows a conchoidal fracture in which the fracture surface cuts through the quartz grains.

Granitization

Certain gneisses appear to be composed partly of metamorphic rock and partly of igneous rock. These are banded rocks known as *migmatites.* Between bands that appear to be true metamorphic rock, originally of sedimentary origin, are bands of the composition and texture of granite. In some cases, the granite bands may represent magma that forced its way between layers of metamorphic rock. An alternative explanation is that the granitelike layers represent matter imported by slow injection of fluids and the diffusion of ions from a nearby magma body. These imported substances have replaced the original rock, and many of the original constituents have been exported to other locations. Thus, granitelike layers may have been formed by metasomatism.

Conversion of preexisting rock to a granite by metasomatism is called *granitazation.* Melting does not occur during granitization, but the end result is much the same as if the affected rock had been melted and recrystallized by cooling to form a plutonic rock. Since we have already stated that large granite plutons are perhaps best explained by wholesale melting of large masses of crustal rock, this means grantie may from both by metasomatism and by melting. The process of granitization is obscure and its importance has been strongly debated for many decades as part of the general question of the origin of granite.

HYDROTHERMAL ALTERATION AND SERPENTINITE

Hydrothermal solutions that rise from intrusive magma bodies in the final stages of cooling. These hot-water solutions consist of volatiles and carry a wide variety of mineral-forming ions in solution. Under favourable conditions, hydrothermal solutions are capable of altering the mineral composition of large masses of rock, a process known as *Hydrothermal alteration.* Because the temperatures and pressures under which hydrothermal alteraction occurs are generally low, the process is often excluded from the scope of metamorphism. Hydrothermal solutions adjacent to shallow magma bodies range in temperature from as low as 60°C to as high as 500°C. Thus, in terms of temperature alone, the process can fall well within the range found in regional metamorphism. However, the hydrothermal process usually operates at pressures under 1 kb and at depths of only 1 to 3 km. This is much less than the pressure and depth required for regional metamorphism.

Chemical changes brought about by hydrothermal alteration are those in which water combines with the rock-forming minerals to form new hydrous minerals. Perhaps the most important of these minerals is serpentine, $Mg_3Si_2O_5\ (OH)_4$. It can be formed through the altertion of olivine, $(Mg,Fe)_2SiO_4$, and other mafic silicate minerals rich in magnesium. The resulting rock is called *serpentinite;* the process by which it is formed is *serpentinization.* Serpentinite is a fine-grained, dark green to black rock with massive structure. Bands of lighter-colored minerals fill joint fractures or are deformed into swirling lines (serpentine patterns), giving the rock an ornamental quality similar to that prized in marble.

Geologists are interested in the origin of large bodies of serpentinite exposed in orogenic belts. The rock is generally considered to have been derived from oceanic crust altered in the presence of water. Evidence is now accumulating to show that hydrothermal activity is intense along spreading plate boundaries. Seawater, which is rich in magnesium, penetrates the rifted crust and is highly heated by the underlying rock. Returning to the surface, the heated water alters the mafic minerals of the crustal basalt to form sepentine-rich basalt. Descending seawater is thought to penetrate the crust to depths on the order of 2 to 3 km, but probably no deeper. For serpentinization to occur deeper in the oceanic crust, affecting the entire crust down to the Moho at a depth of 5 to 6 km beneath the ocean floor, a different source of water would he required.

Professor Harry Hess of Princeton University suggested that the lower part of the oceanic crust is formed from rising peridotite rock of the upper mantle. In the axis of the spreading zone, peridotite is altered to serpentinite on a large scale by the addition of water moving upward from the deeper mantle beneath. The serpentinized mantle rock then spreads laterally to form the oceanic crust. This possibility is to some degree credible, because serpentine and peridotite are often found in adjacent masses at the surface in deeply eroded orogenic belts. These rocks may represent parts of the lower oceanic crust and upper mantle that were finally lifted to the surface by overthrust faulting during orogeny. Evidently, serpentinization may be a process that operates under a wide range of environments, including both low and high confining pressures, and with widely different sources of the water required to form the serpentine.

RADIOISOTOPES AND RADIOACTIVITY

Although we have now covered the full scope of the rock transformation cycle, two basic problems of geology connected with that cycle require detailed attention: (1) What is the source of the internal energy that drives plate motions and provides beat for igneous and metamorphic processes? (2) How do geologists determine the ages of the crustal rocks and of the earth itself? Both questions are answered through an understanding of radioactivity, a fundamental process of nuclear physics.

Element-carbon for example-three exist isotopes. Whereas the atomic number (number of protons in the nucleus) is always the same for a given element, the number of neutrons may vary. Thus, the mass number of a given element differs from one isotope to another. In referring to a given kind of isotope in terms of the composition of its nucleus, physicists use the term *nuclide.* All nuclides with the same atomic number answer to the same element name, but as isotopes of that element, they differ from one another because of differences in mass number.

Protons and neutrons in the atomic nucleus are held together by nuclear forces quite different from any kinds of forces with which we are familiar from every-day experience (e.g., gravitational attraction and electromagnetic forces). The nuclear forces are effective only when protons and neutrons are very close together, as they are in the nucleus of the atom. Some nuclides, with a *stable nucleus* that effectively resists natural forces which might tend to break apart the nucleus, are *stable isotopes*. Other nuclides, with an unstable nucleus whose protons and neutrons are not strongly bound and with excess energy, are *unstable isotopes*.

An unstable isotope will undergo a natural process of nuclear disintegration to be transformed into a stable isotope of a different element. This spontaneous process of nuclear disintegration is called *radioactivity,* a term first applied by a French physicist, Henri Becquerel, who observed in 1896 that uranium emitted a mysterious radiant form of energy capable of leaving a photographic image, even under conditions of total darkness. Isotopes that undergo spontaneous disintegration are called *radioisotopes* or *radionuclides,* and the process is referred to as *radioactive decay.*

Radioactive decay can occur in one of three ways: (1) alpha decay, (2) beta decay, and (3) electron capture. In *alpha decay,* the nucleus emits an *alpha particle,* consisting of two protons and two neutrons. (You will recognize the particle as the nucleus of a helium atom.) As a result of *alpha particle* emission, the mass number is decreased by four, the atomic number by two. The original radioisotope is referred to as the parent isotope, the product of decay is a daughter isotope.

In *beta decay,* the nucleus emits a high-speed electron, whose expulsion has the effect of changing one of the neutrons into a proton. As a result, the atomic number is increased by one but the mass number remains unchanged.

In *electron capture,* one of the protons in the nucleus acquries an electron from one of the electron orbitals. The positive charge of the proton is thus neutralized and the protron becomes a neutron. The atomic number decreases by one, but the mass number remains unchanged.

In viewing radioactive decay as an energy-producing mechanism for geologic processes, we must look at the types of emissions associated with the nuclear changes. In alpha decay, the motion of an alpha particle represents a form of kinetic energy. When they alpha particle is lost, the remaining nucleus is in an excited state and becomes stable only by the emission of a *gamma ray,* which is a high-energy photon. Energy contained in both the alpha particle motion and the gamma ray is produced by a very small reduction in the combined mass of the alpha particle and the remaining nucleus. This mass difference is equated to energy by the Einstein formula:

$$E=mc^2$$

where E is energy

m is mass and

c is the speed of light (300,00 km/sec)

From this equation, we see that an extremely small quasntity of mass converts into a comparatively enormous quantity of energy.

In beta deacy, another form of energy, the *beta particle* is produced. It is an electron travelling at high speed, derived from one of the neutrons in the nucleus. In furnishing one electron (one negative unit electrical charge) for the beta particle, the neutron acquires a unit positive charge and becomes a proton. (The neutron also emits a particle called an *antineutrino,* but this is not important in our discussion of energy released by radioactive decay.)

The important point about radioactive decay is that it generates energy which takes the form of heat in the substance surrounding the radioisotope. This form of heat is called *radiogenic heat.* As alpha particles travel outward through the surrounding matter, they lose energy by interacting with electrons in the orbitals of other atoms and by colliding with other atomic nuclei. This lost energy is transformed into heat and raises the temperature of the substance. Both gamma rays and beta particles interact with elecrons in the surrounding matter, also causing a buildup of heat. The total quanity of radiogenic heat produced per unit of time can be exactly calculated for a given quantity of a radioisotope.

We can illustrate radioactive decay and heat production by using the important *decay series* in which the parent radioisotope uranium-238 (U^{238}) eventually ends up as a stable isotope of lead, lead-206 (Pb^{206}). In the figure, arrows show the direction of successive changes. Note that each step in the direction of the arrow to the left is alpha decay; each diagonal step downward to the right is beta decay. Uranium-238 decays to produce thorium-234. This is followed by a succession of isotopes of seven different elements, listed along the bottom of the graph. The disintegration process achieves a steady rate, or equilibrium, with time. In the case of the uranium-238-lead-206 series, each gram of uranium produces 0.71 calories of heat per year.

Other important heat-producing decay sequences in the rocks of the earth are those of uranium-235, thorium-232, and potassiurn-40. Uranium-235 produces 4.3 calories of heat per gram per year, thorium-232 produces 0.20 calorie, and potassium-40 produces only 0.000027 calorie.

Of great importance in both the early history of the earth and the dating of geologic events is a physical law that governs the rate of decay of radioisotopes. Once an equilibrium has been reached in the process of radioactive disintegration, the ratio of decrease in the number of atoms of the parent isotope with each unit of time as a constant.

Take for example, postassium-40, which decay to the stable isotopes calcium-40 and argon-40. We can start at any point in time. Let the number of atoms of potassium-40 at tune zero be designated by unity (1.0). After 1.31 billion years have elapsed, the number of atoms of potassium-40 will have been reduced to half the initial number, designated as 0.5 on the vertical scale. The span of time of 1.31 billion years is designated as the half-life. In a second elapsed span of 1.31 billion years, the number of atoms of potassium-40 will again be halved, reducing the remaining quantity to 0.25 on the vertical scale. Notice that the ratio of reduction is always the same, i.e., one-half. Such a schedule of decrease in quantity with time is known as exponential decay. This schedule applies to all radioactive decay, but the ratio of change, and hence the value of the half-life, is different from one isotope to another.

It should be noted that calcium-40 is produced about 71/3 times more rapidly than argon-40, hence the curve of calcium-40 rises more steeply. We shall refer to the ratios of parent isotopes to daughter isotopes in our discussion of methods of rock dating.

In projecting the production of radiogenic heat back into earliest geologic time, these differences will be highly important. Isotopes with the shorter half-lives were then present in very much large quantities than today, in contrast with isotopes having extremely long half-lives.

DISTRIBUTION OF RADIOGENIC HEAT

Heat flows continuously upward from depths of the earth toward the surface. The increase in temperature with depth, or geothermal gradient, is well known from observations in deep mines and bore holes and has a value of about 3 C° per 100 m. The flow of heat because of this thermal gradient averages about 1.4 microcalories (0.0000014 calories) per square centimetre per second. At this rate, the total heat flow in one year is about 50 calories per square centimetres, enough to melt an ice layer 6 mm thick. This quantity of heat is extremely small compared with that received by one square centimeter of the earth's surface from solar radiation. Therefore the earth's heat flow from depth is of no significance in the earth's surface heat balance or in powering the atmospheric and oceanic circulation systems.

The rate of temperature increase with depth falls off very rapidly after the first 200 km or so. The flattering of the temperature curve in the lower mantle and core expresses the very low thermal gradient that is postulated to exist within the deep interior. If the thermal gradient decreases rapidly with depth, a logical interpretation is that the rate of production of radiogenic beat is greatest near the earth's surface and decreases rapidly with depth. It has been concluded that the concentration of radioisotopes is greatest in the rocks of the crust, but falls off rapidly in the mantle rocks and is very small in the lower mantle and core.

Both the concentrations and rates of heat production of the radioisotopes of uranium, thorium, and potassium in each of three classes of rocks. These rates are based upon chemical analyses of samples of igneous rocks collected at the earth's surface. If we project these figures to the assumed corresponding rocks of the crust and mantle, gas shown in figure we see that the most rapid production of radiogenic heat is by felsic (granitic) rocks of the upper zone of the continental crust. The ultramatic mantle rock produces very little heat per unit of weight. It has been estimated that about one-half of all radiogenic heat is produced above a depth of 35 km in the continental crust.

Some support for the conclusion that the iron core of the earth produces almost no radiogenic heat is found in the analysis of iron meteorites. These fragments of matter are thought to represent the distrupted cores of plantetary objects of origin similar to the earth, and they show radioactive minerals in only very small quantities.

EARLY THERMAL HISTORY OF THE EARTH

Modern hypotheses of the earth's origin, favour the process of accretion of the earth and other planets through the condensation of a hot interstellar cloud of gases and dust as it collision and gravitational attraction. Once formed, solid masses would have grown by the in fall of solid bodies of many sizes, perhaps including objects similar to the asteroids of our present-day solar system. It is generally supposed that at the time planetary accretion was largely complete, the earth's interior temperature had not risen to the melting point, although local areas may have become molten from the energy of impacts. Modern thinking is thus along quite different lines from that of early scholars, who postualted that hot nebular gases condensed to the molten state and finally cooled to the solid state.

The discovery of natural radioactivity by Henri Becquerel in 1896, followed by the isolation of radium by Marine and Pierre Curie in 1898, radically altered all scientific thinking about the earth's internal heat John Joly in 1909 applied the new knowledge of radioactivity to recalculations of the earth's thermal history. Moreover, Joly brought forward the underlying principle that radiogenic heat provides the prime energy source for volcanism, igneous intrusion, and deformation of the earth's crust into mountain belts. Today radiogenic heat is usually regarded as the basic source of energy for lithospheric plate motions.

We must first accept as a premise that the earth's supply of radioisotopes was furnished, along with all other elements, at the time the earth was formed. It is most unlikely that the earth, at the time of its formation as a planet about 4.5 billion years ago, contained the same quantity and distribution of radioisotopes that we find today. The obvious reason is that radioactive decay progressively reduces the initial store of radiosotopes. We conclude that the total production of radiogenic heat within the earth was at the maximum level at the time of the earth's formation and has diminished even since. The relative rates of decay of uranium, thorium, and potassium isotopes are not the same, but are well established for each isotope.

A graph in which time is plotted on the horizontal axis starting at an arbitrary zero point at the assumed time of formation of the earth. Total planetary radiogenic heat production per year is given on the vertical scale. Curves have been plotted for the major radioisotopes of uranium, thorium, and potassium individually, while the total production is shown in a separate curve. Note that uranium-235 and potassium-40 have short half-lives in comparison with these of uranium-238 and thorium-232. Referring to the total curve, it is obvious that total radiogenic heat production was vastly greater when the earth was first formed than it is at present, roughly by a factor of six. The implications of such a history are of great consequence.

First, assume that at the time of the earth's formation by accretion the radioisotopes were uniformly distributed throughout the entire earth. (There is no reason to think otherwise.) Silicate minerals and iron were also more or less uniformly mixed. As radiogenic beat accumulated at great depths, the temperature of the solid rock would have been raised to a level close to the melting point. At a certain point in time, which may have been about one billion years after planetary accretion was completed, the melting point of iron would have been exceeded. This event probably occurred first in a depth zone ranging from 400 to 800 km. The silicate minerals remained crystalline, forming a spongy mass through which droplets of molten iron could filter down under the force of gravity. As molten iron accumulated in the central core region, the silicate minerals were gradually displaced upward.

At this point, a new factor would have come into play to raise the earth's internal temperature. As the iron sank toward the earth's centre, its potential energy would have been converted into kinetic energy of molecular motion in the form of sensible heat. It seems likely that the additional heating was sufficient to cause melting or partial melting of a large proportion of the entire earth. At various times and places, melted rock would have risen as magma toward the surface, bringing radioisotopes up with it. Minerals containing the radioactive elements are largely of felsic composition and tend to remain in the liquid state at temperatures lower than the mafic minerals. Cooling and crystallization of the mafic minerals would have been accompanied by a sinking of those mineral crystals (which are denser), leaving the less mafic liquid fraction to solidify closer to the earth's surface. Although such a process of differentiation is speculative, it offers a mechanism for the selective removal of the radioisotopes from the inner earth and their eventual concentration near the surface.

So we see that during this great thermal event, the density layering of the earth came into existence, resulting in the concentration of metallic iron in the core, a less dense ultramafic rock mantle above it, and a maficfelsic crust at the top. During and after the segregation process, the rate of radiogenic heat production was steadily falling. Consequently, along with redistribution of the heat-generating isotopes, the final episodes of deep melting must have become fewer and eventually ceased. Today the earth is thermally stable, in the sense that melting and movement of magma are limited to an extremely shallow layer compared with the earth's total diameter. The inner core and much of the mantle are no longer subject to melting through the accumulation of excess heat.

It is fortunate, indeed, that the chemistry of the radioactive elements is such that they would tend to rise towards the earth's surface throughout its history. If, on the other hand, they had tended to sink and collect near its center, the heat produced by their concentrated activity would have repeatedly melted the earth. Under such conditions, no planetary stability would have been possible throughout geologic history. As we find conditions today, radiogenic heat production in the core is negligible, while the rate of surfaceward flow of heat from the upper mantle and crust closely balances the rate of heat production. Consequently, the mantle remains for the most part at a temperature lower than its melting point. Only in the soft layer of the mantle (the asthenosphere) is melting on a large scale a likely occurrence.

RADIOMETRIC METHODS OF DATING THE EARTH

Radioactivity

The radioactivity discovered by Henri Becquerel was a consequence of the fact that some elements, such as uranium and thorium, are unstable. Such elements will decay to form other elements or other isotopes of the same element. To understand what is meant by "decay", let us consider what happens to a radioactive element like uranium 238. Uranium 238 has an atomic weight of 238. The "238" represents the sum of the atom's protons and neutrons (each proton and neutron having a "weight" of 1). Uranium has an atomic number (number of protons) of 92. Such atoms with specification number and weight are sometimes temed *nuclides*. Sooner or later (and entirely spontaneously) the uranium 238 atoms will fire off a particle from the nucleus called an alpha particle. Alpha particles are positively charged ions of helium. They have an atomic weight of 4 and an atomic number of 2. Thus, when the alpha particle is emitted, the new atom will have an atomic weight of 234 and an atomic number of 90. From the decay of the parent nuclide, uranium 238, the daughter of the nuclide, thorium 234, is obtained. A shorthand equation for this change is written:

$$^{238}_{92}\text{U} \rightarrow {}^{234}_{90}\text{Th} + {}^{4}_{2}\text{He}$$

This change is not, however, the end of the Process, for the nucleus of thorium 234 is not stable. It eventually emits a beta particle (an electron discharged from the nucleus when a neutron splits into a proton and an electron). There is now an extra proton in the nucleus but no loss of atomic weight because electrons are essentially weightless. Thus, from 238 90 Th the daughter element 234 91 Pa (protactinium) is formed. In this case, the atomic number has been increased by one. in other instances, the beta particle may be captured by the nucleus, where it combines with a proton to form a neutron. The loss of the proton would decrease the atomic number by one.

A third kind of emission in the radioactive decay process is called *gamma radiation*. It consists of a form of invisible electromagnetic waves having even shorter wavelengths than do x-rays.

As alpha and beta particles, as well as gamma radiation, move through the surrounding materials, their energy is transformed into increased activity of the electrons in the atoms of the surrounding medium. The result is heat. This radiogenic source of heat was the unknown entity in Lord Kelvin's calculations of the earth's thermal history.

The Clocks in the Rocks (Radiometric Dating)

Nuclear adjustments such as those described previously occur many times before a final, stable daughter element, such as lead is formed. The rate at which the steps in the process take place is unaffected by changes in temperature, pressure, or the chemical environment, since these do not involve the nucleus. Indeed, one can confidently assume that the rate of decay of long-lived isotopes has not varied since the earth came into existence. Therefore, once a quantity of radioactive nuclides has been incorporated into a growing mineral crystal, that quantity will begin to decay at a steady rate with a definite percentage of the radiogenic atoms undergoing decay in each increment of time. Each radioactive element has a particular mode of decay and a unique decay rate. As time passes, the quantity of the parent nuclide diminishes and the amount of daughter atoms increases, thereby indicating how much time has elapsed since the clock began its time-keeping. The "beginning," or "time zero," for any mineral containing radioactive nuclides would be the moment when the radioactive parent atoms because part of a mineral from which daughter elements could not escape. The retention of daughter elements is essential, for they must be counted to determine the original quantity of the parent nuclide.

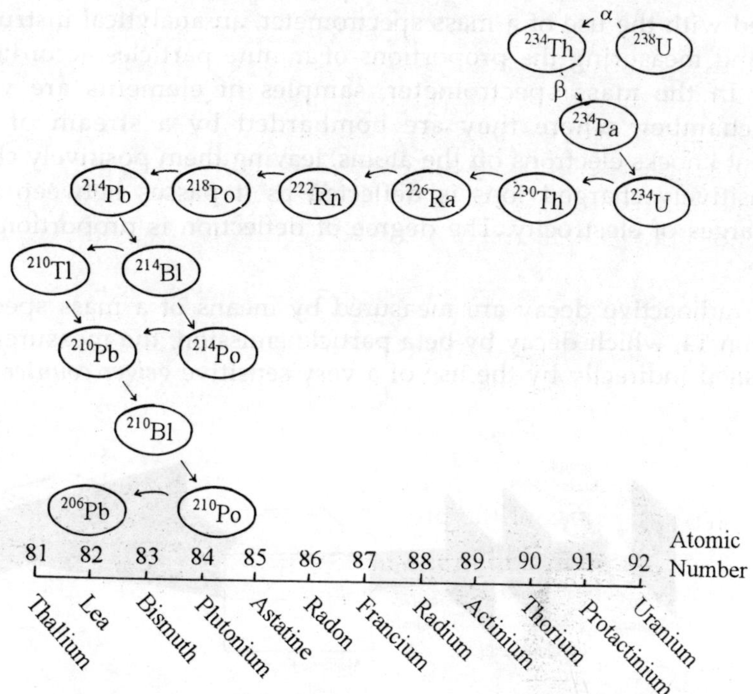

Fig. 2.12: Radioactive decay series of uranium 238 to lead 206.

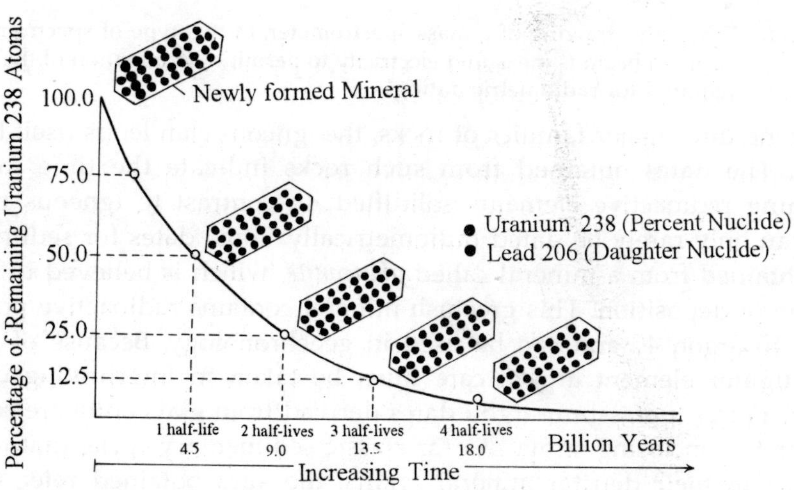

Fig. 2.13: Rate of radioactive decay of uranium 238 to read 206. During each half-life, one half of the remaining amount of the radioactive element decays to its daughter element. In this simlified diagram only the parent and daughter nuclides are shown and the assumption is made that there was no contamination by daughter nuclides at the time the mineral formed.

The determination of the ratio of parent to daughter nuclides is usually accomplished with the use of a mass spectrometer, an analytical instrument capable of separating and measuring the proportions of minute particles according to their mass differences. In the mass spectrometer, samples of elements are vaporized in an evacuated chamber, where they are bombarded by a stream of electrons. This bombardment knocks electrons off the atoms, leaving them positively charged. A stream of these positively charged ions is deflected as it passes between plates that bear opposite charges of electrocity. The degree of deflection is proportional to the masses of the atoms.

Not all radioactive decay are measured by means of a mass spectrometer. In the case of carbon 14, which decay by beta particle emission, the measurement of nuclides is accomplished indirectly by the use of a very sensitive *geiger counter*

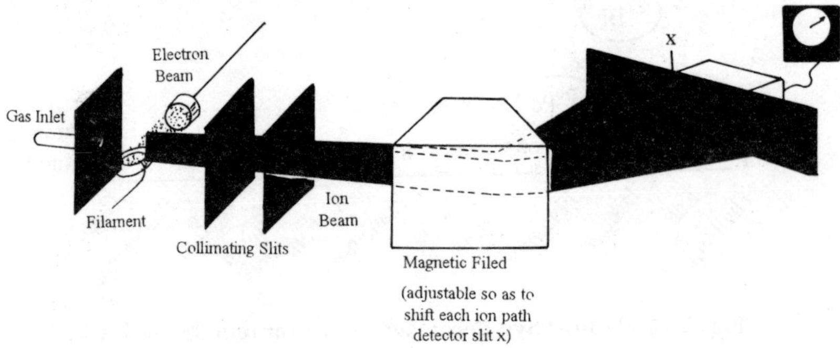

Fig. 2.14: Schematic drawing of a mass spectrometer. In this type of spectrometer the intensity of each ion beam is measured electricity to permit determination of the isotopic abundances required for radiometric dating.

Of the three major families of rocks, the igneous clan lends itself best to radiometric daitng. The dates obtained from such rocks indicate the time that a silicate melt containing radioactive elements solidified. In contrast to igneous rocks, sedimentary rocks can only rarely be dated radiometrically. Some dates for sedimentary strata have been obtained from a mineral called *glauconite,* which is believed to form "in place" at the time of deposition. This greenish mineral contains radioactive potassium 40, which decays to argon 40 and can be used in geochronology. Because of possible, losses in the daughter element argon, care must be taken in interpreting dates, however, in most instances, potassium-argon dates derived from glauconite are considered minimal ages for the enclosing strata. As for classic sedimentary rocks that contain radioactive elements in their detrital mineral grains, the ages obtained refer to the parent rock that was eroded and is older than the sedimentary layer.

Dates obtained from metamorphic rocks may also require special care in interpretation. The age of a particular mineral may record the time the rock first formed or any one of a number of subsequent metamorphic recrystallizations.

There are many other problems that can affect the validity of a radiometric age. If some of the daughter products are removed from the sample by weathering or leaching, its age would be under-estimated. If the element being analysed was a gas, some of that gas (as with argon in glauconite) might have diffused out of the rock. The heat accompanying burial or mountain building might enhance such losses. There is also the possibility that at a later time older rocks may be partially remelted so that the age would be that of the second, rather than the initial, melting event. Clearly, great care must be taken in understanding the field relationships of the rock masses under investigation and in selecting samples.

Once an age has been determined for a particular rock unit, it is sometimes possible to use that date to approximate the age of adjacent rock masses. A shale tying below a lava flow that is 110 million years old and above another flow dated at 180 million years old must be between 110 and 180 million years of age. Fossils within the shale might permit one to assign it to a particular geologic system or series and, by correlation, to extend the age data around the world.

Half-Life

There is no way that one can predict with certainty the moment of disintegration for any individual radioactive atom in a mineral. We do know that it would take an infinitely long time for all of the atoms in a quantity of radioactive elements to be entirely transformed to stable daughter products. Experiments have also shown that the decline in the number of atoms is rapid in the early stages but becomes progressively slower in the later stages. One can statitically forecast what percentage of a large population of atoms will decay in a certain amount of time.

Because of these features of radioactivity, it is convenient to consider the number of years needed for half of the original quantity of atoms to decay. This span of years is termed the half-life. Thus, at the end of the years constituting one half-life, one half of the original quanity of radioactive element still has not undergone decay. After another half-life, one half of what was left is halved, so that one fourth of the original quantity remains. After a third half-life, only one eighth would remain, and so on.

Every radioactive nuclide has its own unique half-life. (Uranium 235, for example, has a half-life of 704 million years. Thus, if a sample contains 50 per cent of the original amount of uranium 235 and 50 per cent of its daughter product, lead 207, then that sample is 704 million years old. If the analyses indicate 25 per cent of uranium 235 and 75 per cent of lead 207, two half-lives would have elapsed, and the sample would be 1408 million years old.

The Principal Geologic Timekeepers

At one time, there were many more radioactive nuclides present on earth than there are now. Many of these had short half-lives and have long since decayed to undetectable quantities. Fortunately, for those interested in dating the earth's most ancient rocks, there remain a few long-lived radioactive nuclides. The most useful of these are uranium 238, uranium 235, rubidium 87, and potassium 40. There are also a few short-lived radioactive elements that are used for dating more recent events. Carbon 14 is an example of such a short-lived isotope. There are also short-lived nuclides that represent segments of a uranium or thorium decay series.

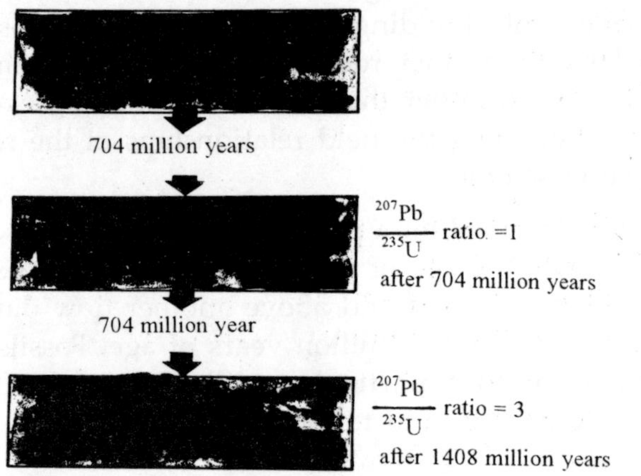

Fig. 2.15: Radioactive decay of uranium 235 to lead 207.

Timekeepers That Produce Lead

Dating methods involving lead require radioactive nuclides of uranium or thorium that were incorporated into the earth's crust when it congealed. To determine the age of a sample of mineral or rock, one must know the original quantity of parent nucildes as well as the quantity remaining at the present time. The original number of parent atoms should be equal to the sum of the present quantity of parent atoms and daughter atoms. This raises the question of whether or not some of the lead may not have already been in the mineral and, if not detected, cause its radiometric age to exceed its true age. Lead 204, which is never produced by decay, provides a means of detecting original lead. All common lead contains a mixture of four lead isotopes. In most minerals used for dating, the proportions of the lead isotopes are nearly constant, so that lead 204 can be used to calculate the quantities of original lead 206 and lead 207. These quantities can then be subtracted from the total to give the amount due to radioactivity.

As we have seen, different isotopes decaly at different rates. Geochronologists take advantage of this fact by simultaneously analysing two or three isotope pairs as a means to crosscheck ages and detect errors. For example, if the $^{235}U/^{207}$ Pb radiometric ages and the $^{238}U/^{206}$ Pb ages agree, then they are said to be *concordant;* there then exists a high probability that the radiometric age is valid. (Uranium-lead ages that vary widely are said to be *discordant*. However, ever the best of concordant dates do not agree perfectly but are expected to vary within reasonable limits. (Inavoidable losses or gains of isotopes by interactions with surrounding solutions or from the beat accompanying geologic processes are the usual causes for variance. Radiometric ages should be considered reasonable approximations of true age.

Radiometric ages that depend upon uranium/lead ratios may also be checked against ages derived from lead 207 to lead 206. Because the half-life of uranium 235 is much less than the half-life of uranium 238, the ratio of lead 207 (produced by the decay of uranium 235) to lead 206 will change regularly with age and can be used as a radioactive timekeeper.

The Potassium-Argon Method

Potassium and argon are another radioactive pair widely used for dating rocks. By means of *electron capture* (causing a proton to be transformed into a neutron), about 11 per cent of the potassium 40 in a mineral decays to argon 40, which may then be retained within the parent mineral. The remaining potassium 40 decays to calcium 40 (by emission of a beta particle from a neutron, thereby transforming it into a proton). The decay of potassium 40 to calcium 40 is not for obtaining radiometric ages, because radiogenic calcium cannot be distinguished from original calcium in a rock. Thus, geochronologists concentrate their efforts on the 11 per cent of potassium 40 atoms that decay to argon. One advantage of using argon is that it is inert-that is, it does not combine chemically with other elements. Argon 40 found in a mineral is very likely to have originated there following the decay of adjacent potassium 40 atoms in the mineral.

Another advantage to the potassium scheme for dating rocks is that potassium 40 is an abundant constituent of many common minerals, including micas, feldspars, and hornblendes. However, like all radiometric methods *potassium-argon dating* is not without its limitations. A sample will yield a valid age only if none of the argon has leaked out of the mineral being analyzed. Leakage may indeed occur if the rock has experienced temperatures above about 125°C. In specific localities, the ages of rock dated by this method reflect the last episode of heating rather than the time of origin of the rock itself. A less serious problem is mechanical entrapment of atmospheric argon in flowing lavas.

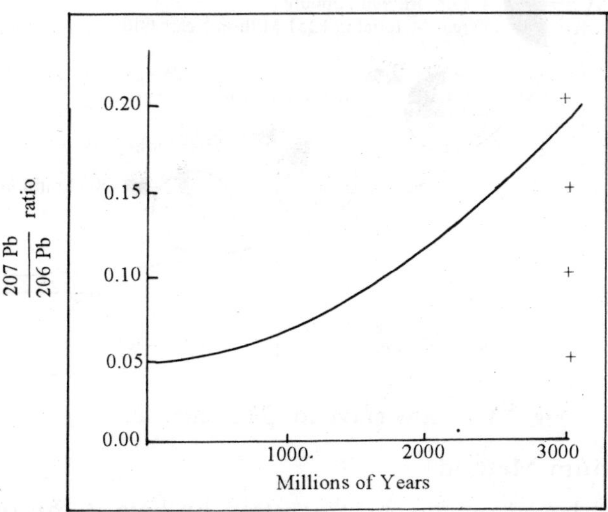

Fig. 2.16: Graph showing how the ratio of lead 207 to lead 206 can be used as a measure of age.

The half-life of potassium 40 is 1251 million years. If the ratio of potassium 40 to origin 40 is found to be 1 to 1, then the age of the sample is 1251 million years. If the ratio is 3 to 1, then yet another half-life has elapsed, and the rock would have a radiogenic age of two half-lives, or 2502 million years.

Potassium-argon is widely used in deciphering various types of geologic problems. Geologists are now using the method in studies relating to sea floor movements. For many years, scientists have been curious about the alignment of the major Hawaiian Islands and the adjacent seamounts. With the advent of the theory of sea floor spreading, scientists developed the concept that these volcanic islands were built over a relatively fixed "hot spot" deep in the upper mantle. Conduits from the "hot spot" brought lavas up to the sea floor, where eruptions periodically occurred. Volcanoes that developed over the "hot spot" were then conveyed along by sea floor movement, and new volcanoes were produced over the vacated position. Geologists reasoned that if this process has taken place in the Hawaiian Islands, then potassium-argon radiometric ages should change in sequence along the island-seamounts chain. The dates do indeed support the theory, and they even suggest that the direction of movement changed from a more northerly trend to northwesterly trend about 40 million years ago.

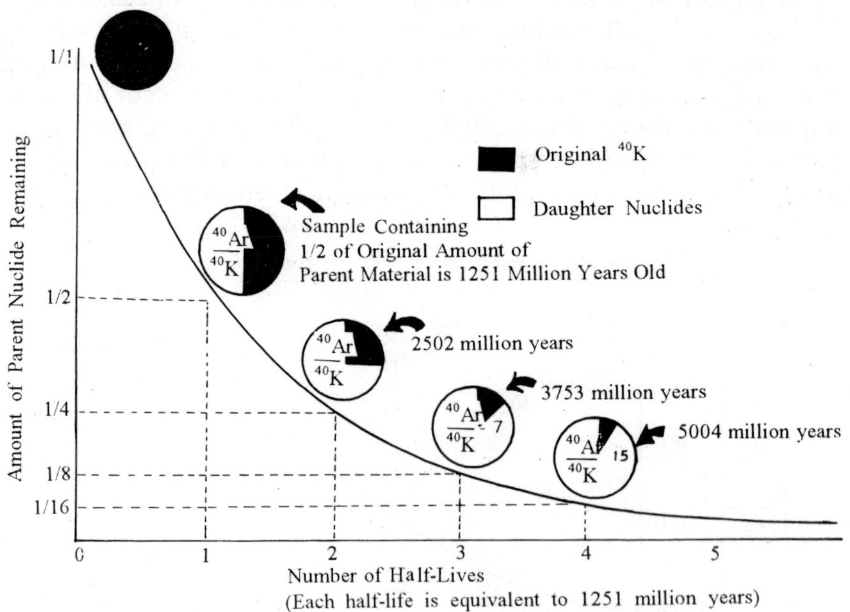

Fig. 2.17: Decay curve for potassium 40.

The Rubidium- Strontium Method

The dating method based on the disintegration by beta decay of rubidium 87 to strontium 87 can sometimes be used as a check on potassium-argon dates, because rubidium and potassium are often found in the same minerals. The rubidiumstrontium scheme has a further advantage in that the strontium daughter nuclide is not diffused by relatively mild heating events, as is the case with argon.

In the rubidium-strontium method, a number of samples are collected from the rock body to be dated. With the aid of the mass spectrometer, the amounts of radioactive rubidium 87, its daughter product strontium 87, and strontium 86 are calculated for each sample. Strontium 86 is an isotope not derived from radioactive decay. A graph is then prepared in which the $^{87}Rb\backslash^{86}$ Sr ratio in each sample is plotted against the $^{87}Rb\backslash^{86}$ Sr ratio. From the points on the graph, a straight line is constructed that is termed an isochron. The slope of the isochron results from the fact that, with the passage of time, there is continuous decay of rubidium 87, which causes the rubidium 87/strontium 86 ratio to decrease. Conversely, the strontium 87/strontium 86 ratio increases as strontium 87 is produced by the decay of rubidium 87. The older the rocks being investigated, the more the original isotope istope ratios will have been changed, and the greater will be the inclination of the isochron. The slope of the isochron permits a computation of the age of the rock.

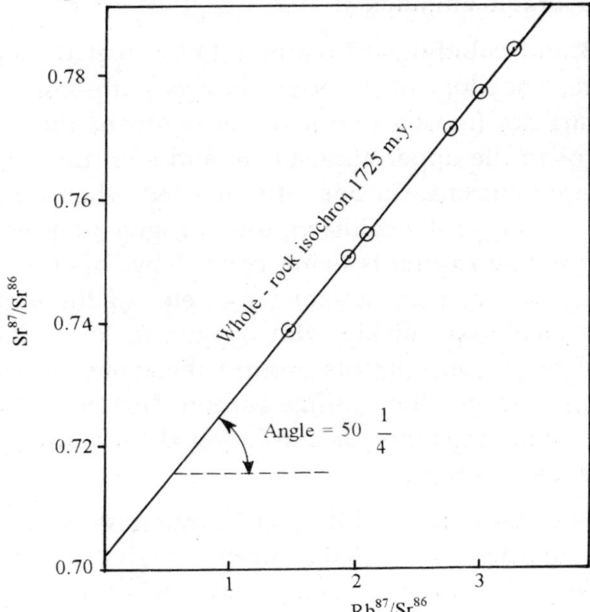

Fig. 2.18: Whole-rock rubidium-strontium isochron for a set of samples of a Precambrian granite body exposed near Sudbury, Ontario.

The rubidium-strontium and potassium-argon methods need not always depend upon the collection of discrete mineral grains containing the required isotopes. Sometimes the rock under investigation is so finely crystalline and the critical minerals so tiny and dispersed that it is difficult or impossible to obtain a suitable collection of minerals. In such instances, large samples of the entire rock may be used for age determination. This method is called *whole-rock analysis*. It is useful not only for fine-grained rocks but also for rocks in which the yield of useful isotopes from mineral separates is too low for analysis. Whole-rock analysis has also been useful in determining the age of rocks that have been so severely metamorphosed that the potassium-argon or rubidium-strontium radiometric clocks of individual minerals have been reset. In such cases, the age obtained from the minerals would be that of the episode of metamorphism, not the total age of the rock itself. The required isotopes and their decay products, however, may have merely moved to nearby locations within the same rock body, and therefore analyses of large chunks of the whole rock may provide valid radiometric age determinations.

The Carbon 14 Method

Techniques for age determination based on content of radiocarbon were first devised by W.F. Libby and his associates at the University of Chicago in 1947. It has become an indispensable aid to archaeologic research and frequently is useful in deciphering the very recent events in geologic history. Because of the short half-life of carbon 14-a mere 5730 years-organic substances older than about 40,000 years no longer contain carbon 14 in measurable amounts.

Unlike uranium 238 and rubidium 87, carbon 14 is created continuously in the earth's upper atmosphere. The story of its origin begins with cosmic rays, which are extremely high-energy particles (mostly protons) that bombard the earth continuously. Such particles strike atoms in the upper atmosphere and split their nuclei into smaller particles, among which are neutrons. Carbon 14 is formed when a neutron strikes an atom of nitrogen 14. As a result of the collision, the nitrogen atom emits a proton and becomes carbon 14. Radioactive carbon is being created by this process at the rate of about two atoms per second for every square centimeter of the earth's surface. The newly created carbon 14 combines quickly with oxygen to form CO_2, which is then distributed by wind and the water currents around the globe. It soon finds its way into photosybthetic plants, because they utilize carbon dioxide from the atmosphere to build tissues. Plants containing carbon 14 are ingested by animals, and the isotope becomes a part of their tissues as well.

Eventually, carbon 14 decays back to nitrogen 14 by emission of a beta particle. A plant removing CO_2 from the atmosphere should receive a share of carbon 14 proportional to that in the atmosphere. A state of equilibrium is reached in which the gain in newly produced carbon 14 is balanced by the decay loss. The rate of production of carbon 14 has varied somewhat over the past several thousand years. As a result, corrections in age calculations must be made. Such corrections are derived from analyses of standards such as wood samples, whose exact age is Known.

The age of some ancient bit of organic material is not determined from the ratio of parent to daughter nuclide, as is done with previously discussed dating schemes. Rather, the age is estimated from the ratio of carbon 14 to all other carbon in the sample. After an animal or plant dies, there can be no further replacement of carbon

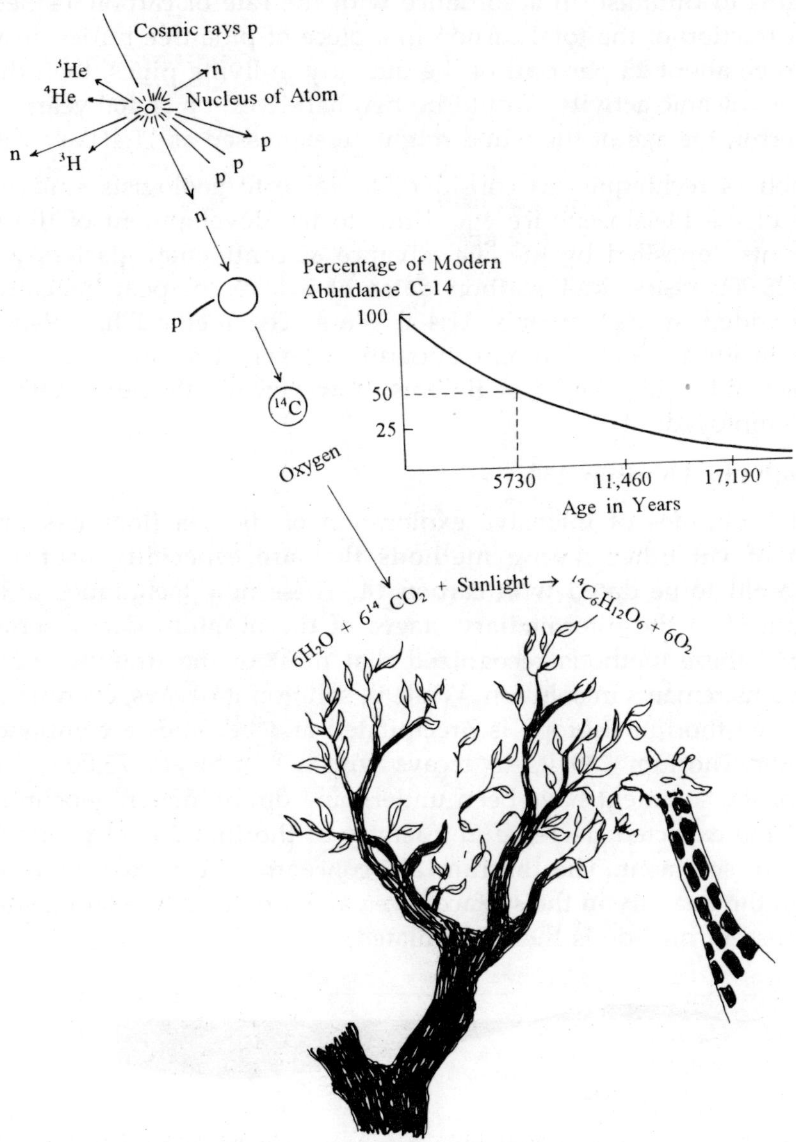

Fig. 2.19: Carbon 14 is formed from nitrogen in the atmosphere. It combines with oxygen to form radioactive carbon dioxide and is then incorporated into all living things.

from atmospheric CO_2, and the amount of carbon 14 already present in the once living organism begins to diminish in accordance with the rate of carbon 14 decay. Thus, if the carbon 14 fraction of the total carbon in a piece of pine tree buried in volcanic ash were found to be about 25 per cent of the quantity in living pines, then the age of the wood (and the volcanic activity) would be two half-lives, or 11,460 years. To allow for unavoidable error, the age of the wood might be expressed as 11,460 +_ 250 years.

The carbon 14 technique has considerable value to geologists studying the most recent events of the Pleistocene ice age. Prior to the development of the method, the age of sediments deposited by the last advance of continental glaciers was surmised to be about 25,000 years. Radiocarbon dates of a layer of peat beneath the glacial sediments provided an age of only 11,400 years. The method has also been found useful in dating the geological recent uppermost layer of sediment on the sea floors. For the deeper and older marine sediments, however, the thorium method described further on is employed.

Methods Involving Thorium 230

The past 2 decades of intensive exploration of the sea floor has prompted the development of yet other dating methods that are especially useful for oceanic sediments too old to be dated with carbon 14. These new techniques utilize isotopes that are produced in the intermediary stages of the uranium decay series. Scientists who developed these methods recognized that most of the uranium brought to the oceans by streams remains in solution. While in solution it decays, eventually producing thorium 230. The thorium isotope is precipitated and becomes a component of ocean floor sediments. Thorium 230 itself decays with a half-life of 75,000 years. Because lower levels of the sediment have been undergoing decay longer, geochronologists are able to detect the expected decrease in quantity of thorium 230 at greater depths in a cored column of sediment. The thorium 230 concentration of each measured interval is compared to the quantity in the surface layer, and a time scale relating age to quantity of remaining parent nuclide is then formulated.

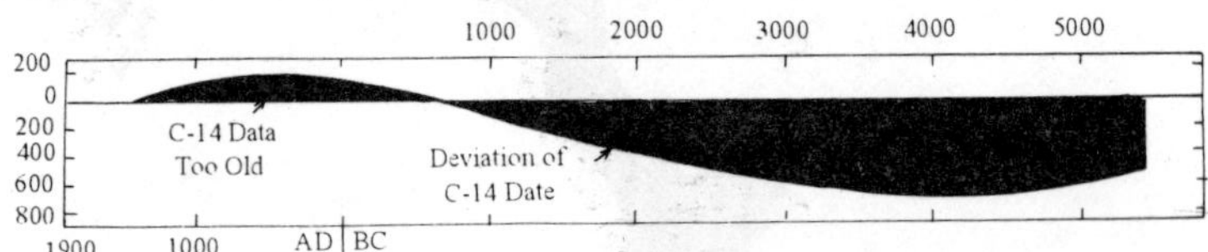

Fig. 2.20: Deviation of carbon 14 ages from true ages from the present back to about 5000BC. Data are obtained from analysis of bristle cone pines from the western U.S. Calculation of ^{14}C deviations are based on half-life of 5,730 years.

Deep sea sediments can also be dated by means of a procedure based upon thorium and protactinium. Thorium 230 has a half-life of 75,000 years and is in the decay series from uranium 238. Protactinium 231 has a half-life of only 34,000 years and is in the line of descent from uranium 235. Both parent nuclide are precipitated in the same proportions, but because of their different rates of decay, the ratio of the two changes regularly with time. Thus, the greater the differences in the quantity of undecayed parent isotopes, the older the sediment.

Nuclear Fission Track Timekeepers

Nuclear particle fission tracks were discovered about 20 years ago when scientists using the electron microscope were able to examine the areas around presumed locations of radioactive particles that were embedded in mica. Closer examination showed that the tracks were really small tunnels-like bullet holes—that were produced when high-energy particles of the nucleus of uranium were fired off in the course of *spontaneous fission* (spontaneous fragmentation of an atom into two or more lighter atoms and nuclear particles). The particles speed through the orderly rows of atoms in the crystal, tearing away electrons from atoms located along the path of trajectory and rendering them positively charged. Their mutual repulsion produces the track. The tracks are only a few atoms in width and are impossible to see without an electron microscope. Therefore, the sample is immersed for a short period of time in a suitable solution (acid or alkaki), which rushes up into the tubes, enlarging the track tunnel so that it can be seen with an ordinary microscope.

The natural rate of track production by uranium atoms is very slow and occurs at a constant rate. For this reason, the tracks can be used to determine the number of years that have elapsed since the uranium-bearing mineral solidified. One first determines the number of uranium atoms that have already disintegrated. This number is obtained with the aid of the microscope by counting the etched tracks. Next, one must find the original number of uranium atoms. This quantity can be determined by bombarding the sample with neutrons in a reactor and thereby causing the remaining uranium to undergo fission. A second count of tracks reveals the original quantity of uranium. Finally, one must know the spontaneous fission decay rate for uranium 238. This information is determined by counting the tracks in a piece of uranium-bearing synthetic glass of known date of manufacture.

Fission track dating is of particular interest to geochronologists because it can potentially be used to date specimens only a few centuries old as well as to date rocks billions of years in age. The method helps to date the period between about 40,000 and 1 million years ago-a period for which neither carbon 14 nor potassium-argon methods are suitable. As with all radiometric techniques, however, there can be problems. If rocks have been subjected to high temperatures, tracks may heal and fade away.

THE AGE OF THE EARTH

Anyone interested in the total age of the earth must decide what event constitutes its "birth". Most geologists assume "year 1" commenced as soon as the earth had collected most of its present mass and had developed a solid crust. (Unfortunately, rocks that date from those earliest years have not been found on earth. They have long since been altered and converted to other rocks by various geologic processes.

The oldest rocks on earth that have been dated thus for include 3.4-billion-year-old granites from the Barberton Mountain Land of South Africa, 3.7-billion-years old granites of southwestern Greenland, and metamorphic rocks of about the same age from Minnesota. These dates permit us to say the planet is at least about 3.7 billion years old.

Meteroites, which many consider to be remants of a disrupted planet that originally formed at about the same time as the earth, have provided uranium-lead and rubidium-strontium ages of about 4.6 billion years. From such data, and from estimates of how long it would take to produce the quantities of various lead isotopes now found on the earth, geochronologists feel that the 4.6-billion-year age for the earth can be accepted with confidence. Substantiating evidence for this conclusion comes from returned moon rocks. The ages of these rocks range from 3.3 to about 4.6 billion years. The older age determinations are derived from rocks collected on the lunar highland, which may represent the original lunar crust. Certainly, the moons and planets of our solar system originated as a result of the same cosmic processes and at about the same time.

Chapter—3

Sedimentary Rocks

SEDIMENTARY ROCKS AND THEIR FORMS OF OCCURRENCE

Rocks are geological bodies composed of mineral aggregates in a certain composition and formed by various geological processes within the crust or on its surface. They may consist of one mineral (monomineralic rocks). Minerals that constitute more than 5 per cent of a rock are called rock-forming, while those present as a slight admixture(less than 5 per cent) are called accessory.

All rocks are divided into three main types according to their origin (genesis): viz., igneous, sedimentary, and metamorphic.

Igneous rocks are formed during the cooling and hardening of magma melts within the Earth or on the surface.

Sedimentary rocks are formed through the destruction of previously formed rocks on the Earth's surface and subsequent accumulation and transformation of the products of that destruction. Atmospheric agents, the hydrosphere, and the organic world are involved in their formation.

Metamorphic rocks are formed from igneous and sedimentary rocks that have been affected by high temperatures and pressures, and chemically active substances within the Earth.

Igneous rocks constitute 65 per cent of the total mass of the rocks composing the crust. Sedimentary and metamorphic rocks constitute the remaining 5 per cent.

Rocks als differ from one another in texture, fabric, and forms of occurrence in the crust, as well as in mineral composition and origin.

The *texture* of rocks is governed by the size, shape, and character of accretion of the mineral grains composing them.

The *fabric* or rocks is governed by the mutural spatial arrangement of the mineral grains composing it and the character of the in-filling of the rock's volume.

FROM SEDIMENT TO SEDIMENTARY ROCK

The most significant factors involved in the origin of sedimentary rocks are *weathering,* which produces sediment, *transportation* and *deposition* of that sediment, and the *liquification* necessary to convert loose particles of sediment into solid rock. Each of these factors may alter the composition or textural features of sediment and thereby provide a varity of different kinds of sedimentary rocks.

SEDIMENT TRANSPORT AND DEPOSITION

Once, weathering products have been formed from preexisting rocks, the next stage in the sequence of events leading to sedimentary rock is the removal and transport of those products. Many denudational agencies, including running water, moving ice, and wind, assist in this removal. The wind is an effective agent in picking up and blowing away the smaller and lighter particles. Glacial ice can move very large pieces of rock and carry an immense load of coarse sediment. Streams are also exceptionally effective in carrying not only solid particles of sediment but invisible dissolved salts as well. Ultimately, sediment-laden streams flow into lakes or the sea and their load of sediment is deposited. It may form sandy beaches, silty floodplains, and sometimes muddy boggy areas of estuaries and deltas.

The solid particles carried by wind or water will be deposited whenever there is insufficient energy to carry them further. For example, if the velocity of dust-laden wind abates, there will be insufficient energy to carry particles of given size, and those particles will be dropped. Similarly, if a stream's velocity is checked, as when entering a standing body of water, the stream also loses energy and is unable to carry the material formerly carried at the higher velocity.

A reduction in a stream's velocity does not, of course, effect the dissolved materials as it does suspended solid particles. Material carried in solution is deposited by a process called *precipitation,* in which dissolved material is changed to a solid and separated from the liquid in which it was formerly dissolved. For example, calcium carbonate, the principal substance in the widespread sedimentary rock known as limestone, may be precipitated from water that contains calcium in solution as indicated below:

Ca^{2+}	+	$2HCO_3$	$\rightleftharpoons$	$CaCO_3$	+	H_2O	+	CO_2
(Dissolved Calcium ions)		(Dissolved bicarbonate ions)		Calcium carbonate		Water		Carbon dioxide

The bicarbonic ions that participate in the above reaction can be derived from ionization of carbonic acid. As indicated by the arrows, the reaction is reversible. If carbon dioxide is removed from sea water, the reaction will proceed toward the right, and calcium carbonate will be precipitated. If, however, carbon dioxide is added to sea water, then the amount of carbonic acid in the water would build, can the reaction would proceed to the left. This would result in a chemical environment not conducive to calcium carbonate precipitation, and one in which existing calcium carbonate might begin to dissolve. As is evident here, the precipitation of calcium carbonate is a complex and delicate process in nature. It is influenced by organisms that utilize of liberate carbon dioxide, by processes that alter the acidity or alkalinity of the water, by the presence of organic compounts, and by ions of sulfure, phosphorus, and magnesium that maybe present.

Lithification

Many changes take place in sediment after it has been deposited. Mineral grains may be dissolved away, some may grow by additions of new mineral matter, and the shapes of particles may be distorted by compaction. The result of some of these changes is to convert sediment into sedimentary rock. The conversion process is called *lithification*. *Cementation, compaction*, and *crystallization* are the principal means by which unconsolidated sediment is lithified.

Cementation involves the precipation of minerals in the pore spaces between larger particles of sediment. The precipitated mineral, which most frequently is either calcium

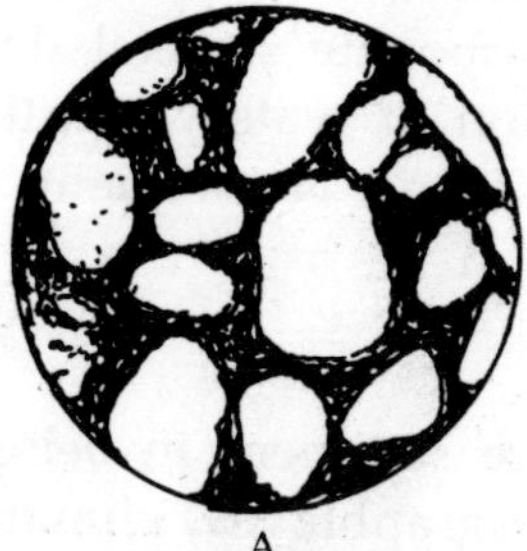

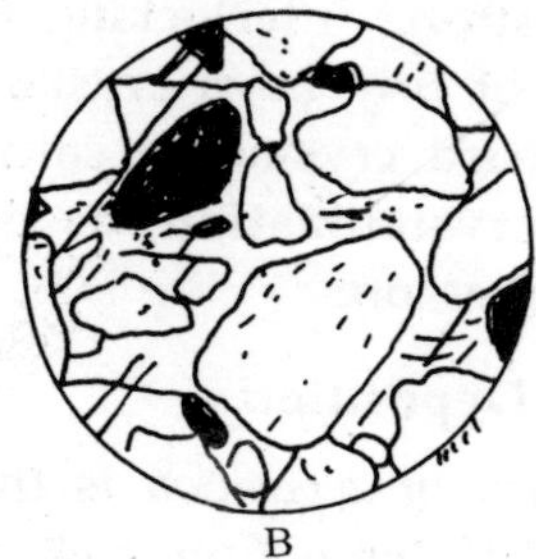

Fig. 3.1: Two common types of cement in sedistones. (A) Quartz sandstone composed of well-sorted rounded quartz grains tightly cemented by quartz (SiO) outgrowths. (B) Sandstone composed of quartz, feldspar, and rock fragments cemented by course sparry calcite. Both are drawing of thin sections as viewed under the microscipe. Diameter of each area is 1.0 mm.

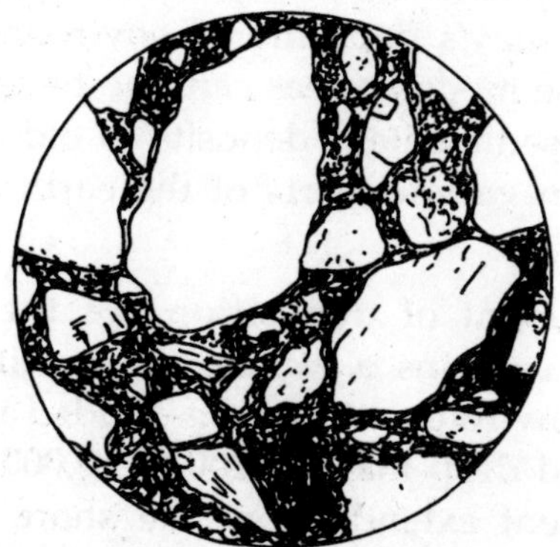

Fig. 3.2: Thin section of sandstone composed of poorly sorted angular grains of quartz (clear), feldspars (thin stripes), rock fragments. This sandstone lacks cement. Spaces between grains are filed with a matrix of clay and slit.

carbonate ($CaCO_3$), silica (SiO2), or ferric iron oxide (Fe_2O_3), called the *cement*. Cement is added to sediment *after* deposition. It differs from a rock's *matrix,* which consists of clastic particles (often clay) that are deposited at the same as the larger grains and help to the grains together.

The reduction in pore spaces in a rock as a result of the pressure of overlying rocks or pressures of earth movements is termed *compaction*. During compaction, individual grains are pressed tightly against one another causing the expulsion of water and rearrangement of particles. The result is often the conversion of loose sediment into hard rock. Compaction is greatest and most important as a lithification process in finer-grained sediments like clay and mud.

Lithification by *crystallization* may begin with an initial chemical precipitate in which the developing crystals grow together to form a crystalline solid. The process of crystallization, however, may also result in a change in the form of grains that have already been deposited. For example, quartz may be precipitated onto rounded quartz grains to form a strong interlocking mosaic of crystals, clay may be converted to a matted aggregate of tiny mica crystals, or calcium carbonate skeletal debris may be reorganized into hard crystalline calcite. The migration of watery solutions through sediment favors crystallization, as does deep burial and consequent increases in temperature and pressure.

Environments of Deposition

The *environment of deposition* is the location where sediment in being deposited. Each environment of deposition is characterized by geographic and climatic conditions that modify or determine the properties of sediment that is deposited within it. Some sediments, such as chemical precipitates in water bodies, are solely the products of their environment of deposition. Their component minerals originated and were deposited at the same place. Other sedimentary accumulations come from a distant source and have experienced particular kinds of transport. Geologists are keenly interested in the sediment of today's depositional environments because the features they find in the deposits of these modern areas can also be seen in ancient sedimentary rocks. Comparing present day sedimentary deposits to old sedimentary rocks permits one to reconstruct conditions in various parts of the earth as they were hundreds of millions of years ago.

The three major environment of deposition are the *marine, continental,* and *transitional environments*. Each contains a number of specific subenvironments. In a very general way, the marine environment may be divided into shallow marine (water depth less than 200 meters) and deep marine (200 to 10,000 meters). As illustrated in, the shallow marine environment extends from the shore to the outer edge of the continental shelves. The kind of sediment deposited in this environment depends on many factors, including climate, parent bedrock, distance from shore, elevation of the adjacent land area, water depth, and the presence or absence of carbonate secreting organisms. Because of wave turbulence and currents, sediment deposited in the shallow marine environment tends to be coarser than material laid down in the deeper realms of the ocean. Sands silts, and clays are common. Where there are few continent-derived sediments and the seas are relatively warm, lime muds of biochemical origin may be the predominant sediment. Coral reefs are also characteristic of warm shallow seas.

In the deep marine environment far from the continents, only very fine clay, volcanic ash, and the calcareous or siliceous remains of microscopic organisms settle to the ocean floor. The exceptions are sporadic occurrence of coarser sediments that are carried down continental slopes into the deeper realms of the ocean by dense masses of mud and silt laden water called *turbidity currents.*

The shoreline of a continent is the transitional zone between marine and nonmarine environments. Here one finds the familiar shoreline accumulations of sand or gravel we call beaches. Mud-covered *tidal flats* that are alternately inundated and drained of water by tides are also found here, as are deltas. *Deltals* form when streams enter bodies of standing water, experience an abrupt loss of velocity, and drop their load of sediment. Provided shoreline currents and waves do not remove the deposits faster than they are supplied, the delta will grow seaward. In general, progressively finder sediment will be deposited in progressively finer sediment will be deposited in progressively deeper water as the current provided by the stream diminishes. At the same time the stream channel is extended over former deposits, periodically chokes in its own debris, and breaks out to form new branches of the delta.

In addition to deltas, the transitional zone in cludes such features as *barrier islands,* which are built parallel to shorelines by wave and current action, and *lagoons,* which are often found between the mainland and a barrier island. *Swamps* are also frequent features of low-lying areas adjacent to the sea.

Continental environments of deposition include river *floodplains,* alluvial fans, lakes, glacial, and eolian environments. The silt, sand, and clay found along the banks, bars, and floodplains of streams are familiar to most of us. In general, stream deposits develop as elongate lenses that are oriented downstream and that grade abruptly from sediment of one particle size to another. In another setting, stream transported materials may accumulate quickly when a rapidly flowing river emerges from a mountainous area onto a flat plain. The result of the abrupt depostion is an *alluvial fan.* Somewhat quieter deposition occurs in lakes, which are ideal traps for sediment. Silt and clay are common lake sediments, although a variety of sediment is possible, depending on water depth, climate, and the character of the surrounding land areas. The *playa lakes* of arid regions are shallow temporary lakes. Periodically they become dry as a result of evaporation. Perhaps the most dramatic environments of deposition are found in areas having active glaciers. Glaciers have the ability to transport and deposit huge volumes and large fragments of rock detritus. Deposits are characteristically unsorted mixtures of boulders, gravel, sand, and clay. Where such materials have been reworked by glacial meltwater, however, they become less chaotic and resemble stream deposits.

Whereas glaciers can move materials of great size, wind is much more selective in the particle size it can transport. Environments where wind is an important agent of sediment transport and deposition are called *eolian environments.* They are characterized by an abundance of sand and silt, little plant cover, and strong winds. These characteristics typify some desert regions.

The Influence of Tectonics on Sedimentation

One final factor, called tectonics, also has an effect on the characteristics of sedimentary rock formed at a particular location. *Tectonics* refers to the crustal behaviour of a part of the earth over a long period of time. For example, a region may be tectonically stable, subsiding, rising gently, or more actively rising to produce mountains and plateaus. Where a source area has recently been compressed and uplifted, an abundance of coarse clastics derived from the rugged upland source area will be supplied to the basin. In the geologic past, such a tectonic setting has resulted in the accumulation of great "clastic wedges" of sediment that thickened and became coarser toward the former mountainous source area. In other tectonic settings of the past, the source area has been stable and topographically more subdued, so that finer particles and dissolved solids became the most abundant components being carried by streams.

The tectonic setting influences not only the size of clastic particles being carried to sites of deposition but also the thickness of the accumulating deposit. For example, if a former marine basin of deposition had been provided with an ample supply of sediment and was experiencing tectonic subsidence, enormous thicknesses of sediments might accumulate. Over a century ago, James Hall, an early New York geologist, recognized that the thick accumulations of shallow-water sedimentary rocks in the Appalachian region required that crustal subsidence accompany deposition. His reasoning was quite straightforward. It was easy to visualize filling a basin that was 12,000 meters deep with 12,000 meters of sediment. However, if the basin was only a few hundred meters deep, the only way to get tens of thousands of meters of sediment into it would be to have subsidence occurring simultaneously with sedimentation.

In a marine basin of deposition that is stable or subsiding very slowly, the surface on which sedimentation is occurring is likely to remain within the zone of wave activity for a long time. Wave action and currents will wear, sort, and distribute the sediment into broad blanket-like layers. If the supply of sediment is small, this type sedimentation will continue indefinitely. Should the supply of sediment become too great for currents and waves to transport, however, the surface of sedimentation would rise above sea level, and deltas would form.

It is possible to view the tectonic frame work of entire continents as well as of particular areas. The principal tectonic elements of a continent are *cratons* and *mobile belts.* The craton consists of the central stable region of continent. Such regions have a generally subdued topography and are composed of ancient igneous and metamorphic rocks that may be either exposed or covered by flat-lying younger sedimentary strata. In contrast to the relative stability of the craton, mobile belts are curving elongate crustal zones of great instability. They are thought to be zones of plate convergence, and are recognized by their high frequency of earthquakes and volcanic eruptions.

COLOR IN SEDIMENTARY ROCKS

We have seen that color in igneous rocks can be used to indicate the approximate amount of ferromagnesian minerals. Color in sedimentary rock can also provide useful clues to identification. For example, carieties of chert can be identified as flint if they are gray of black, or as jasper if they are red. Color is also useful in providing clues to the environment of deposition of sedimentary rocks. Of the sedimentary coloring agents, carbon and the oxides and hydroxides of iron are clearly the most important.

Black Coloration

Black and dark gray coloration in sedimentary rocks-especially shales-usually results from the presence of organic carbon compounds and iron sulfides. The occurrence of an amount of organic carbon sufficient to result in black colouration implies compiles and abundance of organisms in or near the depositional areas as well as environmental circumstances that kept the remains of those organisms from being completely destroyed by oxidation or bacterial action. These circumstances are present in many marine, lake and esturine environments today. In a typical situation, the remains of organisms that lived in or near the depositional basin settle to the bottom and accumulate. In the quiet bottom environment, dissolved oxygen needed by aerobic bacteria to attack and break down organic matter may be lacking. There may also the insufficient oxygen for scavenging bottom-dwellers that might feed on the debris. Thus, organic decay is limited to the slow and incomplete activity of anaerobic bacteria; consequently, in completely decomposed material rich in black carbon tends to accumulate. In such an environment, iron combines with sulfur to form finely divided iron sulfide (FeS_2), which further contributes to the blackish coloration. Such environment of deposition are likely to yield toxic solutions of hydrogen sulfide (H_2S). The lethal solution rise to poison other organisms and thus contribute to the process of accumulation. Black sediments do not always from in restricted basins. They may develop in relatively open areas, provided the rate of accumulation of organic matter exceeds the ability of the environment to cause its decompostion.

Red Coloration

Hues of brown, red, and green are frequently formed in sedimentary rocks as a result of their iron oxide content. Few, if any, sedimentary rocks are free of iron, and less than 0.1 per cent of this metal can color or sediment a deep red. The iron pigments not only are ubiquitous in sediments but also are difficult to remove in most natural solutions.

Iron form two sorts if ions: *ferrous* iron has two positive charges, whereas ferric ion has three positive charges. Thus, iron may form two oxides: FeO (Ferous oxide) and Fe_2O_3 (Ferric oxide). In air, ferrous iron is slowly oxidized to ferric iron. When oxygen is in short supply, ferric iron may be similarly reduced to ferrous iron. Ferric minerals like *hematite* tend to color the rock red, brown, or purple, whereas the ferrous compounds impart hues of gray and green. Hydrous ferric oxide (*Limonite*) is often yellowish in color.

Red Beds

Strata colored in shades of red, brown, or purple by ferric iron are designated red beds by geologists. Oxidizing conditions required for the development of ferric compounds are more typical of nonmarine than marine environments; most red beds are flood plain, alluvial fan, or deltaic deposits. Some, however, are originally reddish sediment carried into the open sea. Electron microscope studies (Walter, 1967), of red beds forming today in Baja, California, indicate that red coloration developed long after the sediment was deposited. After burial, the decary of clastic ferromagnesian minerals released iron that was oxidized by the oxygen in underground water circulating through the pore spaces. Thus, red coloration may be imparted in the subsurface and may be dependent on climate. The interpretations one can draw from red beds should be based to a large degree on the associated rocks and sedimentary structures. Red beds intersperesed with evaporite layers indicate warm and arid conditions.

Although red beds are more likely to represent nonmarine than marine deposition, occasionally the reddish strata are interbedded with fossiliferous marine limestones. In such cases, the color may be inherited from red soils of nearby continental areas. Lands located in warm, humid climates often develop such reddish soils. When the soil particles arrive at the marine depositional site, they will retain their red coloration if there is insufficient organic matter present to reduce the ferric iron to the ferrous state. Otherwise, they will be converted to the gray or green coloration of ferrous compounds.

In summary, sedimentary rocks of red coloration may be a product of the source materials, may have developed after burial as a result of a lengthy period of subsurface alteration, or may be the result of subaerial oxidation. Geologists are suspicious of the last possibility, because most modern desert sediments are not red unless composed of materials from nearby outcroppings of older red beds.

PRINCIPAL TYPES OF SEDIMENTARY ROCK

Detrital sedimentary rocks are formed through the accumulation of products of the mechanical destruction of previously existing rocks. They consist of fragments of various rocks and minerals. Depending on the size of the fragments the following types of rock are distinguished: coarsely fragmented medium fragmented, finely fragmented, and very finely fragmented.

Coarsely fragmented (rudaceous) rocks consist of distinctly visible grains of rocks and minerals, porous or cemented, with particles above one millimeter in size. According to the character and size of the particles and the degree of cementation, friable rocks composed of rounded particles (boulders cobbles, and pebbles) and angular ones (lumps, chippings, rotten stone) are distinguished. Cemented accumulations of angular fragments are known as breccia, and of rounded ones as conglomerate.

Medium fragmented rocks (arenaceous or sandstones) consist of particles of minerals and rocks between 1.0 and 0.1 millimeter in diameter. Sands and sandstones are divided into coarse-grained (1.0-0.5 mm), medium-grained (0.5-0.25 mm.), and fine-grained (0.25-0.1 mm). Arenaceous rocks are also divided into loose (sands) and cemented (sandstones). Depending on the composition of the grains sands may be monomineralic or polymineralic, the former consisting of grains of one mineral and the latter of grains of various minerals. Depending on what mineral is predominant, quartz, micaceous, hornblende, chlorite, feldspar (arkose), glauconite, and other sands are distinguished. Sandstones are classified according to the composition of the minerals forming them, grain size, and the composition of the cement.

Quartz sands and sandstones consist of quartz; feldspars and mica are found as admixtures. The cement is various (siliceous, calcareous, ferruginous, phosphoritic, etc.), and sandstones are classed according to it as siliceous, ferruginous, calcareous, etc. (*e.g.* a coarse-grained calcareous sandstone). Quartz-glauconite sands and sandstones (greensands) consist predominantly of quartz (60-80 per cent), glauconite (40-20 per cent), and mica (as an admixture), and are a yellowish-green. Ferruginous sands and sand-stones are formed by ferruginous minerals and are brown or a rusty orange. Arkose sands and sandstones are composed of quartz and feldspars formed through the breakdown of intrusive igneous rocks (granites, etc.) and are rosy brown. Greywackes consist predominantly of fragments of sedimentary, igneous, and metamorphic rocks, and grains of various minerals, and are a greenish brown or greyish green.

Finely fragmented rocks (argillaceous rocks-mudstones or siltstones) consist of particles of 0.1 to 0.01 mm in diameter.

Finely fragmented rocks (dusts or silts) consist of particles 0.1 to 0.01 mm in diameter. Silts include a big group of rocks of continental origin (sandy loam, loamy clay, and loess). Sandy loams contain up to 25 or 30 per cent of fine sandy particles and between 10 and 20 per cent of clay particles, while loamy clay contains 20 to 25 per cent of clay and 10 to 15 per cent of sandy particles. The latter is formed through the action of atmospheric agents, rivers, and glaciers.

Loess is a pale pinkish yellow or yellowish grey uniform rock consisting of particles of quartz and lime 0.05 to 0.01 mm in diameter with an admixture of clay particles. The calcareous formations often have the form of fine irregular concretions (lime nodules or loess dolls). Carbonates form 6 or 7 per cent of the total mass of the rock, quartz 50 to 80 per cent, and alumina 4 to 20 per cent. Loess is highly porous (up to 55 per cent). It swells rapidly in water and becomes sodden.

Silts cemented by calcareous, siliceous, and other cements are known as shales.

Very finely fragmented (argillaceous) rocks consist of particles finer, than 0.01 mm in diameter. They include clays and mudstones.

Clays consist of particles formed either through the chemical disintegration of bedrock or through its mechanical breakdown. Clay sediments are formed through the depositing of matter from colloidal solutions. Typical clay minerals are kaolinite and montmorillonite. Some clays contain carbonate minerals (marlaceous and calcareous clays), iron pyrites, carbonaceous substances, and bitumen (bituminous clays), gypsum (gypsiferous clays), anhydrite, halite (saliferous clays). Clays are of various colours; reddish brown, yellow, greyish green, and dark grey. Clays readily soften in water. In the dry state they are earthy and grind to a fine powder, when wet they are plastic. Plastic clays consisting of kaolinite, aluminium hydroxides, or micas are called refractory (they do not melt below a temperature of 1700°C). The following clays are distinguished in accordance with their composition: kaolinite, bentonite, and montmorillonite. Clays are also known as meagre (containing a torrigenous admixture of quartz, chalcedony, opal and ferrous oxides) and rich (without terrigenous admixtures).

Mudstones are dense, indurated clay rocks. When enriched with $CaCO_3$, clays become marls.

Biogenic sedimentary rocks are formed through accumulation of products of organisms' vital activity, mainly of the skeletal remains of marine (and less often freshwater) invertebrates. The following biogenic rocks are distinguished according to their composition: carbonate, siliceous, and caustobioliths. This carbonate rocks include limestones and chalk.

Several types of structural fabric are distinguished among biogenic rocks: (1) biomorphic, in which the rock consists of the whole, unbroken skeletal remains of organisms; (2) biomorphicdetrital, in which the biogenic material is both intact and broken skeletal remains; (3) detrital, when the rock consists only of the fragmented skeletal remains of organisms. The structure of the rock may be massive or layered.

Limestones of biogenic origin are formed from the calcareous shells and internal skeletons of various aquatic animals and plants. They consist of calcite ($CaCO_3$,) and are deposited on the bottom of lakes and seas. In appearance they are massive porous or dense rocks. They readily effervesce when reacting with hydrochloric acid and have a comparatively low hardness. Depending on the predominance of certain kinds of remains in their composition the following limestones are distinguished; foraminiferal (nummulitic, consisting of the shells of nummulites; and fusulinite, consisting of the shells of *Fusus*); coral consisting of the skeletons of colonial coral polyps); shelly (consisting almost wholly of the shells of various molluscs). Limestones containing an admixture of organic (bituminous) materials are called bituminous. They are dark grey to black in colour and smell of oil when struck or heated. Siliceous limestones are encountered in which some of the $CaCO_3$, has been replaced by SiO_2.

Chalk is a variety of organic limestone consisting mainly of the very fine remains (coccoliths) of calcareous algae (Coccolithophoreae) and is a soft, fine-grained uniform, usually white, rock that 'boils' vigorously with hydrochloric acid.

Siliceous rocks of organic origin included diatomito and tripoli.

Diatomito is a white cemented or friable rock (diatomaceous earth), consisting mainly of the very fine frustules of diatomic algae (SiO_2nH_2O). It is white in colour. It is very light owing to its microporosity.

Tripoli consists of the very fine opal frustales of diatoms. It is light grey, almost white, in colour, but there are grey and dark grey varieties. It is found both as friable rock and a compact porous mass. Typical features are its capacity to absorb moisture (avidly sticking to the tongue') and its low density.

Chemical sedimentary rocks are formed through the precipitation of matter from true and colloidal solutions. They are precipitated mainly on the bottom of water basins, but are also deposited by underground waters (stalactites, stalagmites, etc.). Chemical sedimentary rocks formed by precipitation from true solutions have a crystalline texture (coarse, medium, fine, and very fine), while those from colloidal solutions are cryptocrystalline. The rocks are mainly of a layered structure but may also be massive (uniform).

These rocks are divided into several groups: carbonates, silicates, ferruginous, halides, sulphates, allitic, and phosphates.

The *carbonate* group includes limestones (crystalline, aphanitic, oolitic), calc tufa (travertine), dolomite, siderite, and stalagmite or dripstone.

Limestones of chemical origin are formed through the precipitation of calcium carbonate ($CaCO_3$) from aqueous solutions. They have a dense, massive structure of a crystalline or cryptocrystalline texture, and effervesce vigorously with hydrochloric acid. Their colour is white, grey, and where there are admixtures of ferrous salts a pinkish brown. Oolitic limestones consist of concentric, shell-like spheres or ooliths, outwardly resembling peas, cemented together by a calcareous cement. Stalagmite or dripstone is formed through precipitation of $CaCO_3$, from percolating ground waters; its typical forms are the stalactites, stalagmites, and incrustations formed in the karst caverns of limestone massifs. They have a crystalline, sometimes 1 a coarse-grained texture.

Calc tufa (or travertine) is a highly porous, spongy or cellular rock with a fine crystalline texture that is formed where underground waters flow out onto the surface; with liberation of carbon dioxide the surplus dissolved calcium carbonate is precipitated. It often includes the shells of terrestrial organisms, or the imprint of leaves and twigs. The colour is predominantly light, while, grey, yellowish, and pink. Dripstones and calc, tufa react vigorously with hydrochloric acid.

Dolomite is a monomineralic rock consisting of the mineral from which it is named $CaMg(CO_3)_2$. It is formed either through joint precipitation of calcium and magnesium salts and subsequent formation of the mineral dolomite in the sediment or through direct, precipitation of dolomite from the water of brackish lagoons. Some dolomites are formed through metasomatic replacement of limestones by the action of underground magnesian solutions (dolomitisation). Dolomites resemble a dense limestone in appearance but differ from it in their weak reaction in powdered form with hydrochloric acid. They are usually light in colour: yellowish, pink, and brownish shades, depending on the various admixtures, and in particular on the presence of clay minerals.

Silicate rocks of chemical origin include siliceous sinter, a hard, sometimes 'weakly porous rock of various light colours, normally white and yellowish white that is formed through the precipitation of amorphous alumina (SiO_2nH_2O) from the waters of hot springs. Siliceous sinter includes the sediments of geysers known as geyserite.

Ferruginous rocks include brown iron ore, which is a mixture of iron hydroxide and clay materials. Iron hydroxides (limonite, goethite, etc.) compose 50 per cent of the rocks. They are laid down both in marine basins and in lakes and bogs. Limonite containing more than 30 or 40 per cent of iron is considered iron ore. Oolitic and pisolitic ores are distinguished; the first consist of ferruginous ooliths, the second of nodules and concretions of ferrous minerals. The rocks are a yellow-brown in colour.

Rocks of a typically Chemical originrock salt, consisting of the mineral halite (NaCl), and sylvinite, composed of sylvine (KCl) and halite-are called *haloids*.

Rock salt is a monomineralic, rock, colourless or a milky white in its pure form. Admixtures give it a yellow, brown, or other hue. The texture is medium or coarse crystalline. The rock has a salty taste, dissolves readily in water, and is very Common. Layers of rock salt sometimes alternate with anhydrite and gypsum.

Sylvinite is a less common rock of a white, grey, yellowish, blue, or yellowred colour. Its texture is medium and coarse crystalline. It has a bitter salty taste and dissolves readily in water.

The group of *sulphate* rocks includes the widespread gypsums and anhydrites. They are formed through precipitation of sulphate salts from aqueous solutions in basins with heightened mineralisation of their waters (shallow lakes and lagoons).

Anhydrite is a rock consisting of the mineral anhydrite (CaSO4). It is usually a bluish white. It forms dense finegrained accumulations.

Gypsum is a rock consisting of the mineral of the same name ($CaSO_4$, $2H_2O$). It is generally layered or massive, of a density varying between a micro-grain and coarse-grained texture. The colour is white, light grey, yellowish, pink, and brown. The massive white and pink, fine-grained varieties are called alabaster, the silvery white and pink parallelfibrous variety selenite. Gypsum is readily identified by hardness, being scratchable by the fingernail.

Allitic rocks include laterites and bauxites, which consist mainly of aluminium and iron hydroxides.

Laterite is a rock formed through the weathering of alumosilicate igneous rocks in regions with a hot, humid climate. The rock is hard and friable, often porous, frequently with an oolitic structure. The colour varies-white, grey, pink, etc. In origin laterites are similar to residual clays because they are residual products of the weathering of igneous rocks rich in alumina.

Bauxite is a rock similar to laterite. It frequently has a pisolitic or oolitic, texture but is a little more uniform. Bauxites are divided by origin into residual (laterite) and sedimentary (redeposited). In the main they are friable, earthy, or harder masses of a greyish white, yellowish brown, brownish red colour.

Phosphate rocks include such common ones as phosphorites (rock phosphate). They are formed in seas and on continents (in lakes, bogs, etc.) and are sedimentary rocks (sands, clays, etc.), heavily enriched with phosphate minerals. The colour of phosphorites is grey and dark grey with violet or brown tinges. Bedded and nodular phosphorites are distinguished. Bedded ones are dense uniform, dark rocks consisting of fine grains (0.01 to 1.0 mm) of weakly crystallised phosphate (apatite) cemented by a phosphate or phosphato-carbonate (siliceous) cement. Pebble phosphates are sandy argillaceous or argillaceous carbonate rocks containing either fine grains of phosphorite (0.1-0.4 mm) or larger phosphorite concretions (0.5 to 5 or 6 mm), composed of phosphate substances mixed with clay or carbonate material.

Sedimentary rocks of mixed origin are complex in structure and contain detrital, biogenic, and chemical material in various proportions. They include calcareous sandstones, marls, siliceous clays, etc.

Calcareous sandstones consist of carbonate material of an organic or chemical origin and detrital sandy material. There are transitional varieties between calcareous sandstones and sandy limestones.

Marl is a calcareous clay rock consisting of calcite and clay minerals. It has a fine-grained texture and a layered, loss often massive (uniform) structure. It is generally light in colour or white, but grey, but grey, brown, and yellow-brown varieties are met. It 'boils' well with hydrochloric, and, forming a dirty patch on the surface of the rock caused by the concentration of clay particles on the spot of the reạction. Marl contains from 20 to 80 per cent $CaCO_3$. With a lower content of carbonate material the rock is called calcareous or marly clay, with a higher content of argillaceous limestone. Marls with an 80 per cent content of $CaCO_3$ and not less than 3 per cent MgO are known as cement clay and are used to produce Portland cement.

Opoka or *gaize* is a porous white, grey, or dark grey rock with conchoidal fracture consisting of argillaceous material and alumina or silicon hydroxide (of organic or chemical origin). It is dense, strong, and when struck produces a characteristic ringing tone.

Caustobioliths (combustible organic rocks) are rocks that are rich in organic matter and are the product of the transformation of the remains of plant and animal organisms in the crust.

Unlike other biogenic rocks they do not consist of the mineral remains of organisms (skeletons) but are the result of transformation of their organic tissues. Caustobioliths are solids (peat, brown coal or lignite, coal, oil shales, asphalt), liquid (oil), and gaseous (natural gas). The most common ones are fossil coal, combustible gases, and oil. They are divided into rocks of the coal and bitumen series.

Rocks of the coal series consist of substances that are insoluble in organic solvents and are formed from the residues of plants (either higher plants or algae) and the simplest planktonic organisms. Those formed from higher plants belong to the humic coals, which are the most common rocks of the coal series; the second are sapropolic rocks. Such formations as peat, lignite (brown coal), coal, and anthracite are distinguished among humic coals, depending on the degree of transformation of the organic matter. Peat corresponds to the initial stage of this transformation and is the initial material for subsequent transformation into lignite, bituminous coal, and anthracite. The processes of coalification, *i.e.* of the transformation of peat into coal and then into anthracite, is controlled by the history of the geological evolution of the sector of the crust concerned and is governed by the depth of burial of the rock and the pressure and temperature at that depth.

Coalification. The formation of coal from peat can he regarded as the diagenesis and katagenesis of sediments represented by an organic, mass. In bog or swamp conditions plant remains sinking to the bottom are partially disintegrated. Their incomplete breakdown is due to a lack or complete absence of oxygen. Anaerobic bacteria play a considerable role.

Ultimately the organic residues accumulating at the bottom of the bog are converted into humus, a dark substance of colloidal nature. Humus constitutes a considerable part of peat. Peat is thus a compacted mass of dead plant residues enriched with carbon. It contains carbon (35 to 59 per cent), hydrogen (6 per cent), oxygen (39 per cent), and nitrogen (2.3 per cent). It is a brown, more or less friable, porous rock.

As layers of peat accumulate to a great thickness they are compacted further and the water squeezed out. The peat is finally consolidated beneath a series of overlying sediments. As the pressure increases water and gases continue to he squeezed out, while the peat itself is enriched with carbon. As a result brown coal is formed in which the plant structure is retained.

Brown coal or lignite contains 70 per cent of carbon. It is a dense, dark brown or black rock with an earthy fracture, and a dull lustre (hardness 1 to 1/2); density-1.1 to 1.3 g/cm^3-depends on the clay and other admixtures that govern its ash content. Occurs in the form of strata and lenses. Brown coals are typical of a number of coalfields in Europe and North America.

Subsequent changes in brown coal through the effect of high pressures and temperatures convert it into coal and anthracite. Bituminous coal contains up to 85 per cent of carbon. The rock is black, dense, with a grainy fracture and dull tarry lustre. Its hardness varies from 1/2 to 21/2, and density from 1.1 to 1.8g/cm^3. *Anthracite* is generally hard and brittle, with a peculiar lustre, burns without smoke, and has high calorific value. Deposits of high quality anthracite are found quite often (in) Donbass, South Wales, and other places).

Coal measures formed in continental lakes and swamps and containing mainly brown coal are called *limnic;* those formed in coastal conditions are called *paralic.*

Rocks of the bituminous series consist. of matter easily dissolved by organic solvents. They include oil, solid bitumen, and bituminous oil shales.

Oil is a mixture of liquid and gaseous hydrocarbons. In appearance it is a light, oily brown, dark brown, reddish brown, sometimes white liquid. Its density is 0.7 to 1.0 g/cm^3.

Asphalt is a dense, brownish black bituminous substance with a very tarry lustre and conchoidal fracture. It has a strong odour of bitumen. It melts and burns easily. Its density is 1 to 1.2 g/cm^3.

Oil (bituminous) shales are think layered argillaceous or marly rocks containing up to 60 per cent of bituminous matter. They are of a dark grey or brown colour and give off a strong smell of bitumen when burning. By composition and occurrence they can be classed as rocks of mixed origin.

The most important rock of the bituminous series is oil, which is a very valuable mineral occuring commonly in the sedimentary layer of the crust.

The origin of oil is a very complicated matter and has not yet been fullyexplained. There is no unanimity among scientists on the origin of oil, it must be noted, and there are two hypotheses—the *inorganic and the organic.*

The originator of the inorganic hypothesis, Mendeleev, considered oil to be formed from the gaseous product of magma, which included mathane (CH_4). In his view these gases, in rising, filled, racks and pores in the rocks, where they were converted into hydrocarbons of the petroleum series. Subsequently they migrated further into strata of sedimentary rocks where they filled reservoirs in the form of gas or liquid oil. At present the inorganic, hypothesis has few supporters because it does not explain a number of points connected with the composition and properties of oil, or with its geographical distribution.

Most scientists support the organic hypothesis. Modern views of the origin of oil wore advanced by Gubkin and developed by his pupils. They concluded that oil is formed from organic matter included in the oozes accumulating in depressions of the sea bed. The very small organisms (plankton) living in profusion in the surface waters sink to the bottom when they die, creating an initial sapropelic mass. With a lack of oxygen the initial organic mass is converted into hydrocarbons of the petroleum series occurring in a dispersed state. Thick series of predominantly argillaceous deposits containing a large amount of organic matter in dispersed form are called *oil-source* beds. During the subsidence of sedimentary series they come into conditions of increased pressure and temperature, with the consequence that the hydrocarbons migrate from the very fine subcapillary pores of the argillaceous' deposits to very permeable porous or fractured reservoir rocks, where liquid oil is formed.

The series in which oil or gas is formed through the migration of hydrocarbons to reservoir strata are called oil-and gas producing. For an oil-source bed to be converted into an oil-producing one two conditions are required: (1) the presence of reservoir strata; (2) the presence in that sector of the crust of descending tectonic movement in the course of which the oil-source series comes into conditions of heightened pressure and temperature.

Carbon forms the basis of all caustobioliths-oil, combustible gases, coal, peat, and oil shales. While there is no doubt about the biogrnic origin of peat and coal, and it is easily demonstrated (by the marks of leaves, spores, and pollen), elements had not been found in oil until recently that have been inherited from living matter; in connection with the rise of the new science of palaeobiochemistry, fragments with a very peculiar biogenic molecular structure have been found in the composition of oil that have undoubtedly come down from living matter. The presence of such structures is yet another factor in favour of the organic origin of oil.

THE FORMS OF OCCURRENCES OF SEDIMENTARY ROCKS

The original form of occurrence of sedimentary formations is the layer or stratum.

A *layer or stratum* is a geological body composed of a uniform sedimentary rock bounded by two more or less parallel bedding surfaces, with a uniform thickness and

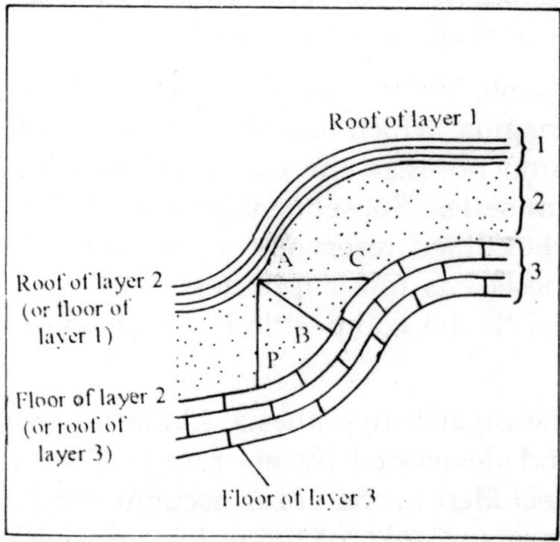

Fig. 3.3: The elements of a bed.

occupying a considerable area. The name given to the layer generally depends on the composition of the rocks forming it, *e.g.* a layer of limestones, or bed of sandstone, etc. The lower surface bounding a layer is called the floor and the upper surface the roof. In a series or suite of beds the roof of the underlying stratum is at the same time the floor of the overlying one.

Every bed or stratum is characterised by thickness. True, vertical, and horizontal thicknesses are usually distinguished. True thickness is the shortest distance between the roof and the floor. Vertical thickness is the distance along the vertical from any point in the roof to the floor. Horizontal thickness is the distance along the horizontal from any point in the roof to the floor of a bed.

The thickness of beds may be relatively constant (sustained) or inconstant (variable). With changes in thickness phenomena of bulging (a marked local increase in thickness) and pinching (marked local thinning of the bed) are observed. Gradual thinning of a bed to its complete disappearance is called petering out. Beds of marine sedimentary series are the most sustained in thickness over large areas. Continental deposits are distinguished by a loss sustained thickness; lens and pocket-like forms of bedding are typical of them.

Apart from the roof, floor, and thickness of a bed its strike and dip are distinguished.

The *strike* is the line of intersection of a bed with the horizontal plane; its position relative to a cardinal points is determined by the azimuth of its bearing.

Dip, i.e. the inclination of a bed to the horizontal plane, is characterised by its direction and the angle it makes with the plane (angle of dip).

The *angle of dip* is the angle between the plane or surface of a bed and the horizontal plane. The direction or azimuth of the dip and the angle of inclination (or dip) are determined by means of a combination compass and clinometer and are measured in degrees. The azimuth of the dip is always perpendicular to the strike of a layer. The strike, dip, and angle of dip are elements of the bedding of a layer that define its position in space.

THE SURVEYOR'S COMPASS AND DETERMINATION OF THE ELEMENTS OF THE BEDDING OF A LAYER

To determine the elements of the bedding of a layer (azimuth of the strike, azimuth of the dip, and angle if dip) a combination compass and clinometer is employed. It differs rather from an ordinary compass in the following ways.

1. The compass is usually fixed to a rectangular plate (castiron or plastic in such a way that the 0°-180° diameter, *i.e.* the north-south direction is parallel to its length.

2. The divisions on the circle from zero to 360° go counterclockwise. The signs for east and west are thus the opposite from the normal compass. This is done so that the value of the azimuth of the strike can be calculated directly from the position of the north-pointing end of the magnetic needle.
3. A clinometer is attached to the compass needle with a half-circle divided from zero to 90°. The angle of the dip is measured by its position on the half-circle.

The elements of the bedding of a layer are determined in the following way. The direction of its strike is first determined on cleared patch on the bed. For that the long side of the compass is applied to the plane of the bed so that the clinometer registers zero. A line drawn along the long side of the plate indicates the direction of the strike. Having determined the strike, the angle of dip is then measured. To do that the compass is turned so that the clinometer gives a maximum reading. A line parallel to the long side of the compass will then indicate the direction of the dip. In all these cases, it must be noted, the line of the bed's dipis always perpendicular to that of the strike. The angle of dip is often determined immediately (when it is small). To determine the azimuth of a bed's dip, the compass is applied to the strike so that the south end is pressed against the bed and the north end is pointed down the dip. The compass is then brought to the horizontal position. When it has been so fixed the locking device is released and the needle allowed to settle; the azimuth of the dip is read from the graduated circle, (according to the direction of the black, north-seeking end of the pointer).

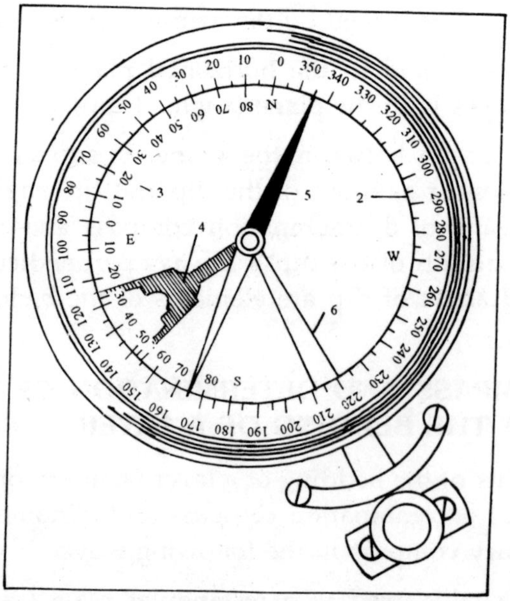

Fig. 3.4: Combination compass-clinometer; I-base plate; 2-azimuth circle; 3-semi-circle; 4-clinometer; 5-magnetic needle; 6-locking device.

Knowing the azimuth of the bed's dip there is no need to measure the azimuth of the strike; it is usually calculated from the measured azimuth of the dip by adding or subtracting 90º; for example, the azimuth of the dip is 300º NW, the strike is 210º SW, and the angle of dip 45º.

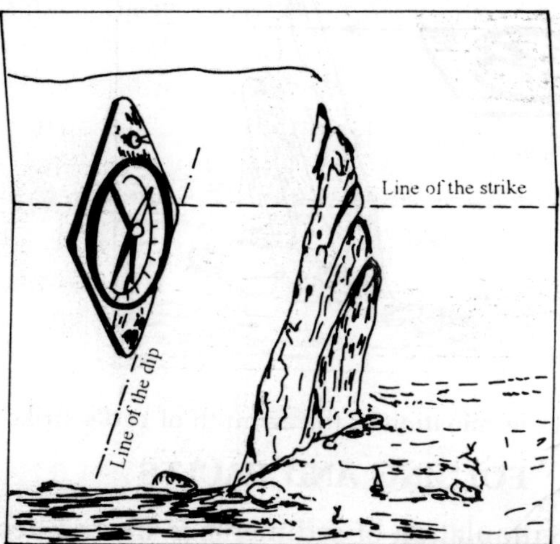

Fig. 3.5: Determination of the dip and angle of dip of a bed.

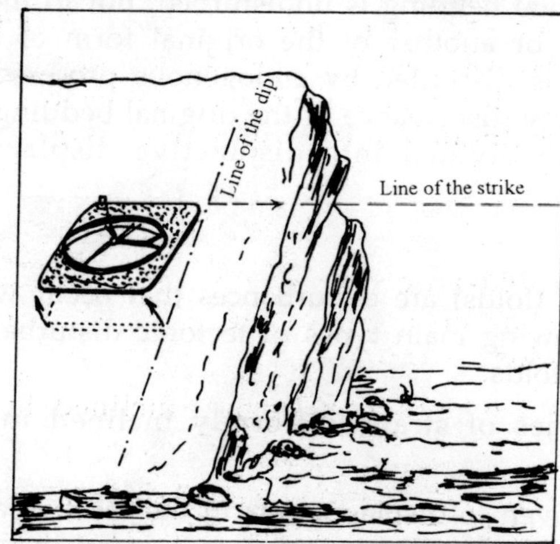

Fig. 3.6: Determination of the azimuth of bed's dip.

If it is necessary to determine the azimuth of the strike by compass, the long side of the latter is applied to the line of the strike, and the azimuth read from the graduated circle. The reverse bearing can be calculated from this reading by adding 180º to the measured azimuth.

The readings obtained for the elements of the bedding of a layer are usually written NE 63ºW48º. Normally the strike and dip are given. When a rock dips vertically, *i.e.* has an angle of dip of 90º. Only the strike is recorded.

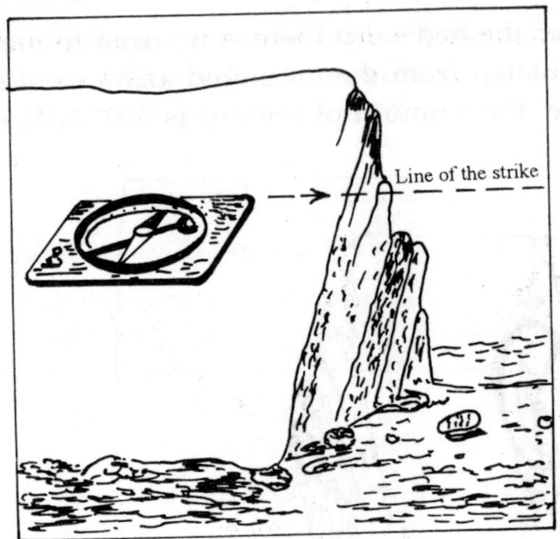

Fig. 3.7: Determination of the azimuth of bed's strike.

FOLDING AND FAULTS

The main area of the accumulation of sediments is the sea bed. There, on a surface planated by erosion, sediments are deposited in the form of nearly parallel horizontal beds. This original horizontal bedding is undisturbed, but in the course of geological development no one area or another of the original form of the bedding of rocks remains undisturbed, but is disrupted by endogenous processes, mainly by tectonic movements of the crust. Any disturbance of the original bedding of rocks is known as a displacement. They are divided into disjunctive displacements and plicated deformations.

Plicated Deformations

Plicated deformations (folds) are disturbances that occur without rupture of the beds' continuity. The following main types of tectonic disturbance are distinguished: monoclines, flexures, and folds.

Monoclines are a series of strata uniformly inclined in one direction for a considerable distance.

Flexure is the name given to displacements in the form of an elbow-like bend, of horizontal and monoclinal beds. They are due to a steep increase in the dip of the beds on a limited sector of a monocline. Upper, lower, and medium or connecting limbs are distinguished. The connecting limb is the part of the series in which the beds have a steep dip and are often thinner.

Folds are the main form of plicated deformations. They exist in two main forms: anticlinal and synclinal. *Anticlinal folds* are convex ones in which the beds dip on the opposite sides, while older rocks are located in the central portions than on the limbs. Synclinal folds are concave ones in which the beds dip toward one another and the rocks at the centre are younger than on the limbs.

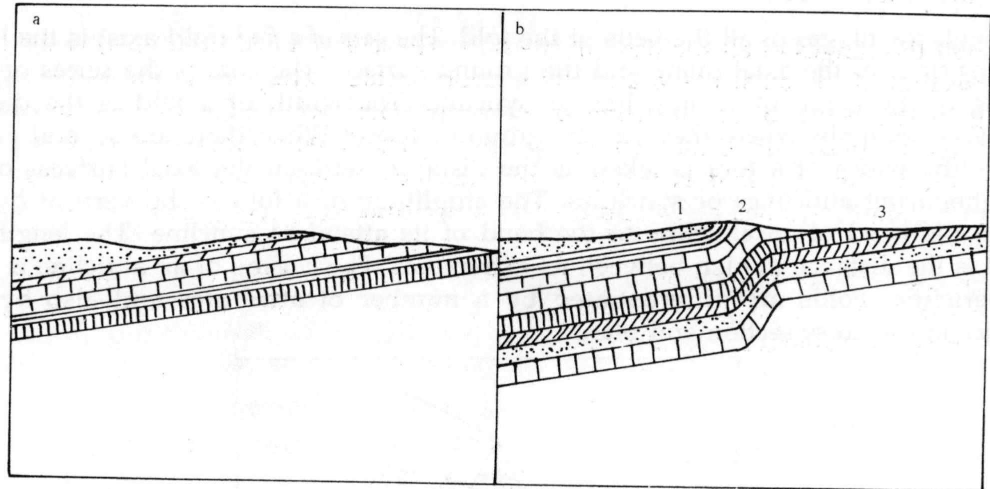

Fig. 3.8: Monocline (a) and flexure (b). Limbs: I-upper: 2-connecting: 3-lower.

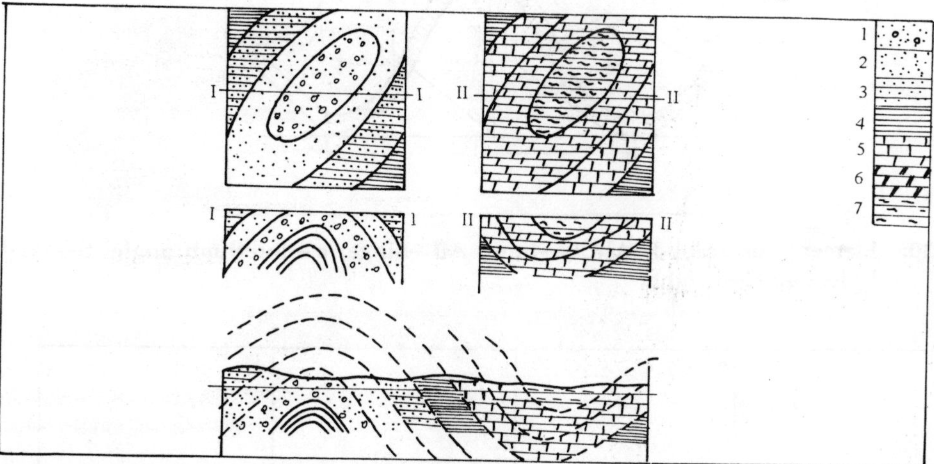

Fig 3.9: Anticlines and synclines in plan and section. a—anticline: b—syncline: C—parallel I—Ordovician: 2—Silurian; 3—Devonian; 4—Carboniferous; 5—Permian; 6—Triassic; 7—Jurassic.

Folds consisting of both concave (synclinal) and convex (anticlinal) sectors are called parallel or concentric.

Anticlines and synclines have the following elements: limbs, crest, hinge, core, axial plane, axis, width, height, and length of the fold. The limbs are the sides of the fold. The hinge is the line running through the points of maximum bending of any of the beds forming the fold. In a longitudinal vertical section the hinge line is often undulating. The crest is the part of a fold in the area of the hinge where the limbs bend. The crest of an anticline is sometimes called a dome and that of a syncline a saddle. The angle of a fold is the angle between the limbs or the interlimb angle, mentally projected to their intersection. The axial plane is the imaginary plane passing through the hinges of all the beds of the fold. The *axis of a fold* (fold axis) is the line of intersection of the axial plane and the ground surface. The core is the series of rocks lying in the bend, of an anticline or syncline. The width of a fold is the distance between its limbs where they cut the ground surface. When there are several parallel folds the width of a fold is taken as the distance between the axial surfaces of two neighbouring anticlines or synclines. The amplitude of a fold is the vertical distance from the bend of an anticline to the bend of its attendant syncline. The length of a fold is the distance in plan between its extremities. The closing of an anticline is called a pericline. Folds are differentiated by a number of attributes and also by their projection in cross-section and plan.

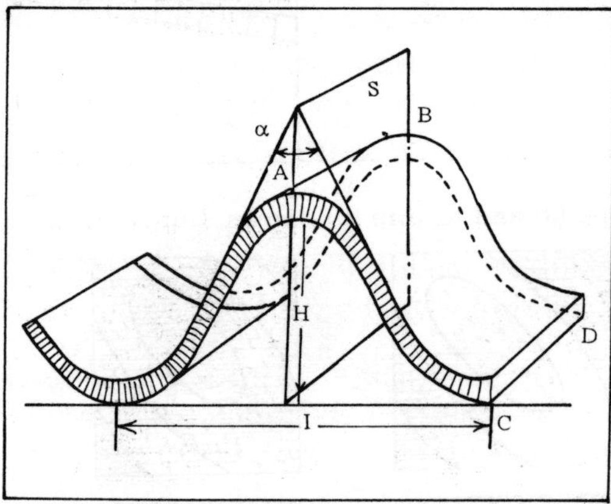

Fig. 3.10: Elements of a solid; ABCD—limo; AB—hinge; α—interlimb angle; S—axial plane; l—length; H—height.

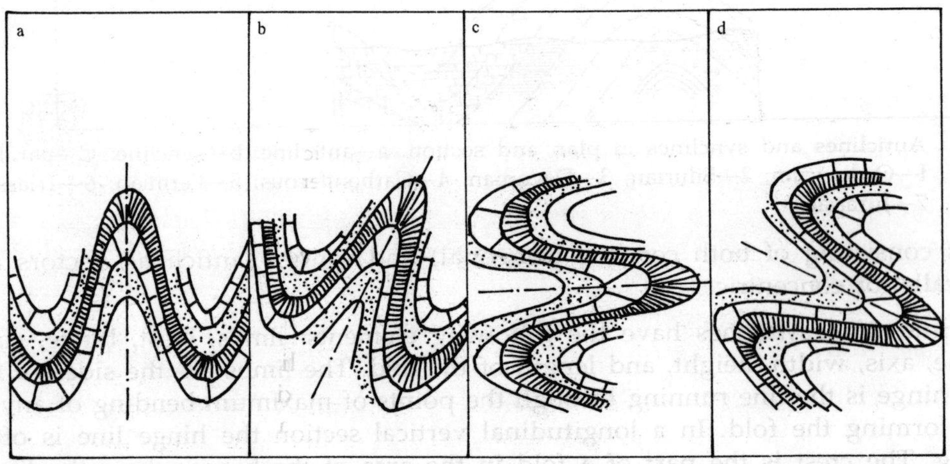

Fig. 3.11: Folds (according to the position of the axial plane and limbs) : a—upright; b—inclined; c—reci, nemt' d—overturned.

The following types of fold are distinguished *by cross-section from the position of the axial plane and limbs:* upright, inclined (reclined), recumbent, and overturned. An *upright* fold has a vertical axial plane and limbs disposed symmetrically in relation to it. An inclined (reclined) fold has an inclined axial plane, with the limbs dipping in different directions. The variety of inclined fold in which the limbs dip in the same direction is known as overturned. A *recumbent* fold is one in which the position of the axial plane is close to the horizontal and the limbs are nearly parallel with one another. An inverted fold is one with a subhorizontal axial plane, the limbs closing downward.

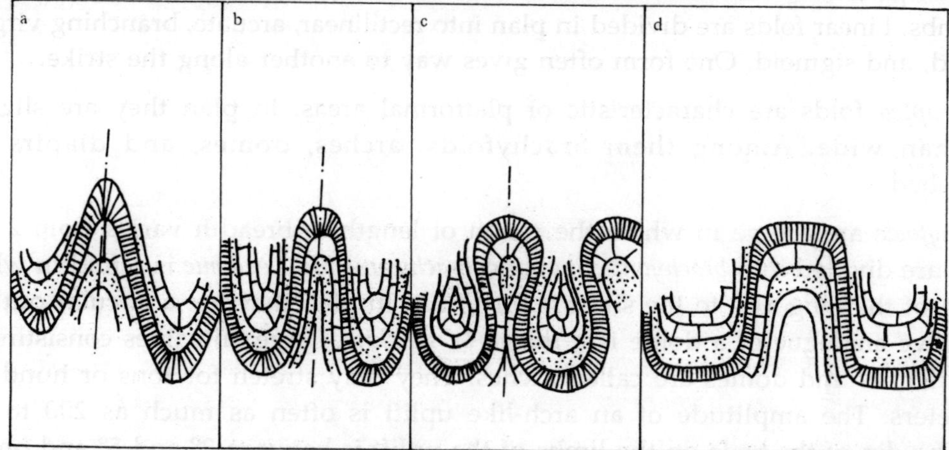

Fig. 3.12: Folds (by character of the disposition of the limbs and shape of the hinge); a—normal: b—isoclinal; c—fanshaped; d—box.

Normal (ridge-shaped), isoclinal, fanshaped, and box folds are distinguished according to the position of the limbs and the shape of the hinge in crossection. In a

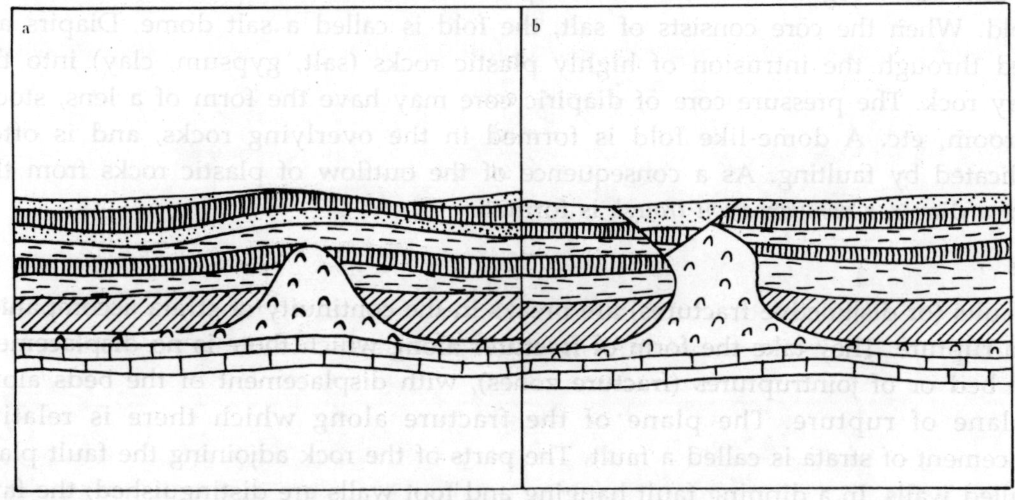

Fig. 3.13: Diapiric fold (salt dome): a—deep seated; b—piercing.

normal (upright) fold the limbs close at an acute angle and the crest has a narrow, sharp shape. *Isoclinal* folds have a sharp crest and parallel limbs. *Fan-shaped* folds have a wide crest, the limbs are located fan-wise and compress the core. Box folds have a broad crest, and relatively steep, almost vertical limbs.

In plan linear and discontinuous folds are distinguished according to the relation of their length and width. Linear folds are characteristic of geosynclinal regions. Their width may be on a fraction their length, the ratio of length to width being 10:1, 20:1, or more. In periclinal and centroclinal structures the slope of the beds is gentler than on the limbs. Linear folds are divided in plan into rectilinear, arcuate, branching virgate, imbricated, and sigmoid. One form often gives way to another along the strike.

Interrupted folds are characteristic of platformal areas. In plan they are slightly longer than wide. Among them brachyfolds, arches, domes, and diapirs are distinguished.

Brachyfolds are a type in which the ration of length to breadth varies from 2:1 to 5:1. They are divided into *brachyanticlines* and *brachysynclines.* A *dome* is a fold in which the ration of the long axis to the short is 1:1 to 2:1. In plan they are a circular outline. The negative analogue of a dome is a *trough or syncline.* Large anticlines consisting of brachyanticlines and domes are called arches. They may stretch for tons or hundreds of kilometers. The amplitude of an arch-like uplift is often as much as 200 to 300 metres. The dip of the beds on the limbs of the uplift is between 3° and 5° and on the downwarp up to 1°.

A special form of dome-like imbricate fold is the *diapir (a dome with an intrusive core).* The presence of plastic rocks (salt, gypsum, clay, etc.) in the core is typical of them, and also a regular increase in the dip of the beds from the limbs to the core of the fold. When the core consists of salt, the fold is called a salt dome. Diapirs are formed through the intrusion of highly plastic rocks (salt, gypsum, clay) into the country rock. The pressure core of diapiric core may have the form of a lens, stock, mushroom, etc. A dome-like fold is formed in the overlying rocks, and is often complicated by faulting. As a consequence of the outflow of plastic rocks from the area of a diapir, a *compensation trough* is formed.

Faults

Faults (or disjunctive fractures) are breaks in the continuity of strata accompanied with a rupture. They take the form of fractures along which there is no displacement of the bed or of jointruptures (fracture zones), with displacement of the beds along the plane of rupture. The plane of the fracture along which there is relative displacement of strata is called a fault. The parts of the rock adjoining the fault plane are called walls. In a dipping fault hanging and foot walls are distinguished; the fault covers the foot wall and lies beneath the hanging wall.

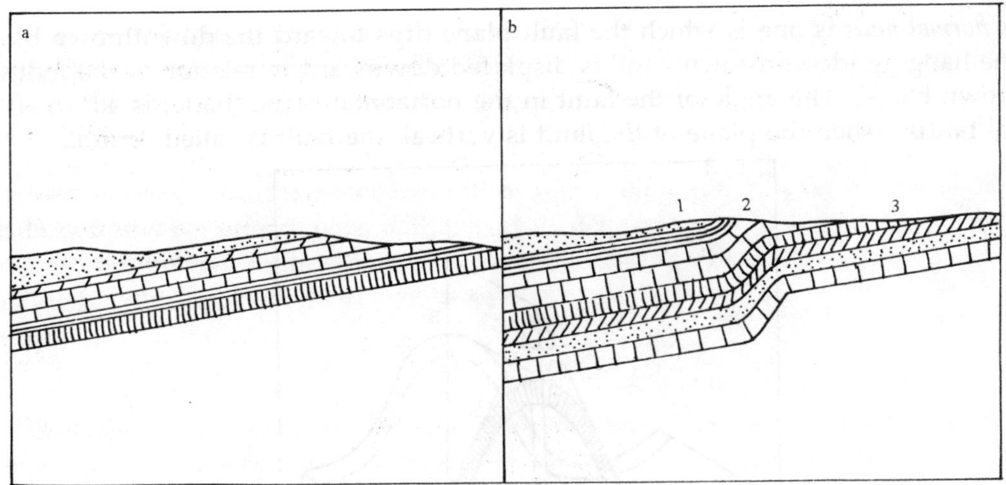

Fig. 3.14: Monocline (a) and flexture (b) Limbs: l. upper; 2, connecting; 3. lower.

The amount of relative displacement of strata along a fault is called its amplitude. The following types are distinguished: true (oblique) amplitude or the distance along the fault plane between the roof or floor of one and the same bed in the hanging and foot walls; vertical amplitude, or heave, the projection of the true amplitude on the vertical plane; horizontal amplitude or throw, the projection of the true amplitude on the horizontal plane; and stratigraphic amplitude, or the distance along the normal between the roof and floor of one and the same bed in the hanging and foot walls.

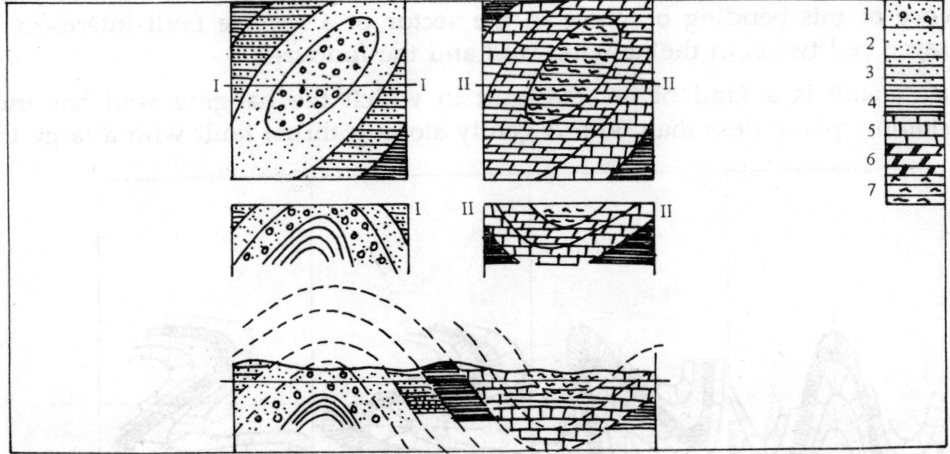

Fig. 3.15: Anticlines and synclines in plan and section, a—anticline; b—syncline; c—parallel; l—ordovician; 2—Silurian; 3—Devonian; 4—Carboniferous; 5—Permian; 6—Triassic; 7—Jurrassic.

Faults are distinguished as normal, reverse, thrust, and tear or wrench (transcurrent) faults.

A *normal fault* is one is which the fault plane dips toward the downthrown block, and the hanging (downthrown) wall is displaced downward in relation to the footwall (upthrown block). The angle of the fault in the horizontal plane (hade) is 40° to 60° in normal faults. When the plane of the fault is vertical, the fault is called *vertical.*

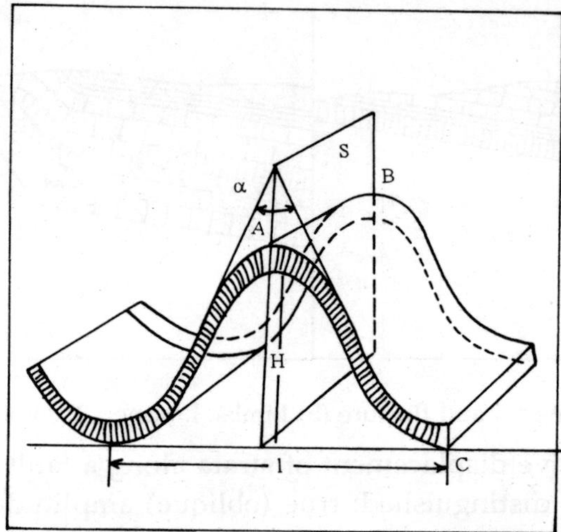

Fig. 3.16: Elements of a fold: ABCD-limb; AB-hinge; α-interlimb angle; S-axial plane; l-length; H-height.

A *reverse fault* is one in which the plane dips toward the upthrown block and the hanging wall (upthrown block) is displaced upward in relation to the footwall (downthrown block) with a steeply dipping fault plane (more than 60°). As a consequence of this bending borehole in the sector of a reverse fault interesects one and the same bed twice, in the hanging wall and the footwall.

A *thrust fault* is a kind of reverse fault in which the hanging wall has moved along a shallow plane (less than 60°). A gently sloping thrust fault with a large throw

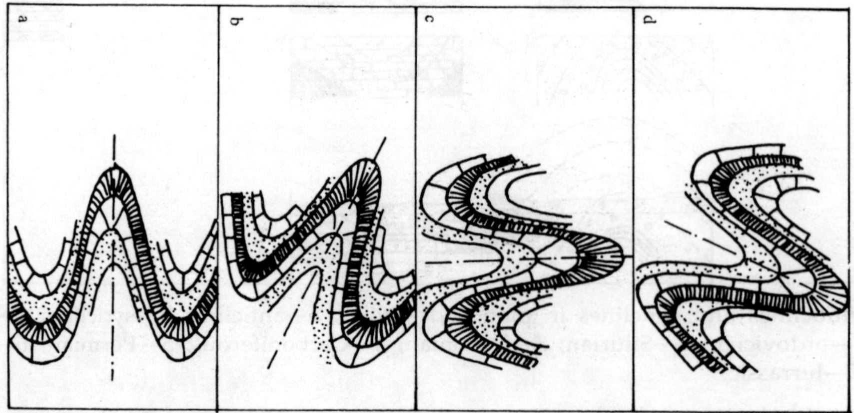

Fig. 3.17: Folds (according to the position of the axial plane and limbs).

and low dip is called a mass overthrust or tectonic sheet. Its horizontal amplitude can be as much as 30 or 40 kilometers.

Tear or wrench faults are fractures whose blocks are predominantly displaced horizontally parallel to the strike of the fault plane. They are frequently combined with normal faults, reverse faults, and thrust faults (strike faults, etc.).

Fractures are usually encountered in groups, forming complex faults (stop faults, grabens and horsts). *Step faults* are a system of normal faults in which each successive block is downthrown in relation to the preceding one. *A graben* is a system of stop faults the central part of which is down-thrown in relation to elevated peripheral blocks. A *horst* is a system, of reverse faults the central part of which is elevated in relation to the peripheral downthrown blocks.

MAPPING THE PAST

Geologic Maps

The study of the geologic history of an area ordinarily begins with the construction of a geologic map. Assume, for a moment, that all the loose material and vegetation were miraculously removed from your home state, so that bedrock world be exposed everywhere imagine, further, that the surfaces of the formations now exposed were each painted a different color and photographed vertically from an airplane. Such a photograph would constitute a simple geologic map, in actual practice a geological map is prepared by locating contact lines between formations in the field and then plotting these contacts on a base map. Symbols are added to the colored areas to indicate formations and lithologic regions, mineral deposits, and structures such as folds and faults. Once the geologic map is completed, a geologist can tell a good deal about the geologic history of an area. The formations depicted represent sequential pages in the geologic record. From the simple geologic map, the geologist is able to reduce that there was an ancient period of compressional folding, that the folds were subsequently faulted, and that an advance of the sea resulted in deposition of younger sedimentary layers above the more ancient folded strata.

Paleogeographic Maps

A map showing the geography of a region area at some specific time in the geologic past is termed a *plaeogeographic map*. Such maps are really interpretations based on all available paleontologic and geologic data. The majority of such maps show the distribution of ancient lands and seas. Palegeographic maps are at best, of limited accuracy, because as seas advance and retreat endlessly through time, the line drawn at the sea's edge may represent an average of several shorline positions. They are nevertheless useful for showing general geographic conditions within regions or continents. To prepare a paleogeographic map, one would plot all occurrences of rocks of a given time interval on a map and enclose the area of occurrence in boundary lines. Areas of nonoccurrence may be places of no deposition or places where deposits once existed but were subsequently eroded away. Nondeposition appears to be the case in the northwestern corner of, for that area is nicely encircled by sandstones that grade outward to shales. If the time-rock unit thins toward areas of nondeposition, this interpretation would be strengthened). With the help of fossils, the nature of the sediments—that is, whether marine or nonmarine—is determined and plotted on the map. The final step is to complete the paleogeographic reconstruction.

Isopach Maps

Isopach maps are prepared by geologists in order to illustrate changes in the thickness of a formation or time-rock unit. The lines on an isopach map connect points at which the unit is of the same thickness. On a base map, the geologists plot the thickness of units as they are revealed in drilling or in measured surface sections. Isopach lines are then drawn to conform to the data points as perfectly as possible. Ordinarily, the upper surface of the unit being mapped is used as the datum from which tickness measurements are made. An isopach map may be very useful in determining the size and shape of a depositional basin, the position of shorelines, and areas of uplift. The map indicates a semicircular center of subsidence in southern New York and Pennysylvania in which over 2000 ft of sediment accumulated. The isopach pattern further indicates a highland source area to the southeast.

Lithofacies Maps

Maps constructed to show areal variations in facies can provide additional details and validity to paleogeographic interpretations. Such graphic representations are called *lithofacies maps*. Geologists first correlate the formations. Then, assuming that the unconformity represents one time plane and the ash bed another, they define the time-rock unit as X. Paleontologic study of the rocks between the time planes confirm the validity of the time-rock unit. Geologists may now prepare the lithofacies map. Time-rock unit X is missing at well number 11; this may be the result of its not being deposited there or, having been deposited, of its being eroded away. It is logical that the sandy facies was deposited adjacent to a north-south trending shoreline.

Figure lithofacies map of rocks deposited over 400 million years ago in the eastern United States. From this map, one can infer the existence of a highland area that existed at that time along our eastern seaboard and that supplied the coarse clastics. Detritial sediments from the source area become fine, and the section thins as one proceeds westward from the source highlands. Finally, as far west as Indian, the map indicates that only carbonate precipitates were laid down. The conglomerates were probably the deposits of great alluvial fans built out from the ancient mountain system. Because the northern border of the time-rock unit cuts across the bands of facies, it is considered to be an erosional border and does not represent the original extent of the unit mapped.

The lithofacies maps just described provide a qualitative interpretation of areal changes in rock bodies. Quantitative lithofacies maps can also be constructed and are frequently used in the study of subsurface formations that are known primarily from well records. By means of contour lines, such maps show the areal distribution of some measurable characteristic of the unit being mapped. For example, contours may be drawn on the percentage of one lithologic component compared to the total unit or on the ratio of one rock type to the others within the unit. An isopach map is ordinarily the base map for any of the quantitative maps, since one must know the total stratigraphic thickness of the unit with which individual components are compared.

Perhaps the most important fact to remember about the mapping techniques used in deciphering earth history is that they are limitless in variety. New and inventive schemes for mapping the past appear regularly in geologic journals as geologists continue to probe into the details of ancient geologic events.

IMPORTANCE OF SEDIMENTARY ROCKS TO SOCIETY TODAY

From caveman to astronaut, humans have learned to use sedimentary rocks to improve their lot. Men and women of antiquity fashioned tools from chert and flint, and they laater learned the art of manufacturing useful containers from clay. Their successors developed the means of extracting copper and iron from the rocks and from these metals made more sophisticated implements. In addition to the metals they sometimes contain, sedimentary formations include coal seams and contain the oil, gas, and groundwater important to our present way of life.

Mineral Fuels

The mineral fuels-petroleum, gas, and coal-are our most important sedimentary resources. They are essential for heat and power and metal refining; they are also the source of many chemicals useful in the manufacturing of plastics and fertilizers. *Coal* is a brownish-black to black combustible rock that forms in beds from a few inches to many feet in thickness and is interbedded with shales, sandstones, and other sedimentary rocks. Extensive coal-bearing sequences characterize the Pennsylvanian and Cretaceous Systems. Pennsylvanian sequences of strata in the Applachians include over 100 individual beds of coal, each composed of the compressed and altered remains of land plants. Coal is abundant in the United States and will become increasingly more important over the next three decades as our reserves of petroleum are consumed.

Commercial accumulations of oil and gas require rather special geologic conditions. These conditions are found almost exclusively in sedimentary rocks. First, there must be a *source rock* for the petroleum. Ordinarily, this is a series of beds rich in the organic remains of unicellular organisms that accumulated along with the particles of sediment. Second, there must be a permeable reservoir rock such as sandstone or porous limestone to provide passage and storage for the gas and oil. The *reservoir rock* is covered by an impermeable *cap rock* to prevent the upward escape of hydrocarbons. Finally, there must be an oil-trapping structure so that the hydrocarbons can be concentrated in one place. Petroleum geologists seek out these structures in many ways. Long before drilling programs are begun, geologists conduct gravity surveys and seismic "reflective shooting" surveys to reveal clues to underground structures. Later, exploratory wells are drilled, and by careful analyses of electrical, lithologic, and paleontologic logs of these wells, a picture of the configuration of the underground layers is gradually developed.

Construction Materials

Other sedimentary materials useful to humans included clays for use in ceramics and bricks, limestones for building stones and cement, sand and gravel for concrete and glass, and evaporites for use in the chemical industry. Some sedimentary strata are rich in iron oxide. Those near Birmingham, Alabama, have been mined and smelted for several decades. Because minerals such as gold, chromite, diamond, cassiterite, and magnetite are heavier than quartz and other common silicates, they tend to accumulate in quiet areas of streams or to be concentrated by waves as placer deposits. Finally, there are sedimentary ores that form in place by the deep chemical decay of parent rock. The most important metal concentrated in this manner is aluminum. Bauxite, an almost pure hydrous aluminum oxide, is the ore mineral of aluminum. Bauxites oped in once humid, tropical regions as a result of the weathering of source rocks rock in aluminium silicates.

Limited Supply of Mineral Resource

Any commentary on mineral resources would be incomplete without noting that ore bodies and oil fields are limited in extent and are the result of unusual associations of geologic conditions. Furthermore, it is not theory but fact that they are exhaustible and irreplaceable. Their effect upon the standards of living and society is so enormous that every citizen should become involved in political decision relating to the management of our natural resources.

Chapter—4

The Oldest Rocks

BENEATH THE CAMBRIAN

When eighteenth century geologists first began to diagram local columns of strata, they frequently found a "basement complex" of igneous and metamorphic rocks beneath the lowest sedimentary layers. Such names as the "Primitive" or "Primary" were applied to these jumbled and often complex rock bodies. In 1835, the geologist *Adam Sedgwick,* while mapping in North Wales, used the name "Cambrian" for the strata lying above the basement rocks. Subsequently, geologists commonly referred to the underlying rocks simply as Precambrian. The term *Cryptozic* ("hidden life") was later applied to the Precambrian time interval, whereas the companion term *Phanerozoic* ("obvious life") was retained for all of subsequent geologic time.

Although early geologists correctly recognized that the Cryptozoic probably lasted several times longer than all the rest of geologic time, they gave rocks of this period only scant attention. Because older Cryptozoic rocks bore few if any fossils and were so contorted and intricately intruded, they were generally thought to be undecipherable. Nevertheless, Precambrian rocks form the very cores of the continents. They are also enormously important source rocks for iron and other metals. For these reasons, it was inevitable that a few intrepid geologists would devote their lives to deciphering the relationships of the tangled masses of schists, slates, and granites.

One of these pioneers of Precambrian geology was *Sir William Logan* of the Canadian Geological Survey. In the middle 1800's, Logan was able to associate groups of Precambrian rocks according to their superpositional and cross-cutting relationships. As a further aid, Logan speculated that those rocks that had suffered the most metamorphism were probably the oldest. Modern geologic investigations clearly show that Logan frequently erred in attempting to correlate degree of metamorphism with age. Older rocks may escape metamorphism, and younger ones may be radically metamorphosed. Nevertheless, in southeastern Canada, where Logan mapped, one could find an older terrain of gneissic rocks as well as younger sequences of less altered metamorphic and sedimentary rocks. For this reason, it seemed reasonable to think of

Precambrian time as divisible into an older *Archem Era* and a younger *Proterozoic Era*. More recent work based on absolute dating techniques that were not available to Logan and his contemporaries has shown that one can continue to use these two terms in a very general way.

Table 4.1:
Some Chronologic and time-rock terms used in Precambrian Historical Geology

TIME IN BILLIONS OF YEARS	CANADIAN CLASSIFICATION			INTERNATIONAL CLASSIFICATION	MAJOR CLASSIFICATION OROGENIES
0.57	CRYPTOZOIC	PROTEROZOIC	Hadrynian	Proterozoic III	
1.0			Helikian	Proterozoic II	Grenvillian (12.9by)
1.5					Elsonian (1.37 by)
				Proterozoic I	Hudsonian (1.73 by)
2.0			Aphebian		
2.5					Kenoran (2.48 by)
3.0		ARCHAEAN		ARCHEAN	
3.5					

* Provisional recommendation of the Subcommission on Precambrian Stratigraphy of the International Union of Geological Sciences.

Today, geologists around the world agree that the only truely reliable basis for chronologic correlation of unfossiliferous Precambrian masses is through isotopic dating techniques. Only after the ages of rocks in a particular region are known can those rocks and the events they record be placed in the proper chapter of Precambrian history. The hundreds of dates already obtained include rocks that are about 3980 million years old, although most near-surface outcrops were formed less than 3600 million years ago.

PRECAMBRIAN SHIELDS

Distribution of Outcrops

Although Precambrian exposures are common place in cores of mountain ranges and canyons of plateaus, the most obvious and largest areas of Precambrian rocks are the erosionally stripped, regionally unwarped, geologically stable regions of the continents. Such regions are called *Precambrian shields* because of their broadly convex shape (reminiscent of a Greco-Roman shield). Shield rocks are exposed over only about 20 per cent of the earth's land surface, yet they represent about 75 per cent of the observable time span accounted for by crustal rocks. Every continent has one or more Precambrian shields bordered by Phanerozoic mobile belts. North America, for example, has the great Canadian Shield, which covers over 3 million square miles. Even where covered by younger layers, shield rocks poke through the cover at structural domes, as in the Ozarks and the Black Hills. Because of the great difficulty in making intercontinental correlations of Precambrian rocks, the terms and groupings used vary for each of the continents. However, as progress continues to be made in establishing the synchrony of outcrops, a standard body of terminology will probably emerge.

Overview of Shield History

Detailed geologic mapping of the Scandinavian and Canadian Shields during the 1930's provided a more complete understanding of Precambrian history. Although the events in any particular shield might differ from those in another, generalities emerged that were applicable to all shields. Radiometric dates for particular outcroppings began to accumulate; they gave geologists their best tool for correlating and deciphering field relationships. The improved data generated new ideas about the way shields developed. One hypothesis for shield growth suggests that at first a few primary continental nuclei might have differentiated from the mantle Linear geosynclinal tracts might next have developed along the edges of these nuclei, experienced depositional and then orogenic phases, and ultimately become welded onto the continental nuclei. Erosion would next begin to wear away the crumpled sediments and volcanies and, aided by occasional uplifts, would strip the highlands down to the intensely metamorphosed and intruded roots of the old mountains. After repeated episodes of uplift and erosion, these once deeply buried stumps of mountains would come into gravitational balance with underlying rocks. They would become stable segments or provinces of the shield. One or more new geosynclines might then develop along the margins of the previously formed tract, and the continent (and shield) would grow by accretion of yet another marginal belt.

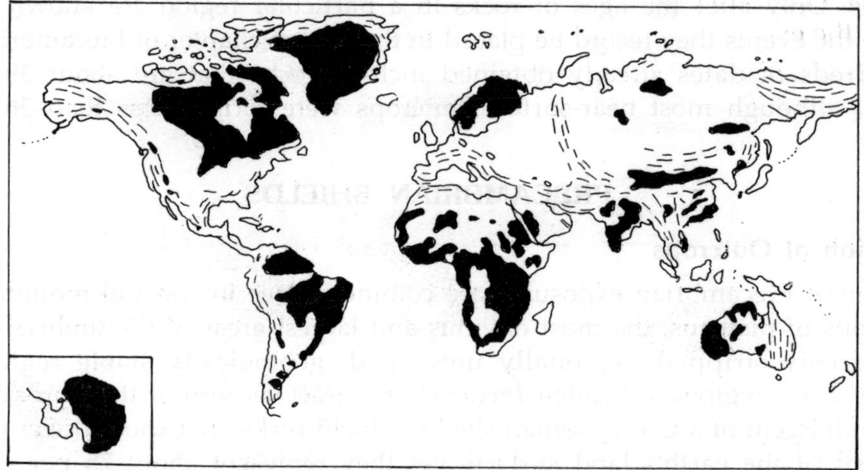

Fig. 4.1: Major areas of exposed Precambrian rocks.

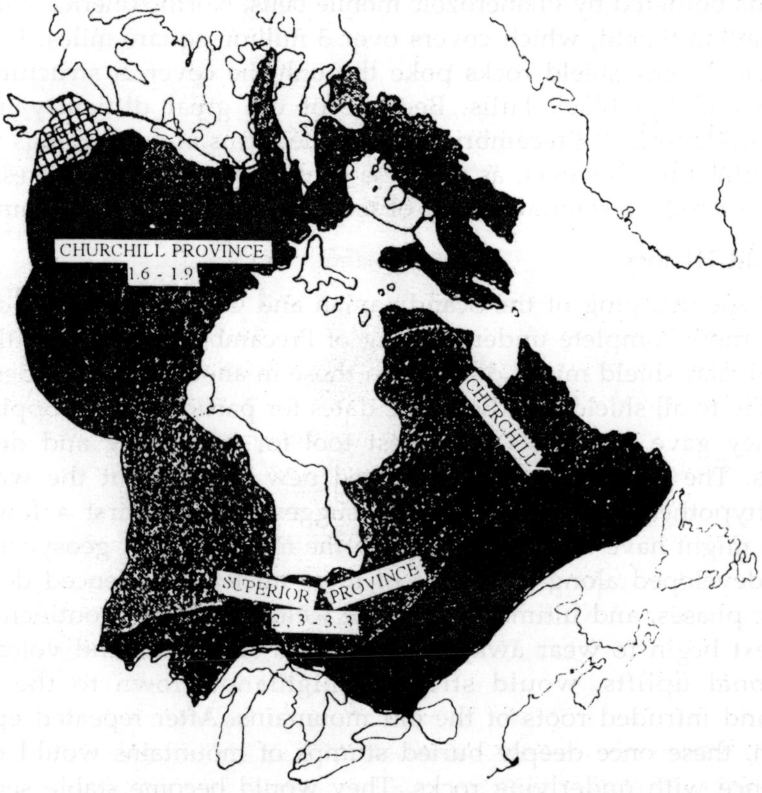

Fig. 4.2: Tectonic provinces of the Canadian Shield.

What is the evidence for this view of shield history? When locations of dated Precambrian rocks are plotted on a map one finds that rocks of about the same age tend to occupy distinct belts. Each belt or Precambrian province is inferred to represent an ancient orogenic belt in which rocks were crystallized during metamorphism and granitic intrusion. Uranium bearing minerals formed at the time of orogeny provide the materials for age determinations.

When the ages of North American shield rocks are plotted on a graph according to frequency of occurrence, they show peaks of metamorphism and intrusion at 2.5, 1.7, 1.4, and 1.0 billion years ago. This suggests that there were four major orogenic episodes that occurred at different times and places in this region of North America. As indicated in the Kenoran was the most ancient of these great disturbances, the effect of which are evident in the Superior and Slave Provinces of Canada. The next great upheaval was the Hudsonian. It affected primarily the Churchill, Southern, and Bear Provinces. The Hudsonian was followed by the Elsonian Orogeny, which produced the complex metamorphic and igneous terrains of the Nain Province. Finally, the geology of the Greenville Province was largely molded to its present state during the Greenville Orogeny.

The delineation of old orogenic belts is not at all a simple matter. One problem that contributes to the difficulty is that an ancient orogenic belt might be remelted or metamorphosed during the orogenesis of a younger belt. Recrystallization during the younger event might reset the "radioactive clocks," and age determinations would no longer give the date of the older event. Another uncertainty is that igneous rocks may be intruded into an ancient belt long after it has become stable. Dates based on such intrusions would not date the original mountain-building episode.

According to modern views, shields may have begun their history as relatively small continental blocks that grew throughout geologic time by additions of rocks chemically differentiated from the upper mantle and by accretion of new material at or near their margins. The process by which the accretionary growth occurs might be like that described in the previous chapter, wherein oceanic crust is subducted beneath the margin of a continent, partially melts, and gives rise to less dense, more siliceous rocks.

The Canadian Shield

Extending across most of Canada and Greenland and from Baffin Island southward to Minnesota is the great Canadian Shield. This great expanse of Precambrian rocks is mostly a low-lying and of confiers, muskeg, and tundra. The shield is bordered by the Appalachian fold belt on the east and the Rocky Mountains on the west. It extends southward beneath the central United States, where deep bore holes have permitted geologists to map its surface and extent. To the north, the shield is bounded by the Franklinian fold belt of the Arctic.

As a result of field work that began in the midnineteenth century, it became evident that the Canadian Shield could be divided into a number of structural provinces, each recognized by characteristic structural patterns, lithologic assemblages, mineralization, and geologic age. Subsequent isotopic dating, mainly by potassium-argon methods, indicated that the rocks of each province were emplaced over well-defined intervals of time. Thus, the shield may be regarded as a patchwork of provinces, each of which represents a cycle or part of a geologic cycle of sedimentation, orogeny, and erosion. Some of the provinces are mere remnants of once larger complexes. The older provinces might have been growing protocontinents that merged to form the single, large, stable craton characterizing the Proterozoic.

Although the old division of Precambrian rocks into an older Archaeozoic and younger Proterozoic is still broadly appropriate for the Canadian Shield, Canadian geologists now utilize a second set of terms (*Archean, Aphebian, Helikian, and Hadrynian*) that better serve to delineate the time encompassed in each geologic cycle.

The Archean

Although Archean rocks are the most ancient rocks of the Canadian Shield, they rarely yield ages older than 2.5 billion years. A small patch of rocks 3.6 billion years old has been found in Minnesota; in Greenland, Archean rocks as old as 3.8 billion years have been discovered. The Kenoran orogenic activity at the close of the Archean undoubtedly metamorphosed, reworked, and melted older rocks, and this may account for the sparsity of exceptionally ancient terrains. The Archean of Canada is best developed in the Superior Province. This province was once part of a much larger continental mass that included regions now occupied by the Nain and Slave Provinces.

In general, there are two large categories of rock assemblages in the Archean of Canada. One of these consists of rocks dominated by enormous amounts of iron and magnesium-rich volcanics that tend to occur along numerous elongate tracts known as "*greenstone belts*." The rocks of greenstone belts often show a very low grade of metamorphism in which such greenish minerals as chlorite, epidote, and actinolite (an amphibole) are well developed. Greenstones are the products of metamorphism of basaltic lavas and other pyrocastic materials. Pillow-lavas (pillowy structures formed during the submarine extrusion of basalt), as well as chemical evidence, indicate some of the lava solidified under water. Metasediments are also present in greenstone belts and include poorly sorted classic rocks that suggest rapid deposition in unstable basins. Iron-rich cherts, thought to have derived their silica from volcanic ash, are also prevalent. The combination of pillow-lava and other volcanics, graywackes and chart beds has led geologists to interpret greenstone belts as remnants of elongate volcanic troughs. Perhaps these arcs were developed parallel to terrains of crystalline granitic rocks, because the conglomerates of the greenstone belts do contain pebbles of granite. Greenstone belts show intricate and complicated patterns of folding and can be traced over several hundred kilometers. They appear to be "supracrustal" features developed on older Archean basement rocks.

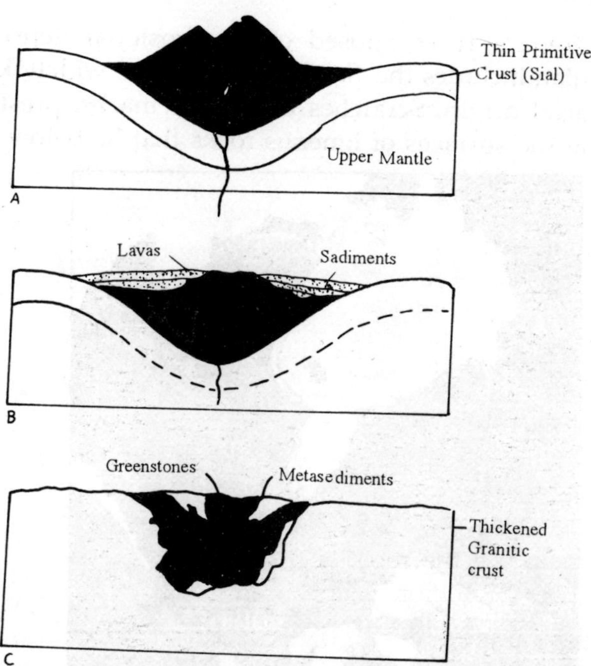

Fig. 4.3: Model for the origin of Precambrian green-stone belts. (A) Development of downwards on an unstable thin primitive crust and accumulation of lavas. (B) Erosion of volcanic fill (and adjacent grantic areas) produces sediment accumulations. (C) Massive invasion by younger grantic melts obliterates primitive crust, but the deep accumulations of lava in the elongate troughs are preserved and metamorphosed to form the greenstone belts.

The second large category of Archean rocks consists primarily of granites and gneisses that occur in the broader areas between greenstone belts. Most of these granitic terrains are products of widespread igneous intrusive activities associated with the *Kenoran Orogeny*. The Laurentian and Algoman granites of the Superior Province are representatives of this Archean assemblage.

Following the Kenoran Orogeny near the end of the Archean, the Superior and Slave provinces were extensively invaded by swarms of basaltic dikes and sills. The arm of the Superior province that extends into Wyoming and Montana experienced the intrusion of a 5-km thick sheet of chromite-bearing gabbroic rocks that composes part of the Stillwater Complex. The layer has since been dated to an almost vertical position and provides excellent exposures for studies dealing with gravity setting of minerals during crystallization.

The Aphebian Era (Early Proterozoic)

Rocks of the Aphebian Era are best exposed along the southern and northeastern perimeter of Superior Province. In the region surrounding Sudbury, Ontario, rocks of Aphebian age are called the *Huronian Succession*. They consist primarily of coarse, clastic sediments that are frequently composed of the erosional debris from the Archean basement. The Huronian includes the *Gowgonda Formation,* widely known for its boulder beds of probable glacial origin. Scratches and scour marks, presumably produced by glaciers, are found on the surfaces of igneous rocks that lie below the conglomerates.

Fig. 4.4: Principal outcrop regions for rocks of the Early Proterozoic, or "Aphebian Era."

Aphebian rocks surrounding the western shores of Lake Superior constitute the *"Animikean System"*. The group is world-famous for the bonanza iron ores it contains. The ores are in the form of hematite (Fe_2O_3) and indicate that by about 1.9 billion years ago free oxygen had begun to accumulate in the atmosphere and to combine with dissolved iron, forming iron oxide. The formation of this compound also implies an abundance of photosynthetic organisms in the world's oceans to generate the oxygen that has oxidized the iron. In the Thunder Bay District of Ontario, the Animikean includes the Gunflint Formation, which contains fossil remains of early forms of life.

South of Lake Superior, the Aphebian section thickens and becomes more severely folded. In this region, shallow-water sandstones, carbonates, and slates constitute the lower part of the section and are overlain by siliceous iron beds, slates, graywackes, and basaltic volcanics.

Another interesting tract of outcropping Aphebian rocks can be found along the northeastern border of the Superior Province. The succession in this far northern region bears a resemblance to the classic model of a geosyncline. Sedimentation occurred within two parallel troughs. The western tract accumulated quartz sandstones and carbonates and appears to have been miogeosynclinal. Pillow-lavas are prevalent in the eastern tract as well as other volcanics, cherts, and graywackes. The rocks of this "eugeosynclinal" area have been severely folded and metamorphosed.

Aphebian orogenic activity continued sporadically in various parts of Superior Province until general province stability was achieved about 1.4 billion years ago. A large central craton was then well established, and subsequent orogenic belts tended to develop around its borders.

The Helikian Era

Among the rocks deposited or emplaced during the Helikian Era were those of the Keweenawan succession. These layers rest upon either the crystalline basement or the Anikikian strata and extend from the Lake Superior region southward beneath a cover of Phanerozoic rocks for hundreds of kilometers. Keweenawan rocks consist of clean quartz sandstones and conglomerates as well as basaltic volcanics. Keweenawan lava flows are well known for their content of native copper. Holes, originally formed as gas bubbles in the solidifying rock, provide the voids in which the metal was deposited.

The Keweenawan lavas accumulated to thickness of over 10,000 meters. Even so, much of the supply of mafic magma was not tapped but remained beneath the surface, where it crystallized to form the 12,000-meter thick and 160-km long mass of the Duluth Gabbro. Basalts, as we have seen, are rock types more characteristic of oceanic areas. When such large quantities of mafic materials come to the surface within the stable parts of continental areas, the cause may be the breaking apart of the continent. The place where the break occurs is characterized by a system of tensional faults that form a rift zone. Along the rift zone the mafic magma may rise toward the surface to form Keweenawan-like accumulations. Evidence from gravity and magnetic surveys, as well as samples from deep drill holes, indicates that the rift zones associated with Keweenawam volcanism extended from Lak Superior southward into the United States. It has been suggested that the rift along which Keweenawan laws ere extruded may possibility have joined with two other arms of a triple-rift system having its junction at the southern margin of ancestral North America. Faulting and volcanism along the Keweenawan and adjoining rifts may have been associated with an early separation of Gondwanaland about 1.15 billion years ago.

The Greenville Province is composed of rocks that can also be included in the Helikian. This Late Proterozoic province has an exposed tract east of the Superior Province. Under a cover of Phanerozoic rocks, it can be traced southwestward almost to Mexico. Because the Girenville parallels the Appalachian belt closely, parts of its have been obliterated by Paleozoic orogenic events in the eastern united States. Typically, Grenville surface rocks consist of highly metamorphosed carbonates and quartz sandstones that have been severely deformed and extensively invaded by intrusive rocks during the Grenvillian Orogeny. Grenvillian deformation produced an impressive mountain range along the eastern margin of the continent. Paleomagnetic studies suggest that this great compressional orogenic episode was the result of a collision with another continent. Indeed, the eastward movement of the Greenville tract to the collision zone may have caused some of the crustal tension to the west

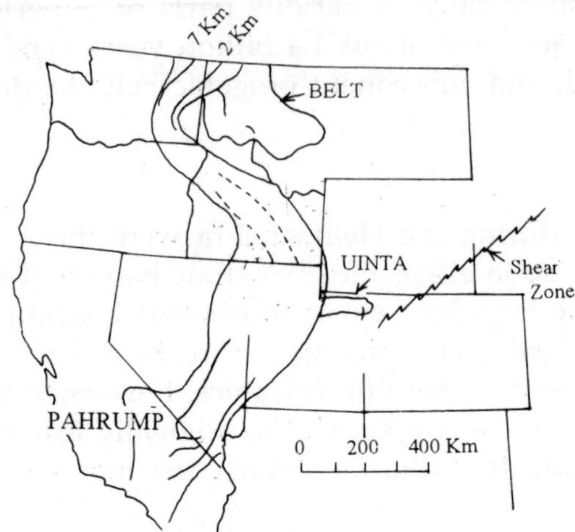

Fig. 4.5: Elements of western United States Precambrian geologic history. Giray are represents the Cordilleran miogeosyncline with generalized isopachs for late Precambrian to early Cambrian continental shelf sediments. Orange patterns are Belt, Uinta, and Pahrump sequences, 0.85 to 1.4 billion year old.

that resulted in the Keweenawan rifts. Clearly, the earth's tectonic plates were actively shifting about during the Late Proterozoic.

The Hadrynian Era

The final segment of Precambrian time has been named the Hadrynian. It is represented in North American by sedimentary rocks and some volcanics that were deposited in the geosynclines that rimmed the continent and were deformed in Paleozoic orogenic events. Rocks of the Hadrynian Era range in age from about 600 to 900 million years old. Of particular interest is the occurrence of unsorted boulder beds considered by many geologists to be glacial deposits. In North America, they have been described in Utah, Nevada, Western Canada, Alaska, and Greenland. If these conglomerates are indeed tillites, they indicate a second major Precambrian continental glaciation about 700 million years ago.

Precambrian Rocks South of the Canadian Shield

Although outcrops are not as extensive as in Canada, Precambrian rocks also occur south and west of the Canadian Shield. Particularly impressive successions are exposed in the Rocky Mountains and Colorado Plateau regions. These rocks have had a complex history that began over 2.5 billion years ago with the evolution of an Archean terrain composed of strongly deformed and metamorphosed granite rocks. Largely on the basis of the study of remnant patches of volcanic rocks and greenstones in Wyoming, it appears likely that this old Archean mass collided with one or more island are terrains about 1.7 or 1.8 billion years ago. The line of collision is believed to be marked by a shear zone in southern Wyoming characterized by severely crushed and brecciated rocks. The next event was a widespread episode of rift development about 0.85 to 1.4 billion years ago. Large fault-controlled depressions were formed in which thick sequences of Late Precambrian sediments accumulated. They include the Uinta Series of central utah, the Pahrump Group of southeastern California, and the Belt Supergroup of Montana, Idaho, and British Columbia. Because of the scenic features with which they are associated, the Belt rocks are of particular interest. In places these rocks are over 12,000 metres thick. Especially impressive are the massive cliff-forming limestones that can be viewed at Waterton Lakes and Glacier National Park. Although exceptionally think, the Belt rocks display features such as ripple marks and algal structures that indicate they were deposited in relatively shallow water.

The last event in the Late Precambrian history of the western United States was the development of a north-south trending geosyncline. At this time the North American plate was moving away from a spreading centre located somewhere to the west, and the geosyncline marked the continent's trailing edge. The continental shelf was the site of the miogeosyncline. A westward thickening wedge of latest Precambrian

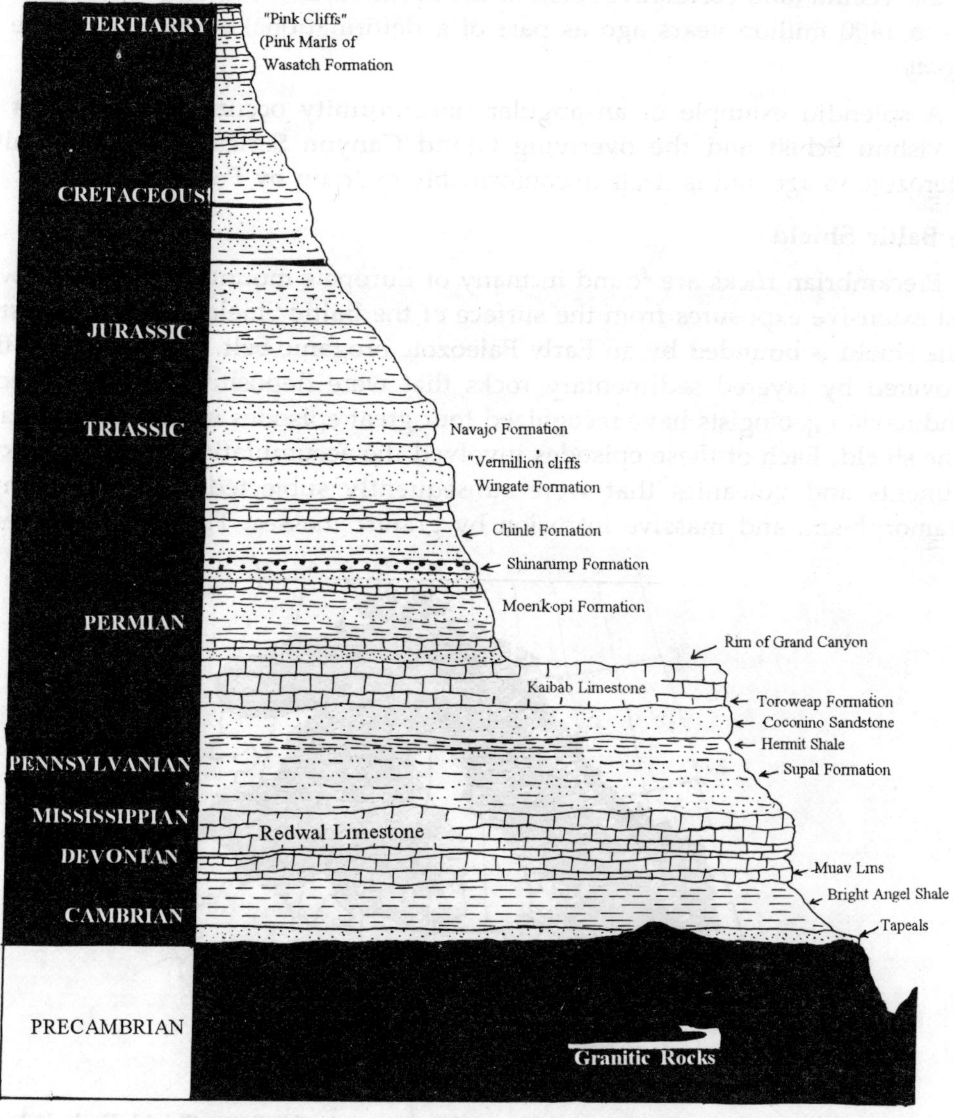

Fig. 4.6: Generalized statigraphic section for the western part of the Colorado Pleateau.

and Cambrian sediments was deposited on this shelf. The distribution of these sediments was not at all affected by the fault-controlled basins of the earlier episode of deposition.

Precambrian rocks of the Grand Canyon region consist of two distinct units. The lower and older unit is the Vishnu Schist, and the upper unit is the Grand Canyon Series. The Vishnu is a complex body of metamorphosed sediments and gneisses that have been intensely folded and invaded by granites. The granitic intrusions that cut into the Vishnu (and correlative rocks of the southwestern United States) were emplaced 1300 to 1400 million years ago as part of a deformational event named the *Mazatzal Orogeny*.

A splendid example of an angular unconformity occurs at the contact between the Vishnu Schist and the overlying Grand Canyon Series. The latter unit is Late Proterozoic in age and is itself unconformably overlain by Paleozoic rocks.

The Baltic Shield

Precambrian rocks are found in many of Europe's mountain ranges. However, the most extensive exposures from the surface of the Baltic Shield. The northwestern edge of the shield is bounded by an Early Paleozoic oroganic belt. The southeast, the shield is covered by layered sedimentary rocks that were deposited after the Precambrian. Scandinavian geologists have recognized four great episodes in the Precambrian history of the shield. Each of these episodes involved the accumulation of great thicknesses of sediments and volcanics that were subsequently subjected to severe compression, metamorphism, and massive intrusion by granitc masses. The rocks representing the

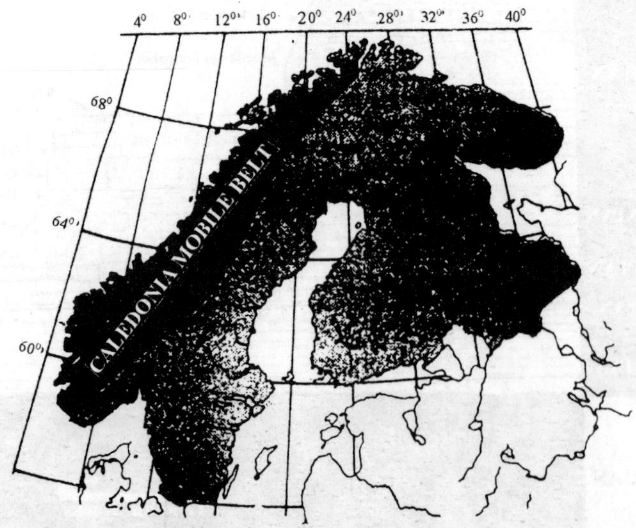

Fig. 4.7: Region over which Precambrian rocks outcrop in the Battic Shield. Early Paleozoic rocks deposited in the Caledonian Geosyncline are shown.

two oldest episodes are mostly schists and gneisses that were once sandstones and vol canies. Conglomerates, ripplemarked quartz sandstones, dolostones, greenstones, slates, and iron ores characterize the third cycle. The topmost Precambrian rocks managed to escape severe metamorphism. These youngest shield rocks consist of lavas and sandstones that were later intruded by granite.

The Angaran Shield

The Angaran Shield is more difficult to study than either the Canadian or the Baltic Shield. These Precambrian rocks are not exposed over a vast region. Rather, they are revealed as small, exposed patches that elsewhere are covered by younger sedimentary rocks of the great Siberian Platform. Covered or not, however, the shield is important as the very nucleus of Asia. To the east and south of this region lay the mobile belts that were later to be fashioned into the Ural and Himalayan Mountains. Not unlike the Precambrian of other shields, the Angaran has an older complex of gneisses, schists, and granites that provides evidence of a long and complicated history. There is also a younger sequence of sedimentary rocks, volcanics, and granitic intrusives that dates from about 1600 millions years ago. Russian geologists have named these rocks *Rhiphaean*. They include the sandstones and limestones of Precambrian shallow seas. Many of the limestones include moundike growths called *stromatolites*. Such structures were produced by algae and formed impressive reef structure along coastal areas.

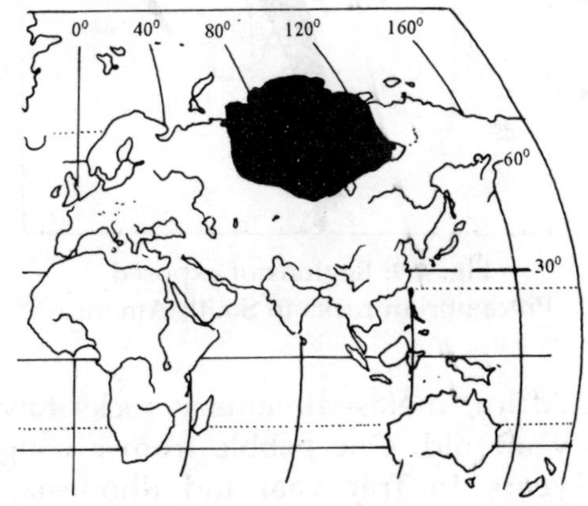

Fig. 4.8: The Angaran Shield. Precambrian rocks ove mass of this vast egic by Phamerozoic strata.

The Shields of South America

The number three seems to characterize the Precambrian geology on South America. There are *three* shields the Guianan, the Brazilian, and the Patagonian also, the Precambrian rocks of South. America are usually grouped into *three* divisions: Early, Middle, and Late Precambrian. As might be expected, gneisses and schists form the oldest sequence. The Middle Precambrian consists, in large part, of metamorphosed sediments, volcanics, and intrusions that appear to be the product of geosynclinal evolution. Metamorphic effects are less profound in the Late Precambrian. Slates and phyllites are still in evidence, however, along with quartzites, conglomerates, volcanic flows, and ash beds. Because of a covering of younger sediments over parts of the shields, as well as the dense forests that cover many regions, the Precambrian rocks of South America are still not as well known as those of the relatively barren shields of Canada and Scandinavia.

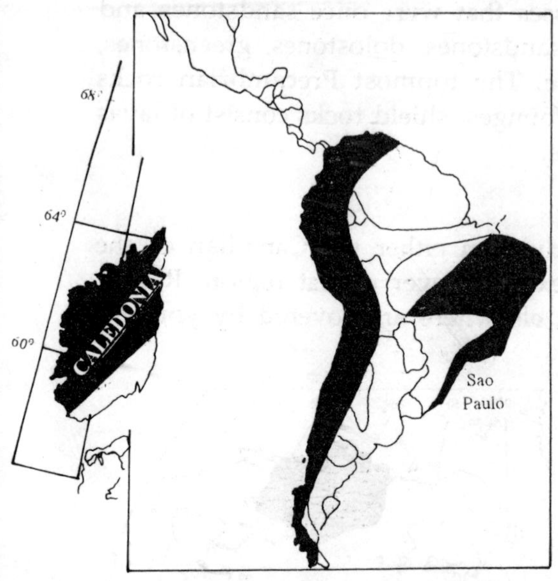

Fig. 4.9: Regions of exposed Precambrian rocks in South America.

Africa

Precambrian rocks outcrop over half the surface of Africa and masswhere lie beneath a veneer of Paleozoic and Mesozoic rocks. Indeed, andica appears to be basically a vast platform of abutting shield segments rips mostly a very stable continent but has along its eastern edge the remarkable faulting and fracturing that has produced the African rift to alleys. Africa's Precambrian rocks are renowned for their treasure of minerals; exploration for gold, diamonds, copper, uranium, chromium, and cobalt has resulted in improved understanding of the continent's very complex geology.

The long Precambrian history of Africa can be divided into an early, middle, and late phase. The Early precasmbrian encompasses events that occurred prior to about 2600 million years ago. Its record consists of some of the oldest rocks in Africa. In the Barberton Moutnain land of South Africa, metasedimentary rocks have been radiometrically dated at 3.0 to 3.6 billion years old. One pebble from a conglomerate yielded an astonishing age of 4.1 billion years. In Transvaal and Rhodesia, the Early Precambrian includes 17,000 meters of

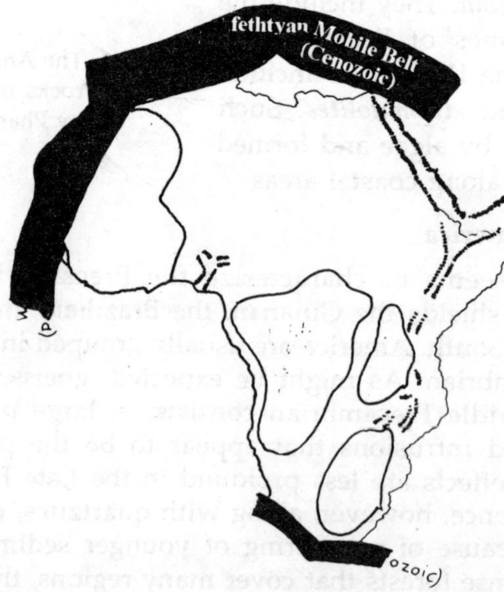

Fig. 4.10: Four large Precambrian cratonic segments formed in Africa by middle Precambrian time.

dense, basic volcanic rocks that appear to be immersed in a "sea" of granite. Geologists speculate that these volcanics were extruded at a time when the crust was very thin and the upper mantle beneath it was extremely hot and possibly event molten. The volcanic fireworks and crustal jarring that accompanied the extrusions must have been spectacular.

More extensive than the basalts in Africa are the massive intrusions of old granties that were emplaced between 2.6 and 3.1 billion years ago. These constitute the "crystalline basement" of South Africa. A great unconformity separates these batholithic masses from younger sequences of the Middle Precambrian.

The African Middle Precambrian probably began about 2.6 billion years ago and ended 1.1 to 1.8 billion years ago. By this time, four stable segments of the evolving African crust had been formed. One large segment now constitutes much of the west African bulge, two large elements were established in central Africa, and one was formed in South Africa. Quartzites, shales, conglomerates, and lavas characteririze the rocks of the Middle precambrian. Except for occasional volcanic eruptions, it was a quieter time than the Early Precambrian had been. The crustal segments were relatively stable.

The Late Precambrian of Africa is characterized by the deposition of mentary rocks and by widespread orogenies in the regions that considered the old crustal segments. Strips of metamorphosed granitic crust med between the segments and welded them into a unified shield. The genic activity continued until about 400 million years ago and resulted continent's taking on its present-day outlines.

India

From a geologic point of view, India can be separated into two distinctly different regions. To the north is the Himalayan orogenic belt at developed long after the Precambrian, whereas to the south lies the dian Shield. It is now believed that the shield was brought into extaposition with the Himalayan region at a comparatively recent date northward drifting of what was once an Indian "island continent."

As is the case in North America, geologists have been able to recognize a number of separate provinces within the Indian Shield that can be distinguished by

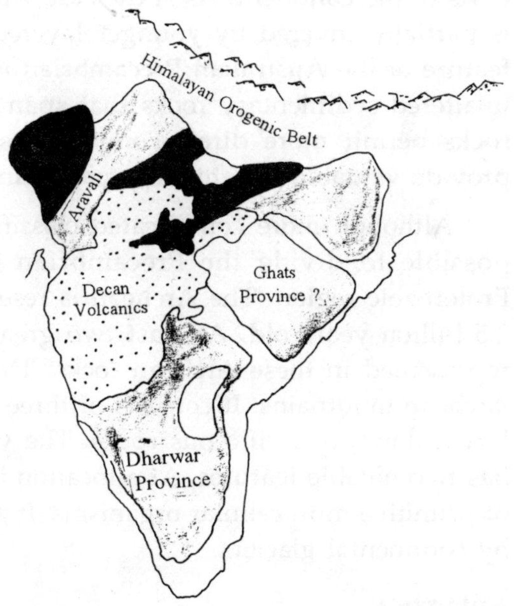

Fig. 4.11: Precambrian geologic segments of India.

the approximate dates at which they had experienced orogeny and by their unique structural characteristics. Altogether, there are five of these provinces. The oldest is the Dharwar Province, which contains rocks 2.4 or more billion years old and probably acted as a nucleus around which other provinces became attached. The granites, gneisses, and volcanics of the Dharwar are exposed along the eastern side of the Indian peninsula. Northeastward lies the 1.6-billion-year-old Ghats Province, which is composed of generally similar rocks. Over 500,000 sq km of the Indian peninsula southwest of Delhi are covered by Late Cretaceous and Early Cenozoic basalts. These flood basalts blanket large areas of the shield. At their northeastern edge, geologists have found evidence of a third province, the Satpura. The Satpura probably stablized about 100 million years ago. Northwest of the Satpura Province is the Aravalli Province, which experienced its culminating orogenies about 750 million years ago. The fifth province contains stratified shales, sandstones, and limestones that rest unconformable on older rocks. They are called the Vindhyan Group and are considered Late Precambrian by the Geological Survey of India. However, part of the Upper Vindhyan strata may be Cambrian in age. These uppermost beds are of particular interest, because they contain small, dislike structures that may be fossils of multicellular animals.

Australia

Structurally, Australia is composed of mobile belts along the eastern third of the continent and a vast Precambrian shield that occupies most of the central and western parts of the continent. As is the case with many other shields, the Precambrian surface is partially covered by younger layered sedimentary rocks. one especially interesting feature of the Australian Precambrian is the presence of a thick sequence of relatively unaltered sedimentary rocks that span a time interval of over 1.5 billion years. Such rocks permit more direct comparisons with modern sedimentational processes and provide valuable insights into paleoclimatic conditions.

Although more complicated classifications have been proposed, it is nevertheless possible to divide the Precambrian of Australia into an Archean region and a Proterozoic region. The Archean is reserved for igneous and metamorphic rocks over 2.5 billion years old. At least two great cataclysms of mountain building seem to be represented in these Archean rocks. The Proterozoic rests upon the worn roots of the Archean moutnains. It consists of three great systems of quartzitic sandstones, basaltic larval flows, and siliceous rocks. The youngest of these rocks, called the Adelaidean, has two notable features. At a location in Edicara Hills, the Adelaidean contains fossils of primitive multicellular organisms. It also contains rocks that were probably deposited by continental glaciers.

Antarctica

Beneath the frozen surface of Antarctica lies a Precambrian shield that was poorly understood until the advent of modern techniques for geophysical exploration. Over

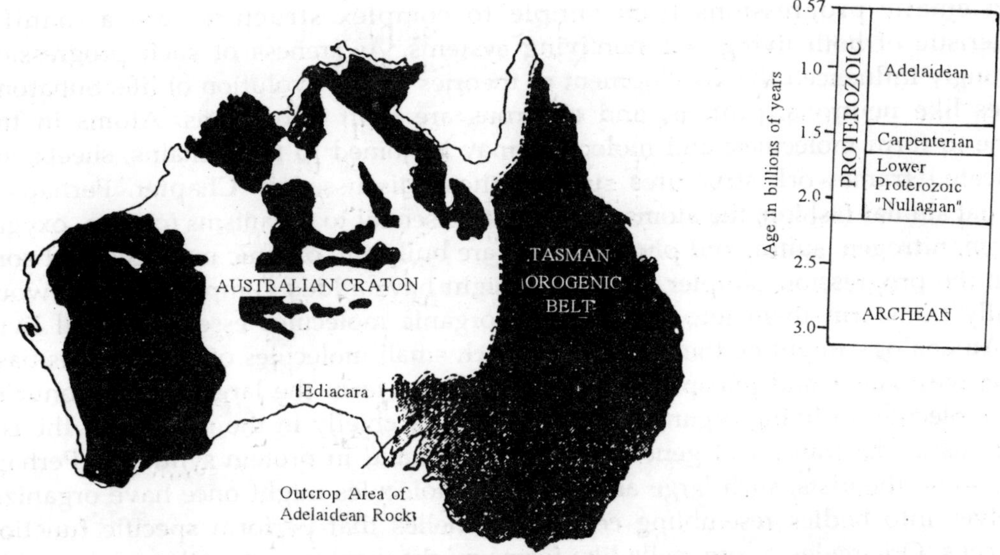

Fig. 4.12: Central and western Australia consists of a vast cratonic region composed of Precambrian rocks exposed as large patches wherever the cover of mostly horizontal Phanerozoic rocks is absent. The eastern third of the continent contains folded Paleozoic strata of the Tasman Orogenic Belt. Inset indicates Australian precambrian time classification.

the past 3 decades, geologists have had an opportunity to examine the bedrock exposed around the edges of the glacier and in moutnain ranges that pierce the ice cover. Most of the rounded eastern half of Antarctica consists of Precambrian shield. The oldest rocks, sometimes referred to as the Archean, occupy the most extreme eastern edge. Rocks of the Proterozoic occupy the remainder of the shield.

Most of the western part of Antarctica consists of fold belts of Late Cretaceous and younger age. These moutnainous tracts show a marked similarity to the Andean mobile belt of South America.

The Begining of Life

The splendid diversity of life today, and as it is recorded in the Phanerozoic fossil record, is primarily the result of billions of years of chemical and biologic evolution that occurred during the Precambrian. There are few clues to the events that led to the first organism. Hypotheses are based on our understanding of modern biology and on our most rational inferences about the nature of the planet 3 or 4 billion years ago. A coherent story of the stages involved in the origin of life has emerged. However, it is a decidedly theoretic story and should not be viewed as undisputed fact.

From Simple to Complex

Systematic progressions from simple to complex structures are a manifest characteristic of both living and nonliving systems. Awareness of such progressions has strongly influenced the development of theories on the evolution of life. Subatomic particles like neutrons, protons, and electrons are built into atoms. Atoms in turn combine to form molecules; and molecules may be joined to form chains, sheets, and complicated framework structures such as those discussed in Chapter. Perhaps in somewhat similar fashion, the atoms of elements essential to organisms (carbon, oxygen, hydrogen, nitrogen, sulfur, and phosphorous) are built into organic molecules. At some stage in the progression, simpler molecules might have added components that would gradually transform them into the complex organic molecules essential to all living thing. An analogy might be the manner in which small molecules of nitrogenous bases combine with sugar and phosphorous compounds to form the large deoxyribonucleic (DNA) molecule. In living organisms, DNA is found chiefly in the nucleus of the cell. It functions in the transfer of genetic characteristics and in protein synthesis. Perhaps, suggest some theorists, such large and complex molecules might once have organized themselves into bodies resembling certain organelles that perform specific functions within cells. Organelles or organelle-like forms might then come together and combine to form simple unicellular structures capable of growth, reproduction, and other attributes that characterize living things. Could a progression even remotely similar to this have actually occurred in the early years of the Precambrian? An unequivocal answer is not yet possible, but at least newly available evidence indicates that such a progression toward life is feasible.

Preliminary Considerations

The oldest fossil evidence for life is found in rocks that are 3.2 billion years old. To produce these simple organisms, life must have originated before that time. If our theory for the evolution of the atmosphere is valid, then the earliest organisms developed on a planet deficient in free oxygen. They were *anaerobic* and did not require oxygen for respiration. The environment in which they lived lacked a protective atmospheric shield of ozone, which is derived from oxygen and which absorbs ultraviolet radiation from the sun. Perhaps early life escaped the lethal ultraviolet rays by developing beneath protective ledges of rock or at deeper water levels. Some biologists insist that in the late stages of chemical evolution, ultraviolet radiation may not have been a hindrance at all but rather served to impel the molecules to interact and to produce the more complex structures that were precursors to true organisms.

Prior to 1828, most geologists believed that all organic molecules were products of living organisms. However, in that year, the German chemist *Frederich Wohler* accidentally discovered that by heating the inorganic compound now called ammonium cyanate, he could produce crystals of urea. Urea, an essential component of urine, is a decidedly organic compound. In the 3 decades that followed, other chemists succeeded in producing several other simple organic substances. Scientists began to favour the view that life may have ultimately arisen from inorganic materials. This view continued to be held for a long period even before experimenters found the means to detect the small quantities of complex organic molecules synthesized in their experiments from inorganic substances. It was not until the late 1960's that biochemists were able to create and detect complex protein molecules.

The elements necessary to produce organic compounds are carbon, oxygen, hydrogen, nitrogen, hosphorus, and sulfur. These elements are also the main components of a living cell. They are among the most abundant elements in the solar system. Clearly, the elementary materials for the development of life were present on the primitive earth. What had to be accomplished was the utilization of those materials in the enormously intricate fabrication of a living cell.

Reduced to a fundamental definition, life can be viewed as a system with four basic componens. The first of these is *proteins*. Proteins are essentially stringes of comparatively simple organic molecules called *amino acids*. Proteins act as building materials and as compounds that assist in chemical reactions within the organism. The second of the basic components is *nucleic acids*, such as DNA, mentioned earlier, and ribonucleic acid (RNA). *Organic phosphorous compounds* provide a third component of life. They serve to transform light or chemical fuel into the energy required for cell activities. The fourth essential for life is some sort of container, such as a *cell membrane*. The enclosing membrane provides a relatively isolated chemical system within the cell and keeps the various components in close proximity so that they may interact.

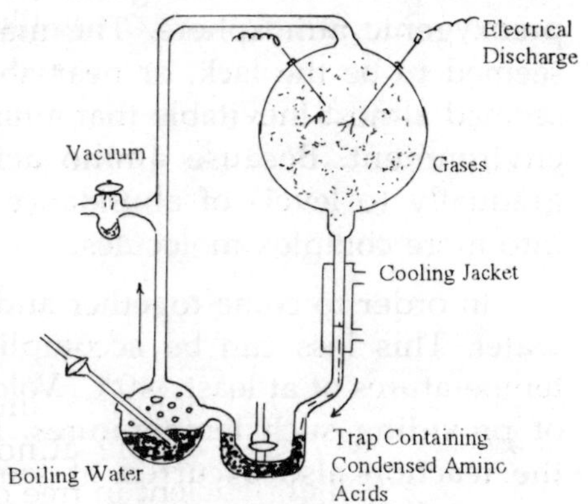

Fig. 4.13: Diagram of the apparatus used by S. Miller in experiments of prebiotic synthesis. The mixture of gases (methane, hydrogen, and ammonia) are circulated through the apparatus together with water vapour by boiling water in a flask. The mixture is subjected to an electrical discharge in a second flask and is condensed by a colling jacket.

From the previous description, it is apparent that amino acids have an important role in the development of the larger and more complex molecules. They are the building blocks of proteins. Two environmental circumstances in the early years of earth history may have been important in the natural synthesis of amino acids. Prior to the accumulation of the ozone layer in the earth's upper atmosphere, ultraviolet rays bathed the earth's surface. As demonstrated in experiments, ultraviolet radiation is capable of separating the atoms in mixtures of water, ammonia, and hydrocarbons and of recombining those atoms into amino acids. A second form of energy capable of accomplishing this feat is electrical discharge in the form of lighting. Either together or separately, lighting and ultraviolet radiation may have stimulated the production of amino acids in the air, in tidal pools, in the upper levels of the oceans, and wherever suitable environmental conditions existed.

Test Tube Amino Acids and a Step Beyond

As noted earlier, scientists in the mid-nineteenth century had succeeded in manufacturing some relatively simple organic compounds in the laboratory. However, it was not until 1953 that the laboratory systesis of amino acids and molecules of roughly similar complexity was announced. *Stanley Miller,* at the suggestion of *Harold Urey,* performed the now famous experiment. He infused an atmosphere at that time thought to be like that of the earth's earliest atmosphere into an apparatus similar to the one. It was a methane, ammonia, hydrogen, and water vapour atmosphere. As the mixture was circualted through the glass tubes, sparks of electricity (stimulated lighting) were discharged into the mixture. At the end of only 8 days, the condensed water in the apparatus had become turbid and deep red. Analysis of the crimson liquid showed that it contained a bonanza of amino acids as well as somewhat more complicated organic compounds that enter into the composition of all living things. In additional experiments by other biochemists, it was shown that similar organic compounds could also be produced from gases (carbon dioxide, nitrogen, and water vapour) of the preoxygenic atmosphere. The main requirement for the success of the experiments seemed to be the lack, or near absence, of free oxygen. To the experimenters, it now seemed almost inevitable that amino acids would have developed in the earth's prelife environment. Because amino acids are relatively stable, they probably increased gradually to levels of abundance that would enhance their abilities to join together into more complex molecules.

In order to come together and form protein-like molecules, amino acids must lose water. This loss can be accomplished by heating concentrations of amino acids to temperatures of at least 140°C. Volcanic activity on the primitive crust would be capable of providing such temperatures. However, the biochemist *S. W. Fox* discovered that the reaction also occurred at temperatures as low as 70°C if phosphoric acid was present. *Fox* and his coworders were able to produce protein-like chains from a mixture of 18 common amino acids. He termed these structures *proteinoids* and reasoned that billions of years ago they were the transitional structures leading to true proteins. This is not extravagant conjecture, for *Fox* was able to find proteinoids similar to those he created in his laboratory among the lavas and cinders adjacent to the vents of Hawaiian volcanoes. Apparently, amino acids formed in the volcanic vapours and were combined into proteinoids by the heat of escaping gases.

Hot aqueous solutions of proteinoids will, on cooling, form into tiny spheres that show many characteristics common to living cells. These *microspheres,* as they are called, have a filmlike outer wall; are capable of osmotic swelling and shrinking; exhibit budding, as do yeast organisms; and can be observed to divide into "daughter" microspheres. They occasionally aggregate linearly to form filaments, as in some bacteria, and they exhibit a streaming movement of internal particles similar to that observed in living cells.

Although complete long-chain nucleic acids have not yet been experimentally produced under profile conditions, short stretches of ordered sequences of nucleic acid components have now been produced in the laboratory. Indeed, in 1976, *Har Gobind Khurana* and his associates at the Massachusetts Institute of Technology announced that they had made a functioning artificial gene-one of thousands in the DNA spiral molecule of the bacterium *Escherichia coli*.

Speculations about Earliest Life

There are several good reasons to believe that the earliest organisms originated in the sea. The sea contains the salts needed for health and growth. The waters of the oceans serve as universal solvents capable of dissolving a great variety of organic compounds, and currents in the oceans ceaselessly circulate and mix these compounds. Such constant motion would have favoured frequent collisions of the vital molecules and would thereby have increased the probability of their combining into larger bodies. One can imagine the larger bodies adding components, increasing in complexity, and ascending in a thousand infinitesimally small steps from nonliving the transitional things of a biologic Netherlands to the first living organisms. Such a sequence could not occur in an environment like that which exists today. Oxygen and present-day microbial predators would destroy the delicate structures. However, life orginated before there were organisms to cause decay and before there was sufficient free oxygen to be troublesome.

Largely from our knowledge of paleontology and biology, it is believed that the first living things were microscopic in size and were unicellular. It is likely that these earliest forms of life would not have evolved a means of manufacturing their own food but rather assimilated small aggregates of organic molecules that were also present in the surrounding medium. Some, undoubtedly, would have consumed even their developing contemporaries. Today, organisms with this type of nutritional mechanism are termed heterotrophs. The food gathered by the ancestral heterotrophs was externally digested by excreted enzymes before being converted to the energy required for vital function. In the absence of free oxygen, there was ónly one way to accomplish this conversion by *firmentation*. There are many variations of the fermentation process. The most familiar reaction involves the fermentation of sugar by yeast. It is a process by which organisms are able to disassemble organic molecules, rearrange their parts, and derive energy for life functions. A very simple reaction may be written as follows:

$$\underset{\text{Glucose}}{C_6H_{12}O} \rightarrow \underset{\substack{\text{Carbon}\\\text{Dioxide}}}{2CO_2} + \underset{\text{Alcohol}}{2C_2H_5OH} + \text{Energy}$$

Animal cells are also able to ferment sugar in a reaction that yields lactic acid rather than alcohol.

Sooner or alter, the original fermentation organisms were to experience difficult times. By consuming the organic compounds of their environment, they would eventually have created a food shortage; this scarcity, in turn, might have caused selective pressures for evolutionary change. At some point prior to the depletion of the food supply, organisms evolved the ability to synthesize their needs from simple inorganic substances. These were the first autorophic organisms. Unlike the heterotrophs, *autotrophs* were able to manufacture their own food. The organisms that had developed this remarkable ability saved themselves from starvation. Their evolution proceeded in diverse directions. Some manufactured their food from carbon dioxide and hydrogen sulfide and were the probable ancestors of today's sulfur bacteria. Others employed the life scheme of modern nitrifying bacteria by using ammonia as a source of energy and matter.

However, more significant than either of these kinds of autotrophs were the *photoautotrophs,* which were capable of carrying on photosynthesis. Their gift to life was the unique capability of dissociating carbon dioxide into carbon and free oxygen. The carbon was combined with other elements to permit growth, and the oxygen escaped to prepare the environment for the next important step in the evolution of primitive organisms. In simplified form, the reaction for photosynthesis can be written as follows:

$$6CO_2 + 6H_2O \xrightarrow{\text{Sunlight}} 6C_6H_{12}O_6 + 6O_2$$

With the multiplication of the photoautrophs, billions upon billions of tiny living oxygen generators began to change the primeval anoxygenic atmosphere to an oxygenic one. Fortunately, the change was probably gradual, for if oxygen had accumulated too rapidly, it would have been lethal to developing early microorganisms. Some sort of oxygen acceptors were needed to act as safety valves and prevent too rapid a buildup of the gas. Iron in rocks of the continental crust provided suitable oxygen acceptors. Eventually, organisms evolved oxygen-mediating enzymes that permitted them to cope with the new atmosphere.

After the surficial iron on the earth's surface had combined with its capacity of oxygen, the gas began to accumulate in the atmosphere and hydrosphere. Solar radiation acted upon atmospheric oxygen to convert part of it to ozone, and ozone in turn formed an effective shield against harmful ultraviolet radiation. Still-primitive and vulnerable life was thereby protected and could expand into environments that formerly had not been able to harbor life. The stage was set for the appearance of aerobic organisms.

Aerobic organisms use oxygen to convert their food into energy. The reaction, which can be considered a form of cold combustion, provides far more energy in relation to food consumed than does the fermentation reaction. This surplus of energy was an important factor in the evolution of more complex forms of life.

Prokaryotes, Eukaryotes, and Symbiosis

There is no unequivocal evidence to date the transition from chemical or prebiotic evolution. The most accurate statement that can be made in this regard is that the transition occurred at some time prior to 3.2 billion years ago. In rocks of that age, paleontologists have discovered the earliest fossil evidence of life. These oldest fossils belong to a category of organisms called *prokaryotes* and are represented today by bacteria and blue-green algae. Primitive organisms such as these lack definite internal organell es and do not have a membrane-bound nucleus in which genetic material is neatly arranged into discrete chromosomes. Modern, prokaryotes do possess cell walls, and most are able to move about. The blue-green algae and some bacteria are prokaryotes capable of photosyntheis.

Prokaryotes are asexual and thereby restricted in the level of variability they can attain. In sexual reproduction, there is a union of gametes (egg and sperm) to form the nucleus of a single cell, the *zygote*. The formation of the zyote results in recombination of the parental chromosomes and leads to a multitude of gene combinations among the gametes that give rise to the next generation. In the asexually reproducing prokaryotes, in contrast, a cell divides from the parent and becomes an independent individual containing the identical chromosome complement of the parent cell. Unless mutation intervenes, the number and kinds of chromosomes in individuals produced by asexual reproduction are exactly the same as those in the parent. Thus, the possibilities for variation are more limited than in sexually reproducing organisms. It is probably for this reason that prokaryotes have shown little evolutionary change through over 2 billion years of earth history. Nevertheless, such organisms represent an important early step in the history of primordial life.

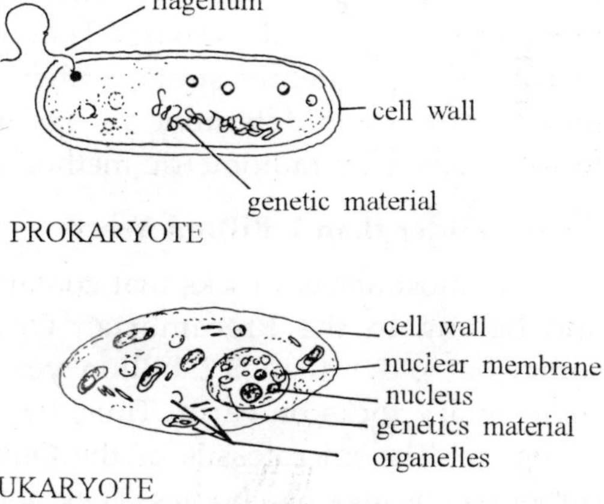

Fig. 4.14: Difference between a eukayote and a prokayote.

Evolution proceeded from the prokaryotes to organisms with a definite nuclear wall, well-defined chromosomes, and the capacity for sexual reproduction. These more advanced forms were *eukaryotes*. Unlike prokaryotes, eukaryotes contain organelles such as chloroplasts (which convert sunlight into energy) and mitochondria (which metabolize carbohydrates and fatty acids to carbon dioxide and water, releasing energy-rich phosphate compounds in the process). Biologists believe that the organelles in eukaryotic cells were once independent microorganisms that entered other cells and then established symbiotic relationships with the primary cell. For example, an anaerobic, heterotrophic prokaryote like a fermentative bacterium might have engulfed a respiratory prokaryote and thereby would have an internal consumers of the oxygen that might otherwise threaten its existence. A nonmotile organisms might acquire mobility by forming a symbiotic association with a whiplike organism like a spirochete.

The resulting cell would appear to have a flagellum for locomotion. Natural selection would clearly favour such advantageous symbiotic relationships.

THE FOSSIL RECORD FOR THE PRECAMBRIAN

Problems in Interpreting the Early Precambrian Fossil Record

Paleontologists seeking traces of the earth's earliest organisms are confronted with rather formidable problems. They must determine not only if a given microstructure is of biologic origin but also if it is truly contemporaneous with the enclosing sediment. The possibility of contamination by microorganisms coming into the rock postdepositionally must be recognized. In cases where chemical traces of life are detected, investigators must evaluate whether the compounds might have leaked into the rocks long after they were deposited or might have formed after deposition by more recent bacterial activity. There are also hazards associated with dating the presumably fossiliferous strata. Rather than directly dating the fossil-bearing formation, investigators often determine its age by correlation and superpositional relationships to rocks dated by radiometric methods.

Fossils Older than 2 Billion Years

The most ancient rocks that contain fossil microorganisms outcrop in South Africa and belong to the Precambrian Onverwacht Series. Onverwacht sediments are somewhat older than 3.2 billion years and are very likely the oldest little-altered sedimentary rocks on earth. Thus, the probability of finding significantly older fossils is remote. The microfossils of the Onverwacht consist of spheroidal and cup-shaped carbonaceous alga-like bodies. They not only are morphologically similar to some living forms of primitive algae but also are closely associated with filamentous structures that contain carbon compounds of biogenic origin.

Approximately 10,000 meters above the Onverwacht beds is another formation that has achieved fame because of its content of primitive fossils. Its name, somehow reminiscent of another story of genesis, is the *Fig. Tree Formation*. The Fig Tree rocks consist of variously colored cherts, slates, ironstones, and tough sandstones. By means of radioactive isotopes, the formation has been dated as 3.1 billion years old. In 1967, samples of the Fig. Tree Formation were studied by Harvard university paleobotanist, *Elso S. Barghoorn,* and his former graduate student, J. William Schopf. With the use of the electron microscope, these scientists were able to find a number of tiny, double-walled, rod-shaped structures that had a striking resemblance to modern bacteria. Barghoorn and Schopf named their find *Eobacterium isolatum*-the "isolated dawn bacteria." In their search for larger fossils, the scientists prepared thin sections of the chert and examined these with the optical microscope. The examination disclosed numerous spheroidal bodies very similar to certain blue-green algae. Because the samples were collected near the town of Barberton, Barghoorn and Schopf named the fossils *Archaeosphaeroides barbertonensis*. The filaments of organic matter and hydrocarbons of organic derivation found in the Fig. Tree Formation (and later in the older Onverwacht rocks) confirmed the existence of a primitive but vigorous flora of microscopic life in the Early Precambrian.

The presence of photosynthetic organisms among the Fig. Tree fossils has not yet been positively established. Like the Onverwacht rocks, the Fig. Tree chert is black and rich in carbon. Biochemists have analyzed the carbonaceous materials extracted from these rocks and have detected organic compounds normally formed during the alteration of a particular component of chlorophyll called the porphyrin-magnesium complex. However, vanadium-porphyrin complexes also result from geochemical alteration of chlorophyll, and these have not yet been detected in the Fig. Tree rocks. Two other organic compounds, phytane and pristane, are also regarded as breakdown products of chlorophyll. Both of these substances were detected in the extracts. However, this evidence has been judged inconclusive because some nonphotosynthetic microorganisms may also be capable of producing phytane and pristane.

Additional evidence for the existence of photosynthetic organisms in the Early Precambrian is provided by structures called stromatolities. Stromatolites are distinctly laminated accumulations of calcium carbonate having rounded, cabbage-like, branching or frondose shapes. Today, the metabolic activities of certain marine colonial blue-green algae result in the formation of similar structures. The fine particles of calcium carbonate settle between the minute filaments of the matlike algal colonies and are temporarily bound within a film of gelatious organic matter. Successive additional layers result in the laminations. The stromatolieis are thought to have had a similar origin, particularly because of the discovery of concentrations of filamentous and spherical blue-green algae in the laminations of Precambrian stromatolities.

Near Bulawayo in southern Rhodesia, stromatolites have been found in limestone that is 2.7 billion years old. The carbon 12/carbon 13 ratio in the Bulawayo rocks suggests the occurrence of biologic fixation of carbon dioxide by photosynthetic organisms. Of equal importance is that it suggests biologically generated atmosphetic oxygen was present about 2.7 billion years ago.

Although stromatolites are found in the Early Precambrian, they do not become common until Middle Precambrian. Late Precambrian stromatolities are sufficiently widespread and abundant to serve as guide fossils. They have formed extensive reeflike structures in Precambrian limestones. In the United States, the most notable of these stromatolitic structures occurs in rocks of the Belt series. They are exceptionally well exposed in Belt strata of Glacier National Park. The photosynthetic activity of stromatolite microorganisms was probably important in causing the precipitation of the calcium carbonate that formed the thick layers of Belt limestone.

Modern stromatolites grow in the intertidal zone with their tops at the high-water mark. In this regard, it is interesting to note that some Late Precambrian stromatolities were approximately 6 meters in height. If these forms were also restricted to the intertidal zone, then Precambrian tides must have been considerably higher than they are today. Such high tides would indicate that the moon was closer to the earth during the Precambrian. More recent stromatolites are not as tall, suggesting the distance between the earth and the moon has been increasing, causing a corresponding decrease in tidal amplitude. It was noted that the length of the day on earth has been increasing. Conservation of angular momentum would indeed require a slowing down of the earth's rotation as the moon increased its distance from the planet. Thus, stromatolite studies are in accord with astronomic observations.

The Gunflimt Flora

Extending eastward from Thunder bay in the northern portion of Lake Superior are outcrops of a rock unit called the *Gunflight Cheri,* Radiometic age determinations indicate that the formation is approximately 1.9 billion years old. It contains a varied and abundant flora of the so-called thread bacteria and nostocalean blue-green algae. Unbranched filamentous forms, some of which are septate, have been given the name *Gunflintia*. Moe finely septate forms, such as *Animikiea*, are remarkably similar in appearance to such living algae as *Oscillatoria* and *Lyngbya*. Other Gunflint fossils, such as Eoastrion (the "dawn star") resemble living iron-and magnesium-reducing bacteria *Kakabekia* and *Eosphaera* are so different from any known microorganism that their classification is uncertain. That these and other Gunflint and stromatolitic organisms were actively producing oxygen and thereby altering the composition of the atmosphere does seem certain. Not only do they resemble living photosynthetic organisms but also their host rock contains phytane and pristane-organic compounds regarded as the breakdown products of chlorophyll.

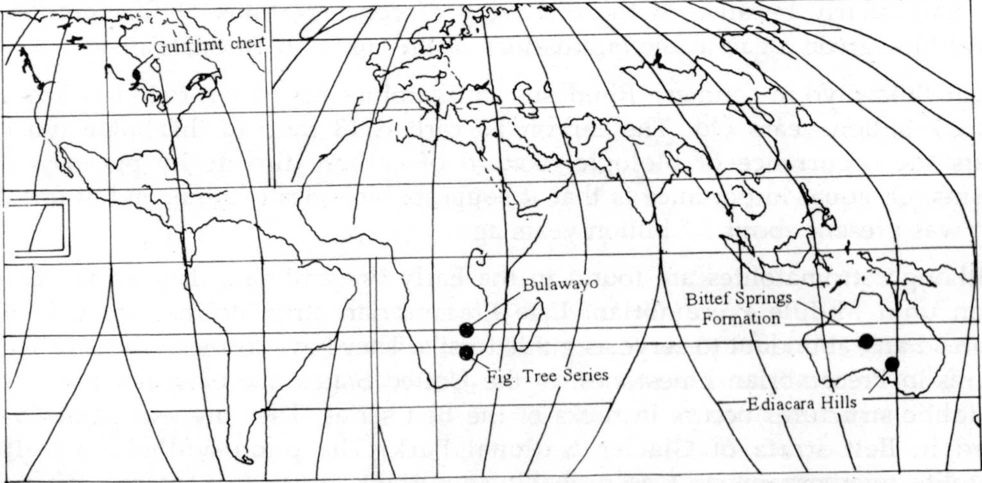

Fig. 4.15: Location of some of the better known fossil-bearing Precambrian formations.

One of the major events in biologic evolution was undoubtedly the origin of the eukaryotic cell, with its enclosed nucleus for the storage and transmission of genetic information. Although the potential for sexual reproduction provided by eukaryotes enormously increased the possibilities for evolutionary change, the fossil record for earliest eukaryotic organisms is ambiguous. This is not surprising, since the first eukaryotes probably were simple microscopic unicellular organisms that in the altered fossil state would be difficult to distinguish from prokaryotes. At the present time, one group of paleobotanists suggests that eukaryotes were present about 0.9 billion years ago. These scientists support their view with microfossils found in the Bitter Springs Formation of central Australia. In addition to a prokaryote assemblage of fossils, the cherts of the Bitter Springs yielded several forms that are the size of eukaryotes and have ghostly internal structures that might be the remnants of nuclei or organelles of eukaryote green algae. If these are indeed the structures they appear to be, then the passage from prokaryotic to eukaryotic life with its capacity for sexual reproduction and genetic variability had been accomplished at least as long ago as 0.9 billion years. However, the identification of some unicellular organisms in the Bitter Springs as eukaryotic is currently being challenged by those who doubt that the internal structures are indeed eukaryotic organelles.

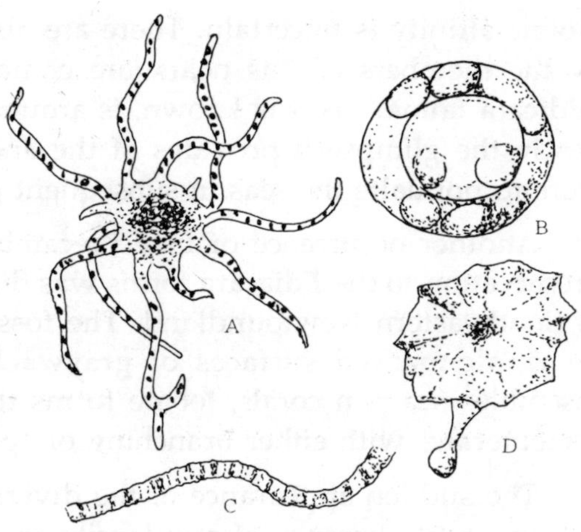

Fig. 4.16: Eoastrion (A). Eosphaera (B). Animikiea (C) and Kakabekia (D) from the Gunfint Cherts. All specimens are drawn to the same scale. Eosphaera is about 30 microns in diameter.

Dawn Animals

As the search for Precambrian fossils continued, it was almost inevitable that some trace of the more advanced forms of life would be discovered. Indeed, since the early 1960's, a fascinating assortment of advanced Precambrian fossils has been found in Australia, Siberia, and Africa. The fossils consist primarily of impressions in sedimentary rocks of the types of animals that can be loosely grouped under the category of *metazoans*. Metazoans are multicellular, possess more than one kind of cell, and have different kinds of cells organized into tissues and organs.

The best known fossils of Precambrian metazoans were found as well-preserved impressions in the consolidated sands of an ancient beach now called the Pound Quartzite. Exposures of the rock in the Ediacara Hills of south Australia have yielded over 600 specimens of soft-bodied animals. The collection includes several different kinds of jellyfish, soft coarls resembling the modern "sea pen," polychaete worms, echinoderms, arthropod-like animals, and a number of unique forms whose precise biolgic affinity is uncertain. There are also large number of tracks and burrows made by the members of this nearshore community. As nearly as can be determined, the "Edicara fauna," as it is known, is around 650 to 700 million years old. Its importance lies in the glimpse it provides of the ancestors to the vast array of invertebrates that were to popualte the seas in subsequent geologic time.

Another occurrence of Late Precambrian meazoan fossils that is rather similar in preservation to the Ediacara forms was discovered in the rocks of the Conception Group of Southeastern Newfoundland. The fossils at that site occur as hundreds of imprints on ripple-marked surfaces of graywackes. They include leaf-shaped animals that resembles sea pen corals, lobate forms that appear to be jellyfish, and other probable coelenterates with either branching or spindle shapes.

The sudden appearance of the diverse Edicara and Conception Group fossils, and the apparent abasence of any fossils or even traces of animal life in rocks older than 700 million years, suggests that metazoans expanded abruptly near the end of the Precambrian. According to *Berkner* and *Marshall* (1964), their rapid expansion may very well correlate with the accumulation of sufficient free oxygen to permit oxidative metabolism in organisms. On the other hand, the Ediacara life may have evolved more gradually from earlier small and naked forms that were essentially incapable of leaving a fossil record. Perhaps, as suggested by *A.G. Fischer* (1965), the ancestral metazoans lived in "oxygen oases" in which marine plants were concentrated. After the atmospheric oxygen content had reached about 1 per cent cent of the present level, these as-yet undiscovered animals would have been free to leave their oases and spread widely in the seas. According to this view, the evolutionary development of metazoans may not have been abrupt, but their dispersal may have been rapid after suitable conditions became prevalent.

The Fig. Tree, Gunflint, and Pound Formations represent only brief and isolated glimpses of the progress of biologic evolution during an almost incomprehensible span of more than 3 billion years. Unlike the fossil record, life is a continuous progression. Other exposures of Precambrian rocks will continue to be scrutinized in the years ahead to find the many missing chapters in the history of Precambrian life. Figure depicts the current state of our knowledge.

ASPECTS OF PRECAMBRIAN CLIMATE AND ENVIRONMENT

Because so much of the Precambrian rock sequence is either alteres or devoid of fossils, interpretations of climate and environment are more difficult than in the rocks of younger eras. This is especially true for the oldest Precambrian rocks. As noted previously, the earth's early atmosphere and hydrosphere were largely devoid of free oxygen. Then about 2 billion years ago, the atmospheric oxygen began to accumulate because of increased plant activity. One result of the oxygen buildup was the accumulation on land of considerable amounts of ferric iron oxide, which stained terrestrial sediments a rust-red color. Such sedimentary rocks are known as *red beds* and are considered a valid indication of the advent of an oxygenic environment. Of course, the oxygen level probably rose slowly and very likely did not approach 10 per cent of present atmospheric levels of free O_2 until the Cambrian Period.

In general, Precambrian rocks provide evidence for a wide range of climatic conditions, but there is no indication that these climates were especially unique in comparison with those of the Phanerozoic Eras. Thick limestones and dolostones with reeflike algal colonies were deposited along the Precambrian equator, where warm tropical conditions prevailed much as they do today. During the Early Proterozoic, the equator lay close to the northern border of North America. Precambrian evaporite deposits in eastern Canada and in Australia suggested that conditions were periodically rather arid during the Middle Proterozoic. In the middle and low latitudes, cliamtes were more severe, as indicated by consolidated deposits of glacial debris (tillites) and glacially straiated basement rocks. The best known of these poorly sorted, thick, boulder-like deposits is the Gowaganda Tillite. Its widespread distribution over the southern portion of the Canadian shield indicates the presence of continental, rather than mountain, glaciers. In deed, geologic mapping suggests that the ice sheet was probably more than 1500 km in diameter and covered most of central Candada. The Gowganda Formation has been dated by radiometric methods as about 2.3 billion years old. Strations on the rock surfaces beneath the tillites suggest that the ice moved toward the Precambrian North Pole, which was located near lat. 22°N, long. 97°W at the time. These Precambrian glaciers ground their way northward into the Arctic ocean from the Canadian and Baltic Shields.

The Gowganda tillites, although impressive, are not the only evidence of glaciation during the precambrian. In Southern Africa, the Chuos Tillite attains thicknesses of over 450 meters and has been recognized across an expanse of over 30,000 sq.km. Tillites of Late protecrozoic age have been found on all continents except Australia, although they are not necessarily synchronous. The ample evidence of Precambrian glaciation around the world clearly indicates that the recent Pleistocene ice age was not at all a unique event in the earth's geologic history.

THE MINERAL WEALTH OF THE PRECAMBRIAN

Precambrian rocks contain a host of metal orea that have been of inestiamble economic importance. Major sources of iron, nicked, gold, silver, chromium, and uranium are derived from precambrian rocks. Iron is most notable of these ores in term of tonnages that have been mined. The world's largest group of iron are localities is in the Canadian Shield. Of these, the most famous are the sedimentary ores of the Lake Superior region. The ore deposits were formed in local areas where the iorn-bearing sedimentary rocks were altered and enriched by removal of nonmetallic constituents. Other major sources of Precambrian iron occur in Sweden, the Ukraine, South Africa, and South America.

About half of the world's gold is mined from Precambrian quartz conglomerates near Johannesburg, South Africa. A Precambrian gabbro intrusion at Sudbury, Ontario, provides 70 per cent of the worls' nickel. Uranium of Precambrian age is mined in lartest quantities in Ontario and South Africa. Enormous amounts of copper were once recovered from Proterozoic rocks along the Keweenaw Peninsula of Lake Superior, however, today the ores are almost exhausted. These abandoned mines provide mute testimony to the fact that every mineral deposit is exhaustible and irreplaceable.

Chapter—5

Fossilisation

Fossils owe their existence to a break in the natural cycle. Normally all of the material used in living organisms is recycled but fossilisation represents a transfer of material from the biosphere to the lithosphere. In the long term, of course, weathering of rocks may return material to the biosphere. The three stages of the process of fossilisation are mortality, biostratinomy, and diagenesis. The whole field is sometimes called taphonomy.

MORTALITY

Here we are concerned with the cause and consequences of death. In nature very few organisms die simply of old age. Far more die prematurely as a result of suffocation, thirst, inadequate light, severe temperature or pressure changes, poisoning, illness, parasites; injury, or predation. The cause of death can only rarely be established in fossils. Examples include encasement in resin or amber, drowning in asphalt (Pleistocene of Rancho La Brea, California) or pitch (Pleistocene of Starunia, West Ukraine) and predation, when the material is preserved in coprolites. In other cases death may have resulted from a number of different causes, for example, an organism buried by sediment might die of asphyxiation, hunger, or pressure of overburden.

An abnormally high number of fossils is often attributed to mass death. At present such events occur where water is polluted by phytoplankton toxins or becomes chilled or oxygen deficient (for example, by heating in tropical lagoons or by upwelling of H_2S-rich water, or by being covered by ice for a long period). Care should be taken, however, when attempting to apply such modern observations to fossil occurrences.

Traces left by death-throes are very scarce but we can sometimes observe the attitude of death and the effects of rigor mortis where the contraction of ligaments and tendons may cause the head to be thrown back in the vertebrates.

BIOSTRATINOMY

Biostratinomy is concerned with the fate of an organism's body from the time of death until it is finally buried by sediment.

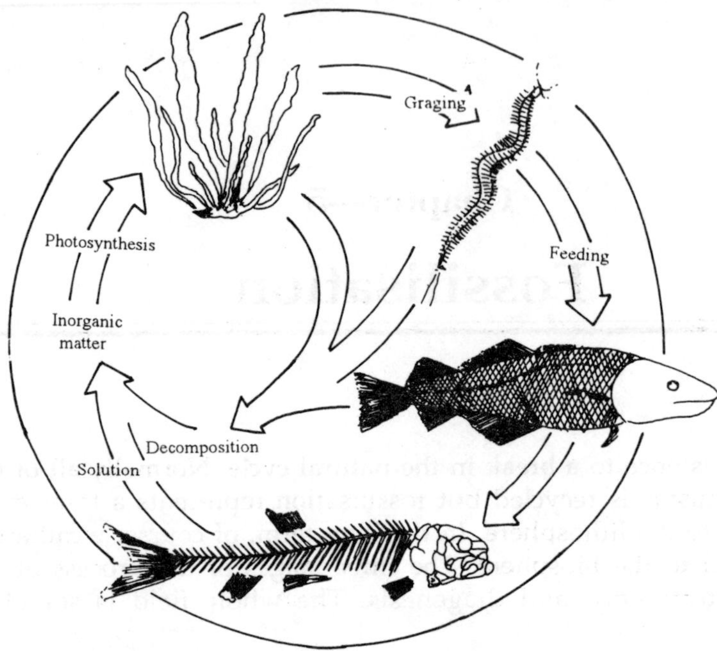

Fig. 5.1: The natural cycle. Decomposition of organic matter replaces the stock of inorganic materials.

Soft Tissue

Among the soft tissue the most important organic components are proteins, fats, and carbohydrates. These begin to break down both chemically and bacterially immediately after death. The cause of this decomposition is controlled by environmental factors. In the presence of water and oxygen the organic substances break down into the simplest inorganic components such as CO_2 and H_2O under the attack of aerobic bacteria. Where oxygen is absent however, in stagnent environments, anaerobic bacteria break down the original molecules using some for their metabolism and converting the remainder into hydrocarbon mixtures of high molecular weight. This process may produce bituminous muds whose organic matter may subsequently convert to bitumens, oil, and natural gas, under the influence of higher temperatures and pressures. Stagnation also permits the preservation of organic matter. True preservation of soft tissue (or mummification) may occur in the absence of water in arid regions or where hygroscopic materials are present. It also results from freezing (mammoths in the Siberian permafrost zone) or incorporation of the body in resin, asphalt, syngenetic silica concretions or other media containing neither water nor oxygen. Often during slow decomposition of soft parts, the infiltration of mineral substances or relatively stable organic materials pseudomorphs the original structure (preservation of saurian skin, soft parts in the Eocene of the Geiseltal near Halle, Germany.)

Skeletal Material

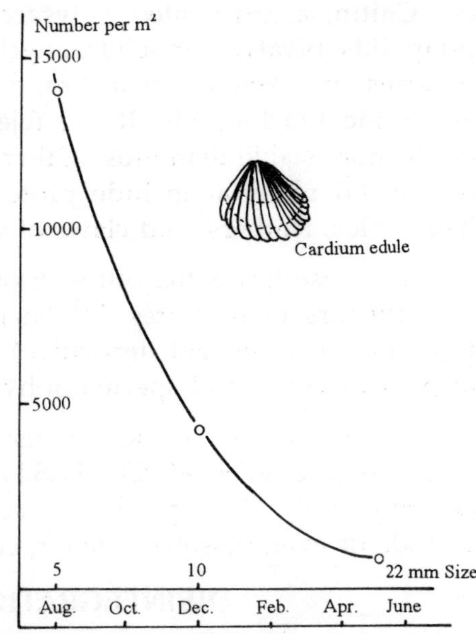

Fig. 5.2: Mortility of juvenile stages as shown by the bivalve Cardium edule from North Sea tidal flats.

Relatively few materials are available to the organic world for the formulation of protective and supporting hard parts. The most readily available are calcium carbonate (calcite and aragonite), calcium phosphate, opal, chitin and cellulose. Agglutinating organisms produce hard parts by cementing together foreign particles with an organic or inorganic cement.

Calcite is the preferred skeletal material of the archaeocyathids, octocorals, bryozoa, brachiopods, some annelids, polychaetes, many bivalves, pectinids, limids, ostreids, sonic cephalopod structures (nautiloid beaks, ammonite aptychi, belemmite rostra) several crustaceans (balanids, ostracods), echinoderms and red algae (Lithothamnia, Solenoporaceae). Calcite may include isomorphous $MgCO_3$ whose abundance increases with temperature. Precipitation of calcite is facilitated by warm, shallow water but by no means confined to that environment. Its solubility in pure water or in alkaline solutions is small (14 mg/l in pure H_2O), but in CO_2 saturated water its solubility increases to 1 gm/1; in acid media calcite is not preserved. In the modern oceans solution of calcite is particularly widespread in the cold depths of the oceans. In sediments, calcite is often leached from permeable strata but more readily preserved in fine grained and impermeable rocks.

Aragonite occurs in the hydrozoa, hexacorals, gastropods, most bivalve and cephalopod shells and the green algae. It is metastable in shallow marine environments but in fresh water it gradually alters to calcite at normal temperature and pressure. Its solubility is some 10% higher than that of calcite.

Calcium phosphate ($Ca_5(PO_4)_3(OH)$) is important to the vertebrates (bone material) and is also found in some brachiopods and arthropods (brilobites, crustacea). It is almost insolube in neutral solutions (0,01 gm/l) and is well preserved in weakly acid or alkaline environments. In more strongly acid environments or where it remains in contact with acid for a long period (as in peat bogs) calcium phosphate dissolves leaving bones fragile and flexible.

Opal forms the test of radiolaria, diatoms, siliceous sponges, and many flagellates. It is weakly soluble in water (river water (river water contains about 10 mg/l of S_1 O_2 in temperate regions and 30 mg/l in the tropics), stable in acids other than HF and slowly soluble in alkalis.

Chitin, a nitrificated polysaccharide occurs in algae, fungi, lichens, cnidarians, priapulids, bivalves, annelids, onychophores, pentastomids, arthropods and tentaculates. It forms the skeleton of arthropods (crustaceans, tribolites, insects, etc.), graptolites, and some brachiopods. It is subject to decomposition like other organic substances but is more stable than most. Other organic substances that are involved in the building of skeletal material include proteins spongin (in some sponges) and keratin (horn, hair, scales, feathers, and claws of vertebrates).

Cellulose forms the cell wall of plants and thereby gives them a certain rigidity. It also appears in tunicates. In the presence of air cellulose breaks down to CO_2 and H_2O. Lignin (a benzol derivative) is an important structural constituent in the wood of pteridophytes and spermatophytes.

Decomposition of the soft tissue affects the preservation of the skeletal material since the gases evolved (CO_2,H_2S,NH_3etc.) vary depending on environmental conditions and alter the pH by producing acids (H_2CO_3,H_2SO_4) or bases (NH_4OH) which may corrode or even dissolve calcium carbonate calcium phosphate, or opal.

DISINTEGRATION OF SKELETAL MATERIAL

Articulated skeletons whose components are held together by organic tissue tend to disintegrate into their individual parts following death of the organism, the pattern depending on the type of skeleton.

Invertebrates

In bivalves, post-mortal relaxation of the adductor muscles allows the valves to spring apart. The ligament decomposes more slowly than the rest of the soft tissue but when it has done so the valves separate. The remainder of the fossilisation process may procede in various ways. Many ammonites possess an aptychus that was probably a mandible aperture. If the organism is in an upright position during decomposition of the soft parts, the structure tends to slip down into the bottom of the body chamber where it is often preserved. When the aptychus consists of two plates these separate along the symphysis only at a late stage. Like most arthropods, the trilobites had a segmented carapace which disintegrated after death. However, since the resulting fragments are identical to those produced by sloughing of the carapace during growth of the trilobite it is difficult to distinguish between the two. In many echinoderm species the test comprises individual elements that are rarely fused together. They are instead held together mainly by ligaments and cartilagenous material. Disintegration begins where this material is most readily accessible, that is, in distal parts of the body such as the tips of the arms.

Fish

Whether a dead fish floats to the surface or sinks to the bottom depends on the amount of gas in its swim bladder at the time of death. Where a fish sinks onto an oxygen-deficient bed, its skeleton may remain intact as it is entombed in sediment.

The same thing may happen to fish stranded on the beach. Gas produced by putrefaction accumulates mainly in the central cavity of the body, the corpse therefore floats belly up with head and tail dangling. The belly later ruptures and as the gas gradually escapes the fish sinks, lands on its head or tail and then usually rolls over onto its side.

Fish with a weak axial skeleton may begin to disintegrate while still afloat. In the ganoids with their overlapping armoured scales the otoliths are dropped first from the fishes ears. The scale-armour then separates from the skull and backbone which also fall apart fairly quickly. Disintegration of the skull begins with solution of the lower jaw. The scale-armour may remain intact for a long period—in Recent ganoids often for months but it then starts to corrode around the periphery and finally dissolves completely. In modern fish (without a scale-armour) disintegration again begins with the otoliths. They also soon lose their scales. Once the abdominal wall ruptures, the link between head and body weakens and they may separate completely. As in the ganoids, decomposition of the skull begins with the lower jaw whose components separate along their symphyses. The jaw joints come apart later as do the upper elements of the skull. The linkage of the tail-fin to the backbone also loosens or fails and finally the individual ribs and vertebrace fall apart. When a fish starts to decompose during drift, large sections of the corpse (head, torso, tail) tend to sink to the bottom where they disintegrate further. Since most parts of a fishes skeleton decompose unusually easily, it is often only the otoliths, teeth, scales, and spines that survive.

Fig. 5.3: Post-mortal decay fish. A-flooting and sinking of the corpse: B and C stages in the breakdown of a floating corpse.

Reptiles and Mammals

Whereas fish possess a weak integument that can hold the corpose together for only a short time, the marine reptiles and mammals have a much stronger one. Moreover they have a large abdominal cavity and the considerable evolution of gas that results from decomposition of the soft tissue can keep them floating for a long time. While it is drifting the corpse loses individual skeletal elements in a regular order. These fragments therefore sink to the bottom in different places. Disintegration begins with the lower jaw, the epidermis and the tip of the tail. The coherence of the skeleton is then gradually lost in the abdominal area, the skull falls away and finally the spinal column and rib-cage disintegrate into individual bones and vertebrae. Complete skeletons are preserved in some littoral or oxygen-deficient environments. Where the rate of disintegration of the soft tissue is greatly retarded, the skeleton may be preserved completely but in a rather tangled jumble. In completely anaerobic environments the skeleton remains intact during burial. Terrestrial reptiles and mammals decompose on the same sequence and if they are covered by fluvial or lacustrine sediments the results are similar to those of marine environments. It is only if they die on dry land that aridity and other subaerial processes may produce a different pattern of decay.

Birds

Because their feathers trap plenty of air and because their bones have air spaces, dead birds float on water for a long time. During this period, their skeletal elements are scattered over the sea floor. The down goes first, then the foot bones, then the head separates from the body although it may remain linked for a time by means of the tough trachea. Later, the pinions and tail feathers become detached from the decomposing periostreum and finally the torso disintegrates. From observations of modern gulls it would appear that complete skeletons are preserved only on land or on the beach, but whether these findings apply to all other Recent and fossil birds remain uncertain.

Burial

The chances of *in situ* preservation depends on mode of life and whereabouts at death. Many benthonic organisms live under the sediment-water interface and are readily preserved in their growth position. Others living on the surface appear simply to have fallen over. In most cases, however, the organism did not live where we find it. A phase of transport intervened between death and burial, and during this period the organic remains were damaged, current-sorted, and aligned.

When organic remains undergo transport either alone or with sediment they may be damaged or destroyed. Shattering is the dominant process affecting large particles in gravel deposits whereas abrasion predominates in sandy sediments with a grain size between about 0.05 and 0.5 mm. Abrasion is produced by other small shell fragments as well as by the sand grains themselves. The slit and clay fractions have little mechanical effect. Where the fossil is being transported with a mass of shifting sand it is equally abraded on all sides whereas facetting is likely to result if the shell is lying on the surface of the sediment or anchored to a bed over which a uni-directional current is flowing. Bowl-shaped objects develop first a round and then a horseshoe shaped notch whose opening faces up-current. If the shell changes position or the current direction fluctuates several facets are developed. The results are similar to those seen in ventifacts. In areas of stronger currents or tidal flow where the fossils tip over and roll, all the projecting parts of the shell are abraded to produce shapes similar to those of shells abraded within moving sediment. Circular and horseshoe shaped pits are again developed. Glide facets are produced when the shell slides along an abrasive bed an has its base eroded.

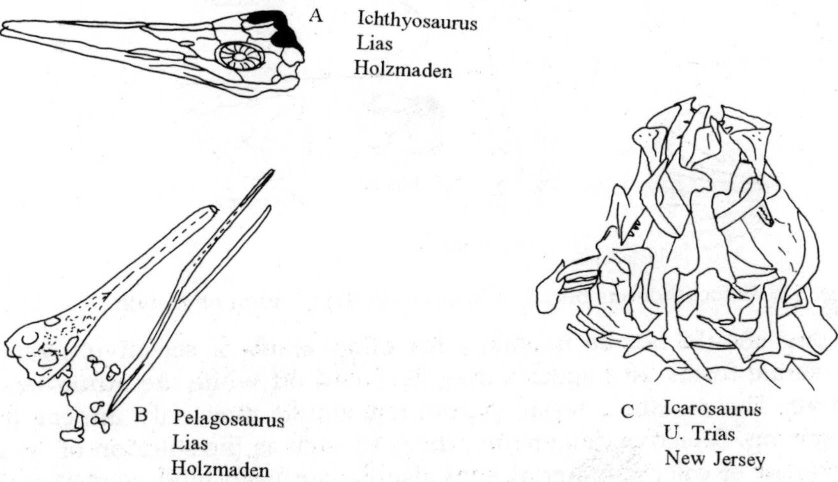

Fig. 5.4: Stages in the breakdown of reptiles skulls.

Shattering and dislocation of skeletal material results not only from purely mechanical processes but also from organic activity—boring microorganisms such as algae and fungi, bivalves, sponges, and other animals all weaken the skeletal structure. Scavenging also plays a considerable part in biogenic disintegration.

The fate of bodies that lack any inherent buoyancy depends greatly on the ease with which they can be carried in suspension. Suspendibility can be calculated from the formula S=0/4G (Where O=surface area in mm^2 and G=weight in gms). The higher the value of S, the more easily is the material transported. Organic debris which is easily suspended is carried largely in the turbulent zone above the bed whereas less easily suspended material is carried in the bed load. There two different modes of transport produce different types of wear. Buoyant bodies experience quite different conditions. They float until they are either washed ashore or lose whatever feature (gas, air chambers, fat droplets, etc.) confers that bouyancy.

During transport, organic debris is sorted according to its size, weight, shape, and resistance to abrasion. The strongest skeletal elements survive longest. The greater the range of suspendibility the more widespread the area over which the material is deposited. From their sorting curve we may determine whether the remains have suffered long or short transport.

The suspendibility of fossil material can be greatly affected by the presence of projections that act as anchors. Flat, disc-shaped bodies such as some gastropod opercula lie tight against the bed. They present only a small surface area to the current and may therefore remain unmoved despite their high suspendibility while other larger objects are carried away. If the area dries out and is then recovered by water, such objects may be held up by surface tension and transported further than other material.

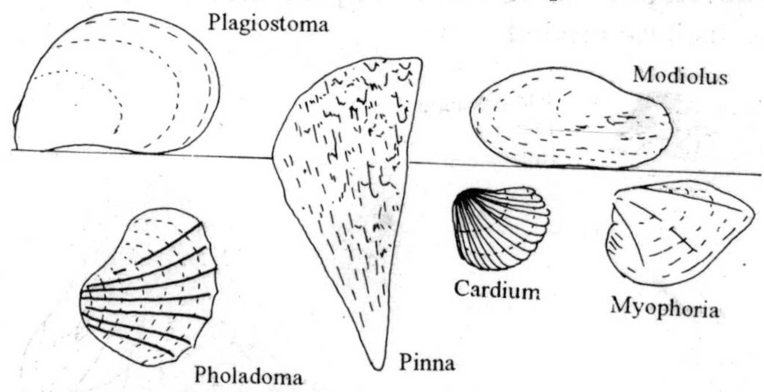

Fig. 5.5: Autochthonous burial of bivalves in the position of growth.

Current sorting should not be mistaken for other kinds of selective, post-mortal processes. All juvenile forms of a species may be killed off while the adults resist the attack and survive. The resulting fossil population would then only appear to have undergone size sorting. Selective diagenetic processes such as the solution of aragonitic tests and preservation of calcitic material may also be confused with current sorting.

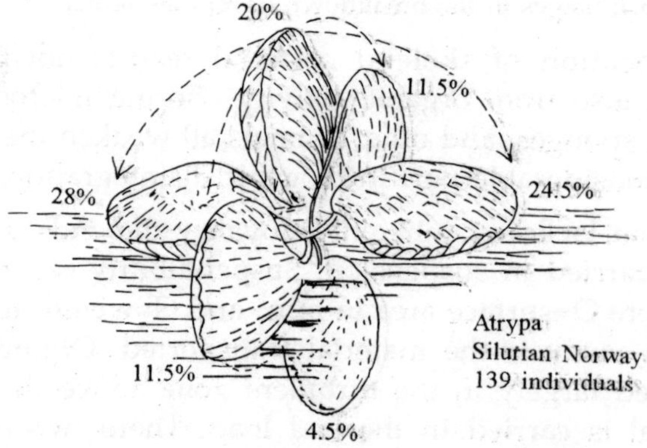

Fig. 5.6: Autochthonous burial of brachiopod colony whose individual members have topples over into different attitudes, and whose relative abundance is shown by percentage values.

Flow conditions in any medium (water, air, ice, sediment) affect objects within or at the surface of that medium. They may result in orientation of shell material. Movement about a horizontal axis may produce overturning or imbrication while rotation about a vertical axis produces alignment.

Moveable objects within a fluid are readily over-turned when they rise or fall under the influence or gravity, buoyancy, and drag. Curved objectics turn convex side down. This happens, for example, when individual valves of bivalves settle through still water, or when they are reworked from bottom sediment then allowed to settle again. When conical objects such as gastropods are free to move within a bottom mud or other sediment whose consistency enables objects to be held in any position, they tend to come to rest point down where buoyant forces predominate but right way up where gravitational forces predominate, provided that a reasonable amount of vertical motion can take place.

Where the body being reoriented lies at the boundary between two media, its freedom of movement is partly inhibited by one of the media; the sediment in the case of a sediment-water interface, the ground in the case of a subaerial land surface. The restrictions imposed have considerable biostratonomic and palaeogcographic significance, because shells or other organic detritus tend to become overturned into the attitude where they offer least resistance to the current flow. This is particularly apparent in the case of cup-shaped shells. Here the rule is that in subaqueous environments with a sufficient current flow velocity, cup-shaped objects will normally come to rest and be buried convex side up. This fact affords an important indicator of the way up of a sedimentary sequence.

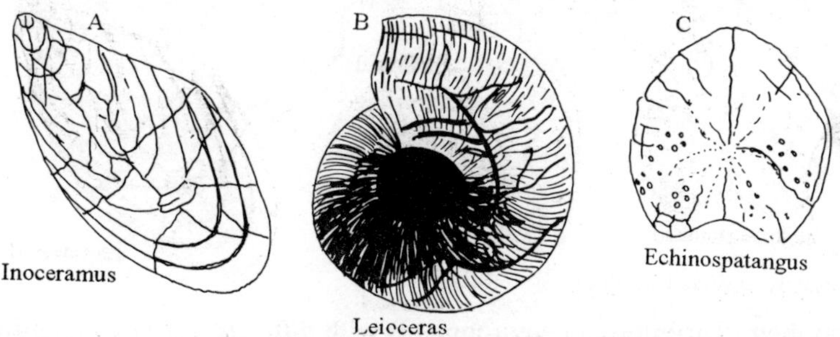

Fig. 5.7: Fracture patterns caused by mechanical stress and overburden pressure.

Exceptions to the rule are common when shells are able to sink into the sediment or are so closely packed that they mutually interfere. In the former case the shells often come to rest concave side up, while the latter situation produces random orientation. Irregularities in shape of the shell may also cause inversion or random orientation.

Where alignment occurs within one medium it may be described as 'free' whereas at the interface between two media it is termed 'limited'. It is best seen in elongate bodies, and most information is available when such bodies have differently shaped ends, for example, if they are elongate cones. When bodies of this shape are lying on a substrate over which a uni-directional current is flowing they roll into a position where they are offering least resistance to the flow. This position occurs when the long axis of the body is parallel to the current and its pointed end is usually directed downstream although with open-ended shells such as turriform gastropods the pointed end may face up current. When such bodies are being rolled around by an oscillating current, for example, waves they lie with their long axis perpendicular to the flow. Spindle-shaped or cylindrical objects roll with their long axes perpendicular to the current. Preferred orientation of the most easily aligned shells required current velocities in excess of 15 cm/s while a flow rate of over 40 cm/s may be needed in the case of large, heavy, or strongly ornamented shells. When the object has a projection that can serve as an anchor, the body pivots about it until the pivot point is directed up-current. A shell without such a projection can be pivoted in the same way if it is partly projecting out of the water or lying on a cohesive substrate.

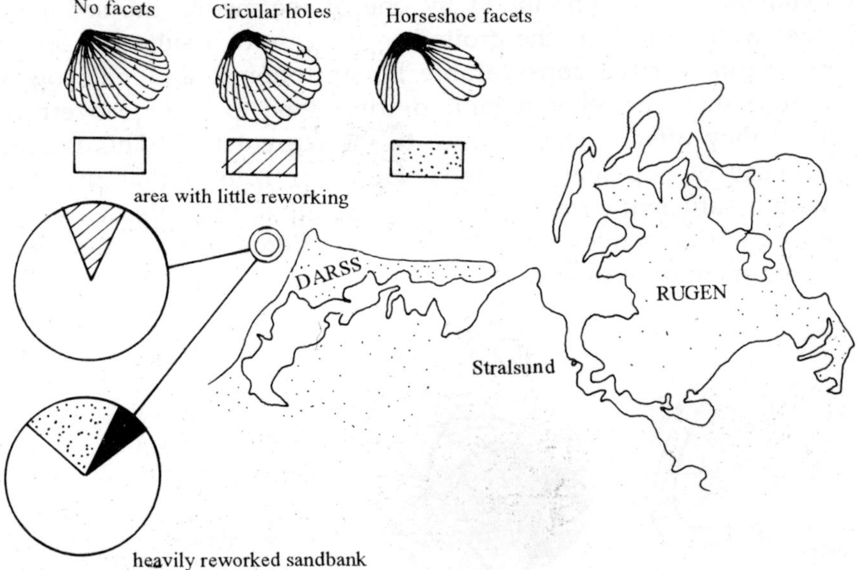

Fig. 5.8: Facets and their distribution in environments with different current strengths neat Darso in the western Baltic.

Preferred orientation of elongate bodies can be analysed by means of a current rose. Symmetrical, bimodal distributions (propellorshaped) suggest orientation perpendicular to the flow. If the shells are conical we can then deduce that the current was an oscillating one. Such a deduction cannot however be made in the case of cylindrical objects. Uni-modal distributions indicate orientation parallel to a uni-directional current with, in the case of pivotable objects, the pivot end up-current. Non-anchored cones may, however, lie with their points upstream.

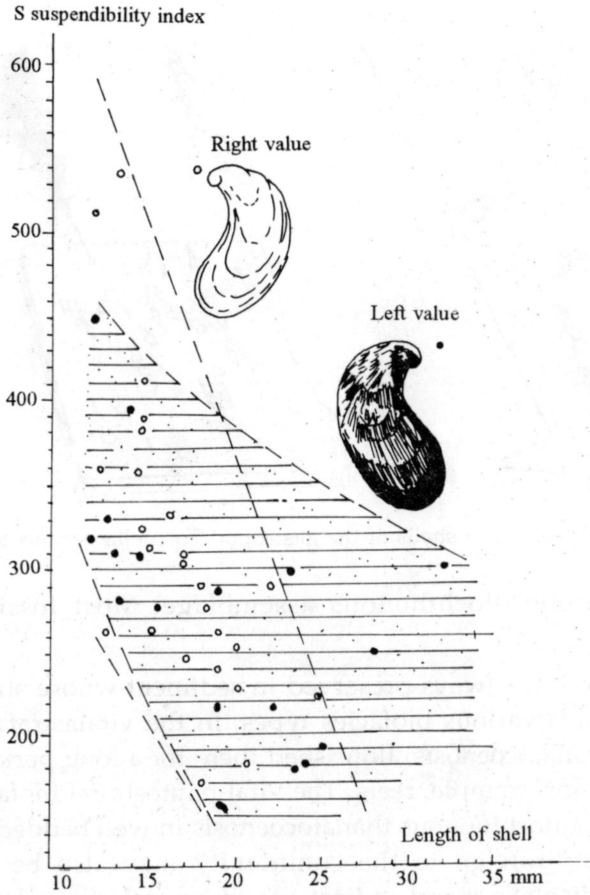

Fig. 5.9: Variation in ease of suspension as a function of shell size as shown by the Jurassic oyster Exogyra virgula.

In addition to these mechanisms, we also find preferred orientation of biogenetic origin caused by the living organism aligning itself in the current in such a way that will allow its various organs to functions most efficiently.

'Sole marks' are imprints on the sediment surface produced physically by the movement of pebbles or, less commonly, by current-drifted hard parts of deal organisms.

Fossil Communities

A 'biocoenosis' is a community of living organisms. It cannot possibly remain unchanged in a fossil state. An assemblage of dead organisms that lived together and have been preserved *in situ* is known as a 'thanatocoenosis' or death assemblage. A 'taphocoenosis' is a fossil assemblage whose members are derived from different habitats. It may comprise either a mixture of autochthonous and allochthonous

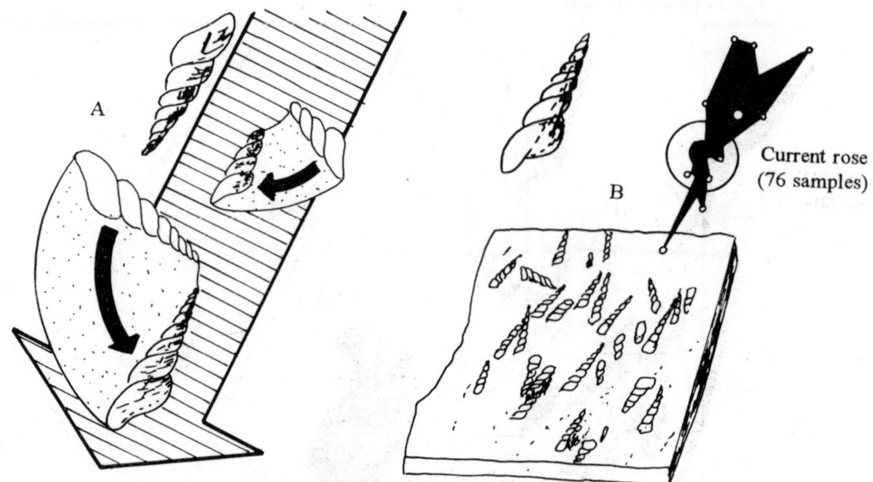

Fig. 5.10: Alignment of turriform shells of the gastropod Turritella parallel to the current direction.

individuals or a purely allochthonous assemblage. Most fossil communities are taphocoenoses.

Fossil assemblages are always preserved in sediment whose structures and texture enable us to distinguish various biofacies types. In the vitalastratal biofacies, a single biocoenosis (now a thanatocoenosis) flourished there for a long period. It forms massive unbedded sediments, for example, reefs. The vital-pantostratal biofacies is characterised by the presence of an autochthonous thanatocoenosis in well bedded sediment. Remains of animals and plants foreign to the community may also be present. The vital-lipostratal biofacies displays many indications of reworking and sedimentary breaks. In it the remains of thanatocoenoses are supplemented by allochthonous taphocoenoses. The same sedimentary characteristics reappear in the lethal-lipostratal biofacies whose fossil assemblage is an allochthonous taphocoenois without any autochthonous organisms. Autochthonous life forms are again missing from the lethal-pantostratal biofacies whose assemblage again consists entirely of allochthonous forms. In this case, however, the bedding is undisturbed as the sediment has not been reworked.

Thanatocoenoses and taphocoenoses turn up in a number of frequently recurring guises the most important of which are shell banks, pavements, sorted pavements and bonebeds. Shell banks (lumachelles) are accumulations of skeletal remains (mostly shells) whose state of preservation ranges from intact through broken to shattered. Most shell banks are allochthonous taphocoenoses. They occur widely on shorelines and beaches, in estuaries, ahead of river deltas, on submarine ridges and on submerine dunes. Shells can also be piled up on the surface of reworked sediment as a result of the activity of burrowing organisms. In this case the assemblage is autochthonous. Shell pavements are bedding planes enriched in shell material. The density of shell

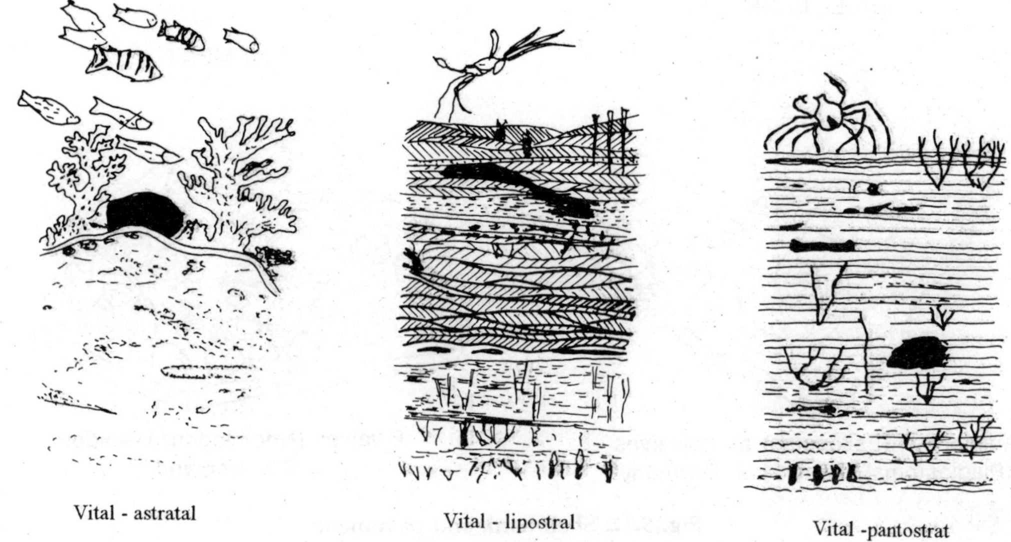

Fig. 5.11: Vital biofacies types.

cover may vary from sparse to almost complete. Pavements form in the surf zone either as allochthonous taphocoenoses or by winnowing of the autochthonous material. Most of the shells always lie convex side up *Lesedecken* are pavements on which the shells have undergone size-and shape-sorting as a result of current activity. The long axes of the shells are commonly oriented parallel to the flow, for example, in belemite 'graveyards'. Bonebeds are tabular enrichments of bones and teeth. Their components are always allochthonous and have often suffered considerable abrasion. Apart from the fissure fillings in karstified areas which also contain allochthonous taphocoenoses, bone beds are the most widespread sites of rich vertebrate assemblages. The famous localities in the lethalpantostratal biofacies, for example, Mansfeld Monte San Georgio, Holzmaden, and Solenhofen are exceptional cases.

Concentration of fossil material on a considerable scale can arise as a result of condensed sequences that form when sedimentation ceases for a time or when sediment is winnowed. The organic remains continue to accumulate at their usual rate during these periods giving enriched sequences that can be recognised by the differing stratigraphic age of the various constituents.

Concentrations may also arise by selective sorting (for example, placer deposits or bone beds) or by being washed into a cavity (for example, fissure fillings). Unusually well-preserved material occurs where decomposition of soft tissue is prevented or greatly retarded. Articulated skeletons whose elements are held together by soft tissue may then survive intact. Such conservation may occur in stagnant oxygen-deficient environments, for example, in bituminous shales, or in preserving media such as resin or peat, or in early diagenetic concretions, or in rapidly deposited sediments.

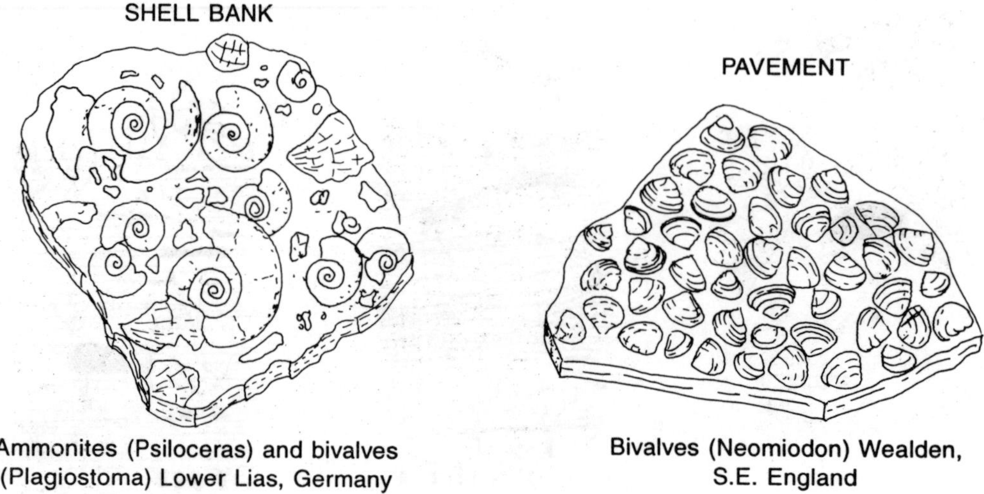

Fig. 5.12: Shell bank and pavement.

DIAGENESIS OF FOSSILS

Diagenesis of fossils covers the fate of organic remains after they are buried in sediment. It is controlled largely by conditions of sedimentation and by the petrography of the sediment.

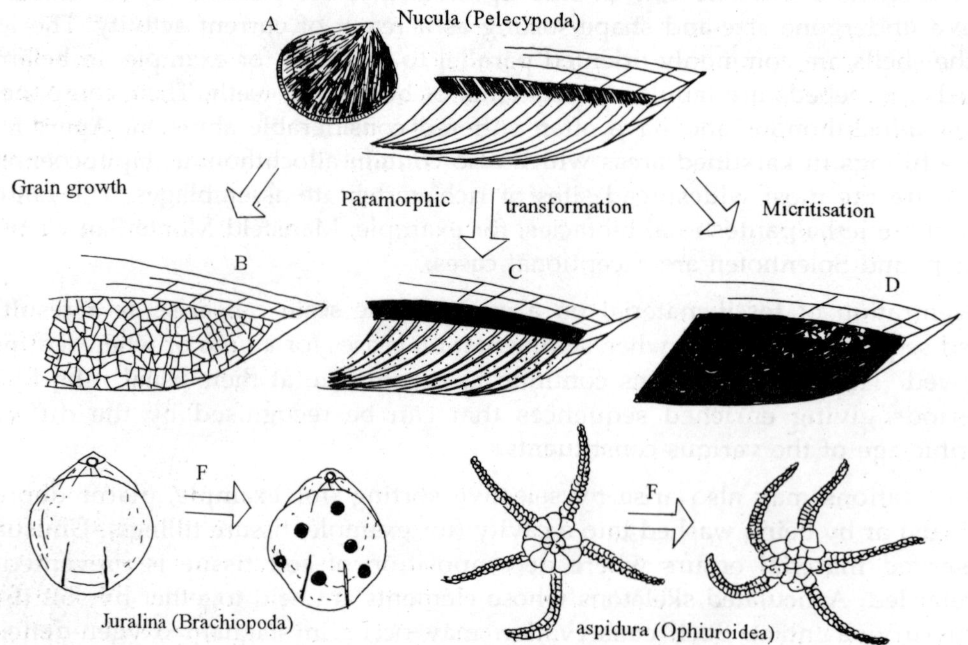

Fig. 5.13: A-D; patterns of recrystallisation shown by the aragonitic shells of the bivalve Nucula E: Grain growth of skeletal elements in echinoderms.

Preservation of Material

In the simplest case, the chemistry and structure of the organic remains does not change. Such fossil remains are fairly common in recent sediments but with increasing age of the material the likelihood of some kind of charge increases. Original material is rarely preserved in Palaeozoic rocks. A sediment in which the circulation of pore water is reduced to a minimum is the main pre-requisite for preservation of such material.

Solution

Organic remains may be dissolved away by circulating ground water especially in course clastic sediments. Solution is assisted by acids and bases in that water. Aragonite is usually the first mineral to be destroyed; calcitic, phosphatic and siliceous materials are more resistant. Thus, we get a selective destruction of fossils depending on the mineralogy of their skeleton.

Recrystallisation

Recrystallisation without change in the chemical composition of skeletal material can arise in a number of ways. Metastable phases of a polymorphic substance are transformed into stable phases in the course to time, for example, aragonite transforms to calcite. The shell structure may survive unaltered during this inversion although the finer details are usually lost. *Grain growth* involves growth of the larger crystals in the skeleton at the expense of the smaller ones and thus produces a completely new texture. The diagenesis of echinoid tests provides a fine example of this process. Here, the individual plates which in life are permeated by a network of cavities are each transformed after death into a single crystal of calcite with almost complete loss of the original structure. Occasionally, crystal growth does not stop at the shell margin

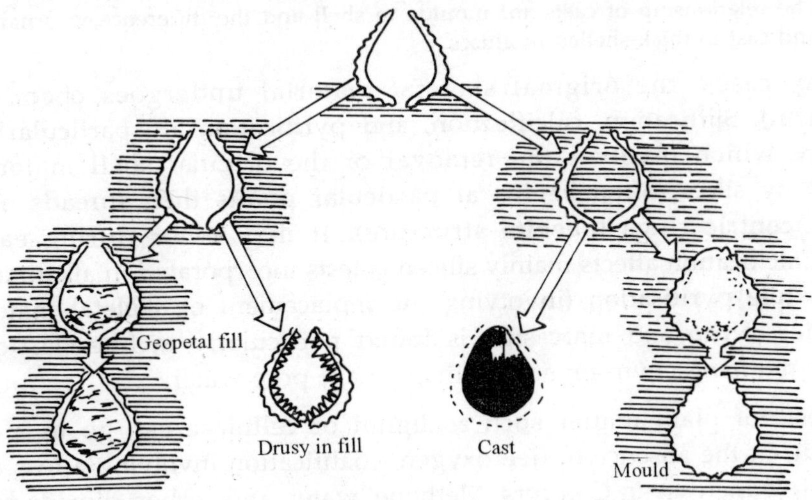

Fig. 5.14: Mode of formation of casts, moulds, cavities, drusty infills and geopetal fills.

but extends out into the sediment. Diagenesis of skeletal opal produces fine-grained quartz with the loss of water of crystallisation. Grain growth may or may not involve mineralogical transformations.

Micritisation involves the early diagenetic breakdown of skeletal material into a cryptocrystalline and usually structureless aggregate. Algae are usually responsible and aragonite is the mineral most affected.

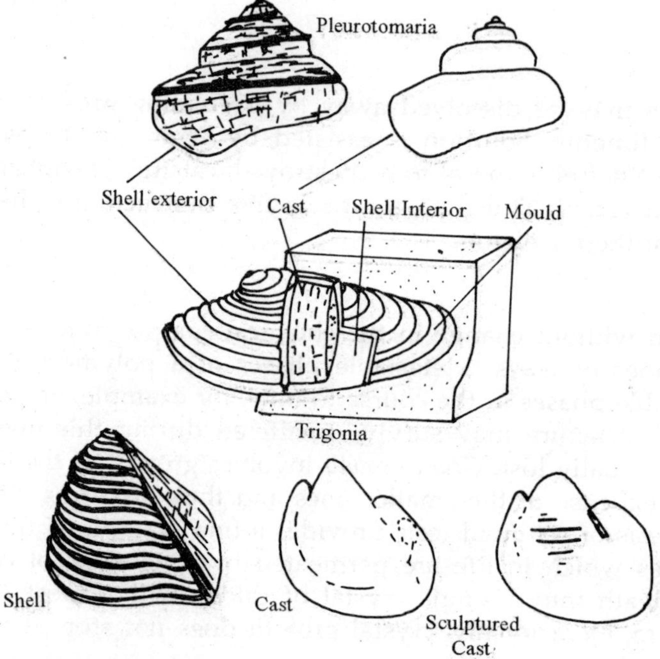

Fig. 5.15: The relationship of casts anf moulds to shell and the difference in ornament of shell and cast in thick-shelled molluscs.

In many cases, the original skeletal material undergoes chemical change (metasomatism). Silification, calcification, and pyritisation are particularly common. Silicification, which involves the removal of the original shell material and its replacement by silica, often begins at particular points then spreads outwards to produce concentric rings (beekite structure). It may occur during early or late diagenesis. Calcification affects mainly siliceous tests incorporated in alkaline, calcareous sediments, while pyritisation (involving the replacement of skeletal material by the iron sulphides pyrite and marcasite) is found particularly in fine clastic sediments deposited in reducing environments with stagnant pore water.

Diagenesis of plant matter such as lignin or cellulose can go in a number of directions but, in the absence of free oxygen, coalification involving a loss of O,N and H and a relative increase in C occurs. Methane, water, and carbon dioxide are released.

$$4C_6H_{10}O_5 \rightarrow 2C_9H_6O + 2CH_4 + 4CO_2 + 10H_2O$$

Under higher temperatures and pressures the reaction proceeds towards the production of almost pure carbon (althracite and graphite).

Impregnation

Minerals are often precipitated in the pore spaces of skeletal material and thereby impregnate them. The minerals involved naturally depend on the chemistry of the ground water. Calcite, silica, and barytes are common.

Encrustation

Skeletal material may become surrounded by crusts, especially of calcite, produced either by algal activity or by chemical precipitation from supersaturated water (cf ooliths). Such material is sometimes incorrectly described as 'mummified'.

Internal Moulds

An internal mould (sometimes called a cast) is a replica of the inner surface of shells such as foraminifera, brachiopods, bivalves, gastropods, cephalopods, and echinoids. The material involved is usually derived from the surrounding sediment, although in the case of tightly closed shells that prevent the ingress of sediment, a crystalline fill (usually calcite) precipted from aqueous solution may be present. When the shell is only partly filled with sediment with the remainder either empty or containing crystalline precipitates the geopetal texture indicates which way up it should be like a 'fossil spirit level'. When no deposition occurs within tightly shut shells, and when the shell material itself is later dissolved after lithification, hollow moulds are produced. These may subsequently have a skin of crystals developed on their walls in the position of the original shell.

Internal moulds consisting of pyrite or marcasite often originate in anaerobic clays. They form in the same way as concretions and often infill only the most inaccessible parts of shells, for example, the innermost chambers of cephalopods or the tip of gastropod shells and so produce an illusion of stunted growth.

In thin-shelled organisms the inner and outer surfaces usually have similar shapes whereas in thick-shelled animals they are often very different. If such a shell (within which an internal mould has already formed) subsequently dissolves in still unconsolidated sediment the imprint of the external surface of the shell may be compaction become impressed on the internal cast to provide a 'sculptured cast'.

In ammonites, sediment can only enter the body chamber because the phragmocone chambers are hermetically scaled by septae. For this reason these chambers are often preserved empty or filled with later crystalline calcite. Since many other ammonite shells do, however, display internal moulds of sediment, the coarser-grained sediment must have entered through fractures in the shell, while fine-grained material appears to have passed in through the siphuncle. Where structural elements such as keels are separated from the interior of the shell by a partition (septate keels) they may remain empty when the internal mould forms to produce hollow keels that contrast with the filled keels produced where such partitions are absent.

Internal moulds are mostly formed during early diagenesis. Moulds on which epifaunas are present must have been reworked and have remained on the sea floor long enough for the epifauna to develop.

Concretions

Concretions result from the localised segregation of originally dispersed material. They may consist of calcite, siderite, silica, pyrite, marcasite, phosphate or other minerals and often contain organic remains. The concretionery mineral frequently occupies only the pore space of the sediment (especially in calcareous concretions) or it may make room for itself by replacement (siliceous concretions) or by physical displacement of the host sediment (marcasite concretions).

Concretions may result either from the decomposition of organic matter or because of changing solubility of salts in rising ground water. Calcareous concretions form when the carbonate material present in most sediments segregates in an alkaline environment. Calcareous shells act as ideal nuclei. Siderite concretions occur where the O Eh level lies just below the sediment-water interface. Below it, iron dissolves in the pore water only to reprecipitate as it approaches the surface. Phosphatic nodules may either be pure or may grade into calcareous and ferriginous varieties. They are commonly found where phosphate-rich upwellings enter areas of slow sedimentation and are closely associated with organic remains and faecal pellets. Iron sulphides develop in acid or neutral-reducing environments by the reaction of iron and bacterially-produced H_2S. Both the stable form (pyrite) and the metastable form (marcasite) are found. Siliceous concretions (opal, flint, chert) derive their material from the dissolution of siliceous tests, or sponge spicules, present in the fine fraction of the sediment or from volcanic SiO_2. They often grow at a very early stage of diagenesis in association with organic remains. Mobilisation of silica is possible only in alkaline environments.

Early diagenetic concretions are very important in the preservation of fossils which become encased and protected before they can be destroyed. Fossils within concretions can be also undergo reworking without suffering damage.

Deformation

Deformation of fossils refers to shape changes that occur after the fossil is embedded in sediment. Such changes are caused by compaction that results from increasing overburden pressure and dewatering. If the fossil consists of articulated elements, these elements move relative to each other. On the other hand, shells, armour, or bones are rigid units that yield by fracturing. Regular radial, concentric and axial fractures appear along which relative displacements occur. Thin sections frequently reveal that shells that appear superficially to be intact are internally shattered.

Within skeletal material whose internal structure is so weakened by solution that individual crystals can move relative to each other, deformation can occur without fracturing. In this way curved shells can be squeezed flat and thick bones can be attenuated. Plastic deformation of internal moulds is a widespread phenomenon (pelomorphic deformation). The individual features of the cast are compressed vertically. The ratio V_2/V_1(V_1=original volume; V_2=final volume) is known as the 'compaction ratio'. The ease with which deformation can be recognised depends on the original attitude of the mould.

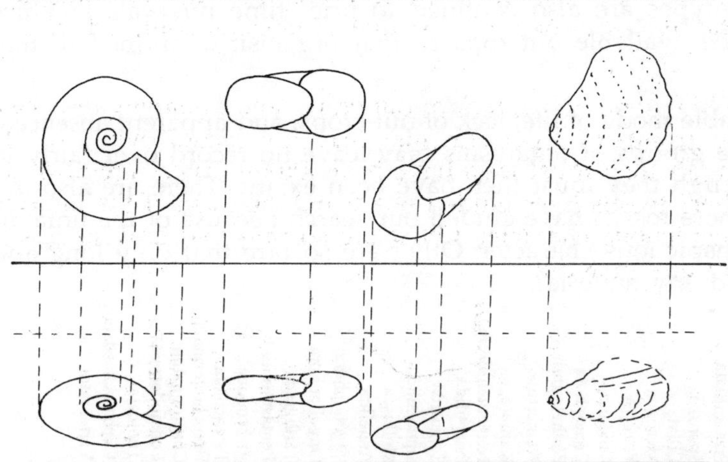

Fig. 5.16: Deformation of fossils resulting from compaction of the enclosing sediment.

Tectonic deformation also causes distortion, extension, and disruption of fossils enclosed in sediment but this is no longer to be regarded as diagenesis.

Corrosion

Corrosion involves partial solution of shells by water, acids (H_2CO_3,H_2SO_4), or alkalis (NH_4OH). It commonly occurs when buried fossils are re-exposed on the sea floor and frequently causes removal of those parts of the shell that project above the new surface. Such planar corrosion on the sea floor is called 'subsolution'.

Corrosion can also occur within the sediment when organic remains in unconsolidated sediment are attacked by rising pore water. In lithified sediments this can produce partings and stylolites. Weathering also produces corrosion mainly because of CO_2 dissolved in rainwater. The carbon dioxide released from plant roots also dissolved carbonate.

GAPS IN THE RECORD

Only a minute fraction (well under 1%) of all organisms are preserved as fossils and different groups fare quite differently. Completely soft-bodied animals and plants survive in only a few luckly cases whereas organisms with skeletal parts stand a better chance of preservation, although even then the record is very fragmentary.

Because of their mode of life, some organisms are hardly ever preserved. For example the plants and animals of the tropical rainforest decompose remarkably quickly after death as do organisms in turbulent seas and flying animals.

Only a very small part of the area over which fossiliferous sediments were once deposited now lies at the Earth's surface. The greater part has either been eroded away or been buried under younger sediments. Moreover the beds at the surface are exposed only is places and only some of the exposures are accessible and have been studied. Many biotypes are also confined to brief time intervals and may not then occur in any of the available outcrops so that organisms confined to these biotypes remain unknown.

Where unsuitable mode of life, lack of out-crops, and apparent absence of a biotope all coincide, whole groups of organisms may leave no record over fairly long periods of time even although they must then have been extant. There are also many gaps in out knowledge where fossils have eluded our search because of the unfavourable size, chemistry or inconspicuous character. Other are so rare that only long and expensive investigations yield any material.

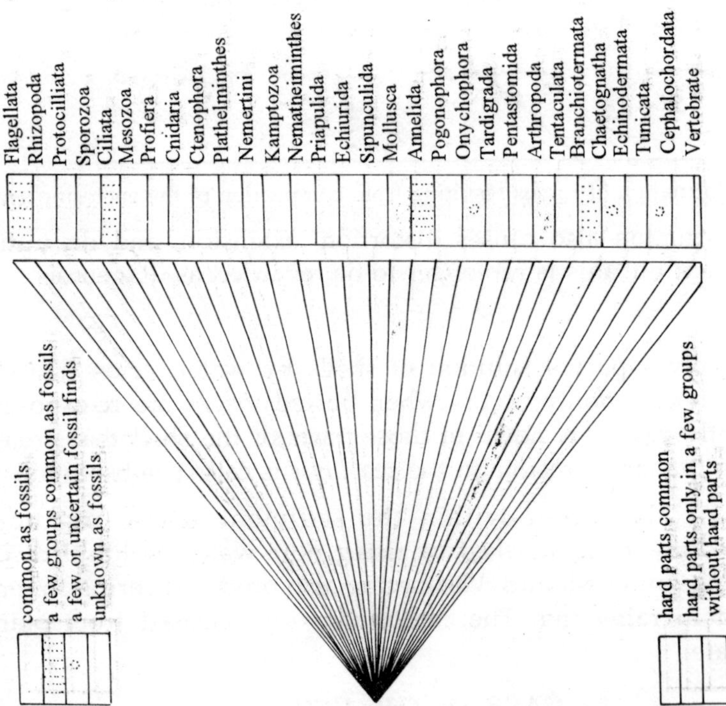

Fig. 5.17: Abundance of hard parts in different animal phyla and its influence on the fossil record because of the very limited preservation of soft-bodied organisms.

The smallness of the percentage of organisms preserved suggests that we can expect to find only the more common types as fossils. Representatives of rarer groups are found only in exceptional circumstances. Thus we observe only the peaks in the frequency curve of past life and it is astonishing how much valuable material they alone have yielded.

Chapter—6

Chemical Fossils

If you ask a child to draw a dinosaur, the chances are that he will produce a recognizable picture of such a creature. His familiarity with an animal that lived 150 million years ago can of course be traced to the intensive studies of paleontologists, who have been able to reconstruct the skeletons of extinct animals from fossilized bones preserved in ancient sediments. Recent chemical research now shows that minute quantities of organic compounds—remnants of the original carbon-containing chemical constituents of the soft parts of the animal—are still present in some fossils and in ancient sediments of all ages, including some measured in billions of years. As a result of this finding organic chemists and geologists have joined in a search for "chemical fossils": organic molecules that have survived unchanged or little altered from their original structure, when they were part of organisms long since vanished.

This kind of search does not require the presence of the usual kind of fossil—a shape or an actual hard form in the rock. The fossil molecules can be extracted and identified even when the organism has completely disintegrated and the organic molecules have diffused into the surrounding material. In fact, the term "biological marker" is now being applied to organic substances that show pronounced resistance to chemical change and whose molecular structure gives a strong indication that they could have been created in significant amounts only by biological processes.

One might liken such resistant compounds to the hard parts of organisms that ordinarily persist after the soft parts have decayed. For example, hydrocarbons, the compounds consisting only of carbon and hydrogen, are comparatively resistant to chemical and biological attack. Unfortunately many other biologically important molecules such as nucleic acids, proteins and polysaccharides contain many bonds that hydrolyze, or cleave, readily; hence these molecules rapidly decompose after an organisms dies. Nevertheless, several groups of workers have reported finding constituents of proteins (amino acids and peptize chains) and even proteins themselves in special well-protected sites, such as between the thin sheets of crystal in fossil shells and bones.

Where complete destruction of the organisms has taken place one cannot, of course, visualize is original shape from the nature of the chemical fossils it has left behind. One may, however, be able to infer the biological class, or perhaps even the species, of organism that gave rise to them. At present such deductions must be extremely tentative because they involve considerable uncertainty. Although the chemistry of living organisms is known in broad outline, biochemists even today have identified the principal constituents of only a few small groups of living things. Studies in comparative biochemistry or chemotaxonomy are thus an essential parallel to organic geochemistry. A second uncertainty involves the question of whether or not the biochemistry of ancient organisms was generally the same as the biochemistry of present day organisms. Finally, little is known of the chemical changes wrought in organic substances when they are entombed for long periods of time in rock or a fossil matrix.

In our work at the University of California at Berkeley and at the University of Glasgow we have gone on the assumption that the best approach to the study of chemical fossils is to analyze geological materials that have had a relatively simple biological and geological history. The search for suitable sediments requires a close collaboration between the geologist and the chemist. The results obtained so far augur well for the future.

Organic chemistry made its first major impact on the earth sciences in 1936, when the German chemist Alfred Treibs isolated metal-containing porphyrins from numerous crude oils and shales. Certain porphyrins are important biological pigments; two of the best-known are chlorophyll, the green pigment of plants, and here, the red pigment of the blood. Treibs deduced that the oils were biological in origin and could not have been subjected to high temperatures, since that would have decomposed some of the porphyrins in them. It is only during the past decade, however, that techniques have been available for the rapid isolation and identification of organic substances present in small amounts in oils and ancient sediments. Further refinements and new methods will be required for detailed study of the tiny amounts of organic substances found in some rocks. The effort should be worth-while, because such techniques for the detection and definition of the specific architecture of organic molecules should not only tell us much more about the origin of life on the earth but also help us to establish whether or not life has developed on other planets. Furthermore, chemical fossils present the organic chemist with a new range of organic compounds to study and may offer the geologist a new tool for determining the environment of the earth in various geological epochs and the conditions subsequently experienced by the sediments laid down in those epochs.

If one could obtain the fossil molecules from a single species of organism, one would be able to make a direct correlation between present day biochemistry and organic geochemistry. For example, one could directly compare the lipids, or fatty compounds, isolated from a living organism with the lipids of its fossil ancestor.

Unfortunately the fossil lipids and other fossil compounds found in sediments almost always represent the chemical debris from many organisms.

The deposition of a compressible fine-gained sediment containing mineral particles and disseminated organic matter takes place in an aquatic environment in which the organic content can be partially preserved; an example would be the bottom of a lake or a delta. The organic matter makes up something less than 1 per cent of many ancient sediments. The small portion of this carbon containing material that is soluble in organic solvents represents a part of the original lipid content, more or less modified, of the organisms that lived and died while the sediment was being deposited.

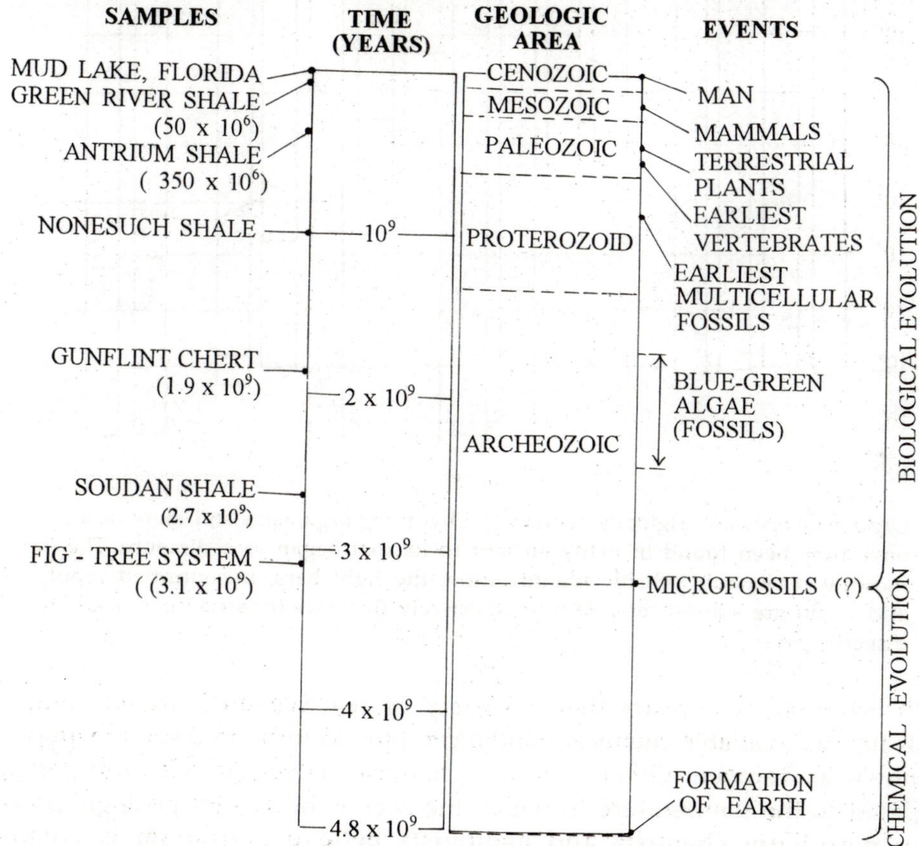

Fig. 6.1: *Geological Time Scale* shows the age of some intensively studied sedimentary rocks (*left*) and the sequence of major steps in the evolution of life (*right*). The stage for biological evolution was set by chemical evolution, but the period of transition is not known.

The organic content presumably consists of varying proportions of the components of organisms-terrestrial as well as aquatic—that have undergone chemical transformation while the sediment was being laid down and compressed. Typical transformations are reduction, which has the effect of removing oxygen from molecules and adding hydrogen, and decarboxylation, which removes the carboxyl radical

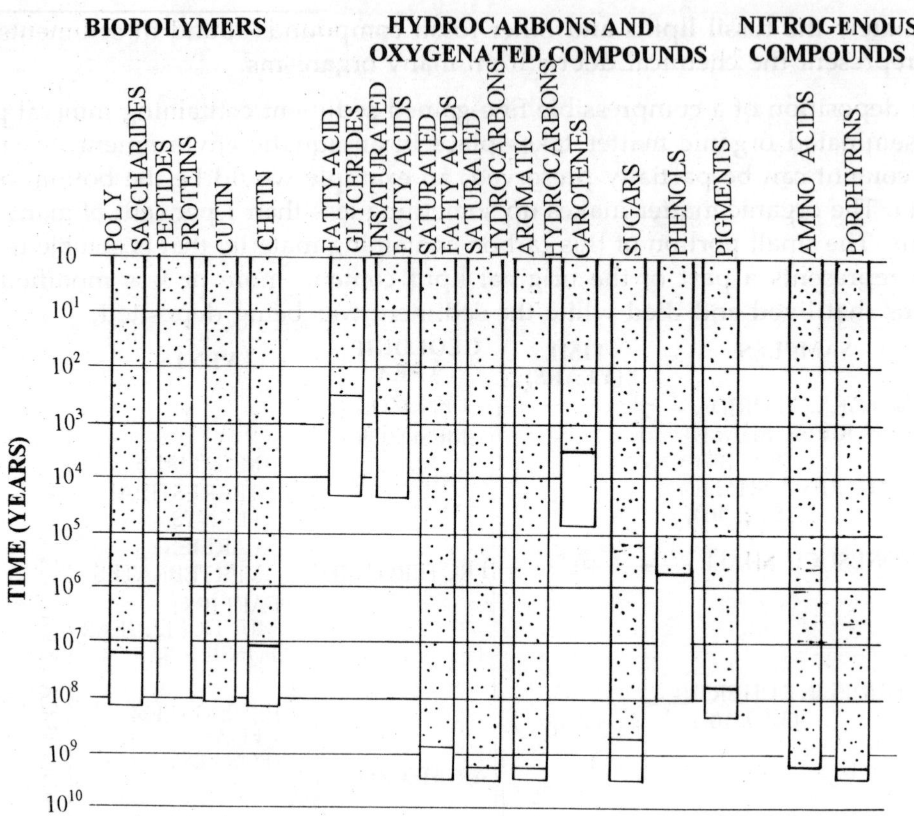

Fig. 6.2: *Organic Compounds* originally synthesized by living organisms and more or less modified have now been found in many ancient rocks that began as sediments. The dark bars indicate reasonably reliable identification; the light bars, unconfirmed reports. Cutin and chitin are substances present respectively in the outer structures of plants and of insects.

(COOH). In addition, it appears that a variety of reactive unsaturated compounds (compounds having available chemical bonds) combine to form an insoluble amorphous material known as kerogen. Other chemical changes that occur with the passage of time are related to the temperature to which the rock is heated by geologic processes. Thus many petroleum chemists and geologists believe petroleum is created by progressive degradation, brought about by heat, of the organic matter that is finely disseminated throughout the original sediment. The organic matter that comes closest in structure to the chains and rings of carbon atoms found in the hydrocarbons of petroleum is the matter present in the lipid fraction of organisms. Another potential source of petroleum hydrocarbons is kerogen itself, presumably formed from a wide variety of organic molecules; it gives off a range of straight-chain, branched-chain and ring-containing hydrocarbons when it is strongly heated in the laboratory. One would also like to know more about the role of bacteria in the early steps of sediment formation. In the upper layers of most newly formed sediments there is strong bacterial activity, which must surely result in extensive alteration of the initially deposited organic matter.

In this article we shall concentrate on the isolation of fossil hydrocarbons. The methods must be capable of dealing with the tiny quantity of material available in most rocks. Our general procedure is as follows.

After cutting off the outer surface of a rock specimen to remove gross contaminants, we clean the remaining block with solvents and pulverize it. We then place the powder in solvents such as benzene and methanol to extract the organic material. Before this step we sometimes dissolve the silicate and carbonate minerals of the rock with hydrofluoric and hydrochloric acids. We separate the organic extract so obtained into acidic, basic and neutral fractions. The compounds in these fractions are converted, when necessary, into derivatives that make them suitable for separation by the technique of chromatography. For the initial separations we use column chromatography, in which a sample in solution is passed through a column packed with alumina or silica. Depending on their nature, compounds in the sample pass through the column at different speeds and can be collected in fractions as they emerge.

In subsequent stages of the analysis finer fractionations are achieved by means of gas-liquid chromatography. In this variation of the technique, the sample is vaporaized into a steam of light gas, usually helium, and brought in contact with a liquid that tends to trap the compounds in the sample in varying degree. The liquid can be supported on an inorganic powder, such as diatomaceous earth, or coated on the inside of a capillary tube. Since the compounds are alternately trapped in the liquid medium and released by the passing stream of gas they progress through the column at varying speeds, with the result that they are separated into distinct fractions as they emerge from the tube. The temperature of the column is raised steadily as the separation proceeds, in order to drive off the more strongly trapped compounds.

The initial chromatographic separation is adjusted to produce fractions that consist of a single class of compound, for example the class of saturated hydrocarbons known as alkanes. Alkane molecules may consist either of straight chains of carbon atoms or of chains that include branches and rings. These subclasses can be separated with the help of molecular sieves: inorganic substances, commonly alumino-silicates, that have a fine honey comb structure. We use a sieve whose mesh is about five angstrom units, or about a thousandth of the wavelength of green light. Straight-chain alkanes, which resemble smooth flexible rods about 4.5 angstroms in diameter, can enter the sieve and are trapped. Chains with branches and rings are too big to enter and so are held back. The straight-chain alkanes can be liberated from the sieve for further analysis by dissolving the sieve in hydrofluoric acid. Other families of molecules can be trapped in special crystalline forms of urea and thiourea, whose crystal lattices provide cavities with diameters of five angstroms and seven angstroms respectively.

The families of molecules isolated in this way are again passed through gas chromatographic columns that separate the molecular species with the family. For example, a typical chromatogram of straight-chain alkanes will show that molecules

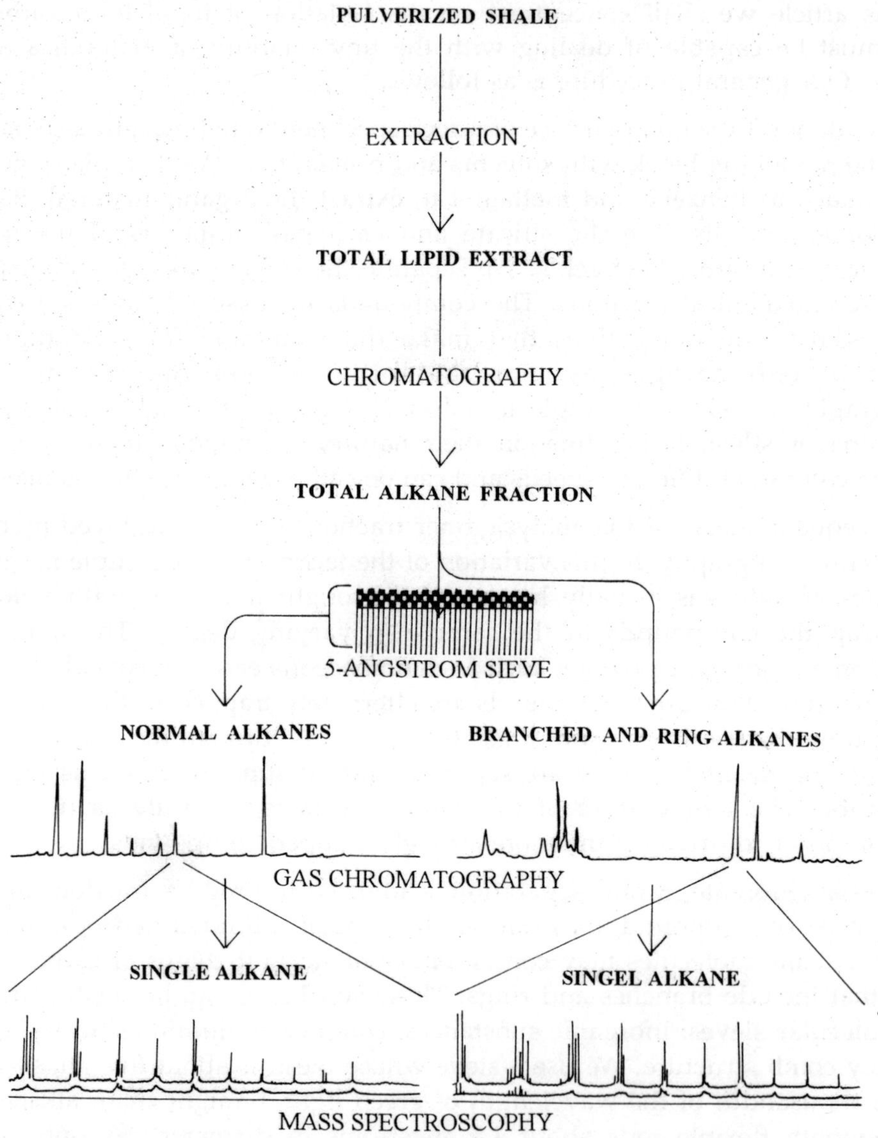

Fig. 6.3: *Analytical Procedure* for identifying chemical fossils begins with the extraction of alkane hydrocarbons from a simple of pulverized shale. In normal alkanes the carbon atoms are arranged in a straight chain. Molecular sieves are used to separate straight-chain alkanes from alkanes with branched chains and rings. The two broad classes are then further fractioned. Compounds responsible for individual peaks in the chromatogram are identified by mass spectrometry and other methods.

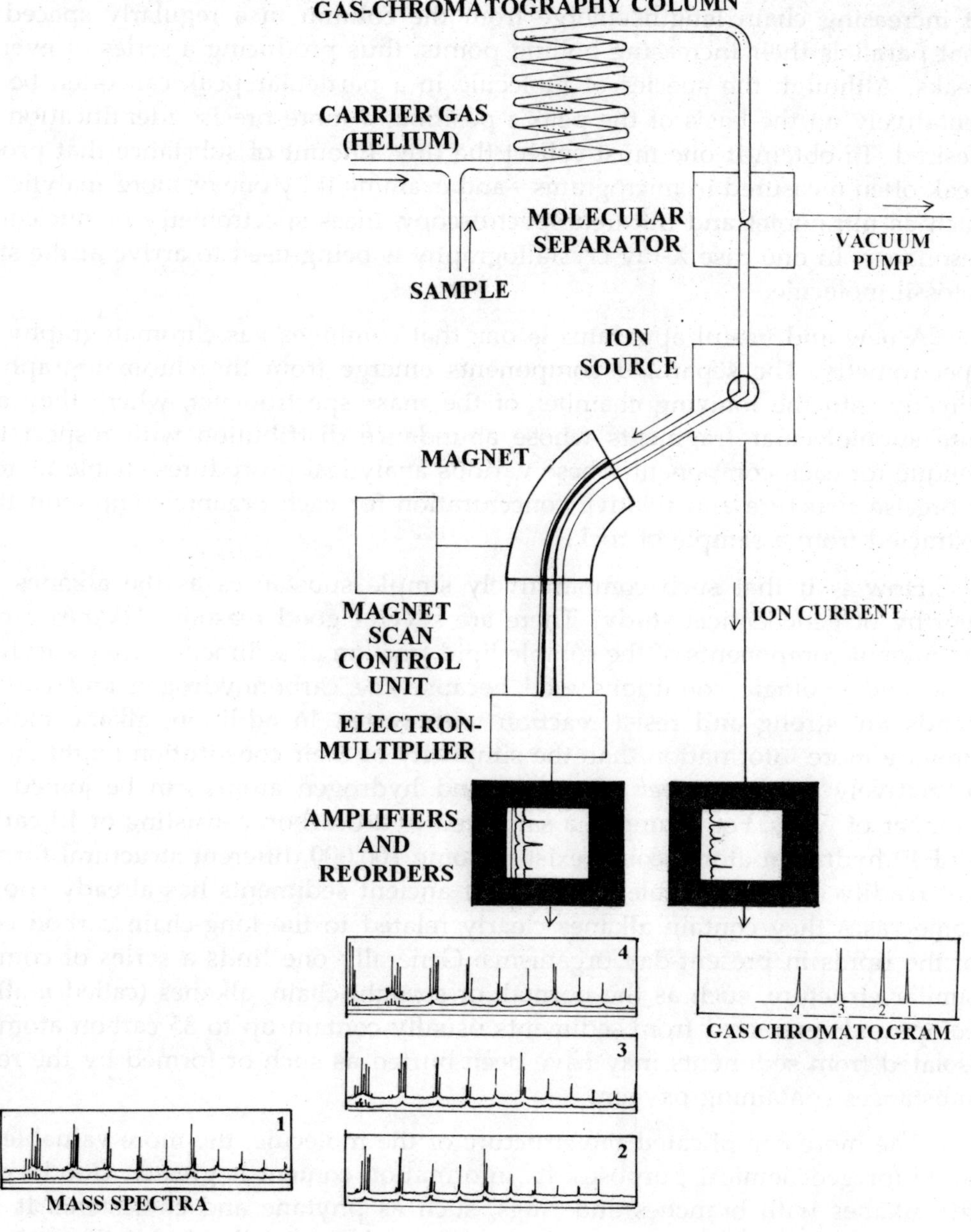

Fig. 6.4: *Combination Instrument* feeds the output of a gas chromatography directly into a mass spectrometer. As hydrocarbon molecules emerge in sequence from the chromatograph and enter the spectrometer, they are ionized, or broken into charged fragments. The size of the ionization current is proportional to the amount of material present at each instant and can be converted into a chromatogram. In the spectrometer the charged fragments are directed through a magnetic field, which separates them according to mass. Each species of molecule produces a unique mass-distribution pattern.

of increasing chain length emerge from the column in a regularly spaced sequence that parallels their increasing boiling points, thus producing a series of evenly spaced peaks. Although the species of molecule in a particular peak can often be identified tentatively on the basis of the peak's position, a more precise identification is usually desired. To obtain it one must collect the tiny amount of substance that produced the peak-often measured in micrograms—and examine it by one or more analytical methods such as ultraviolet and infrared spectroscopy, mass spectrometry or nuclear magnetic resonance. In one case X-ray crystallography is being used to arrive at the structure of a fossil molecule.

A new and useful apparatus is one that combines gas chromatography and mass spectrometry. The separated components emerge from the chromatograph and pass directly into the ionizing chamber of the mass spectrometer, where they are broken into submolecular fragments whose abundance distribution with respect to mass is unique for each component. These various analytical procedures enable us to establish a precise structure and relative concentration for each organic compound that can be extracted from a sample of rock.

How is it that such comparatively simple substances as the alkanes should be worthy of geochemical study? There are several good reasons. Alkanes are generally prominent components of the soluble lipid fraction of sediments. They survive geologic time and geologic conditions well because the carbonhydrogen and carbon-carbon bonds are strong and resist reaction with water. In addition, alkane molecules can provide more information than the simplicity of their constitution might suggest; even a relatively small number of carbon and hydrogen atoms can be joined in a large number of ways. For example, a saturated hydrocarbon consisting of 19 carbon atoms and 40 hydrogen atoms could exist in some 100,000 different structural forms that are not readily interconvertible. Analysis of ancient sediments has already shown that in some cases they contain alkanes clearly related to the long-chain carbon compounds of the lipids in present-day organisms. Generally one finds a series of compounds of similar structure, such as the normal, or straight-chain, alkanes (called *n*-alkanes); the compounds extracted from sediments usually contain up to 35 carbon atoms. Alkanes isolated from sediments may have been buried as such or formed by the reduction of substances containing oxygen.

The more complicated the structure of the molecule, the more valuable it is likely to be for geochemical purposes: its information content is greater. Good examples are the alkanes with branches and rings, such as phytane and cholestane. It is unlikely that these complex alkanes could be built up from small subunits by processes other than biological ones, at least in the proportions found. Hence we are encouraged to look for biological precursors with appropriate preexisting carbon skeletons.

In conducting this kind of search one makes the assumption, at least at the outset, that the overall biochemistry of past organisms was similar to that of present-day organisms. When lipid fractions are isolated directly from modern biological sources, they are generally found to contain a range of hydrocarbons, fatty acids, alcohols, esters and so on. The mixture is diverse but by no means random. The molecules present in such fractions have structures that reflect the chemical reaction pathways systematically followed in biological organisms. There are only a few types of biological molecule wherein long chains of carbon atoms are linked together; two examples are the straight-chain lipids, the end groups of which may include oxygen atoms, and the lipids known as isoprenoids.

The straight-chain lipids are produced by what is called the polyacetate pathway. This pathway leads to a series of fatty acids with an even number of carbon atoms; the odd-numbered molecules are missing. One also finds in nature straight-chain alcohols (*n*-alkanols) that likewise have an even number of carbon atoms, which is to be expected if they are formed by simple reduction of the corresponding fatty acids. In contrast, the straight-chain hydrocarbons (*n*-alkanes) contain an *odd* number of carbon atoms. Such a series would be produced by the decarboxylation of the fatty acids.

The second type of lipid, the isoprenoids, have branched chains consisting of five-carbon units assembled in a regular order. Because these units are assembled in head-to-tail fashion the side-chain methyl groups (CH_3) are attached to every fifth carbon atom. (Tail-to-tail addition occurs less frequently but accounts for several important natural compounds, for example beta-carotene.) When the isoprenoid skeleton is found in a naturally occurring molecule, it is reasonable to assume that the compound has been formed by this particular biological pathway.

Chlorophyll is possible the most widely distributed molecule with an isoprenoid chain; therefore it must make some contribution to the organic matter in sediments. Its fate under conditions of geological sedimentation is not known, but it may decompose into only two or three large fragments. The molecule of chlorophyll a consists of a system of interconnected rings and a phytyl side chain, which is an isoprenoid. When chlorophyll is decomposed, it seems likely that the phytyl chain is split off and converted to phytane (which has the same number of carbon atoms) and pristane (which is shorter by one carbon atom). When both of these branched alkanes are found in a sediment, one has reasonable presumptive evidence that chlorophyll was once present. The chlorophyll ring system very likely gives rise to the metal containing porphyrins that are found in many crude oils and sediments.

Phytane and pristane may actually enter the sediments directly. Max Blumer of the Woods Hole Oceanographic Institution showed in 1965 that certain species of animal plankton that eat the plant plankton containing chlorophyll store quite large quantities of pristane and related hydrocarbons. The animal plankton act in turn as a food supply for bigger marine animals, thereby accounting for the large quantities of pristane in the liver of the shark and other fishes.

An indirect source for the isoprenoid alkanes could be the lipids found in the outer membrane of certain bacteria that live only in strong salt solutions, an

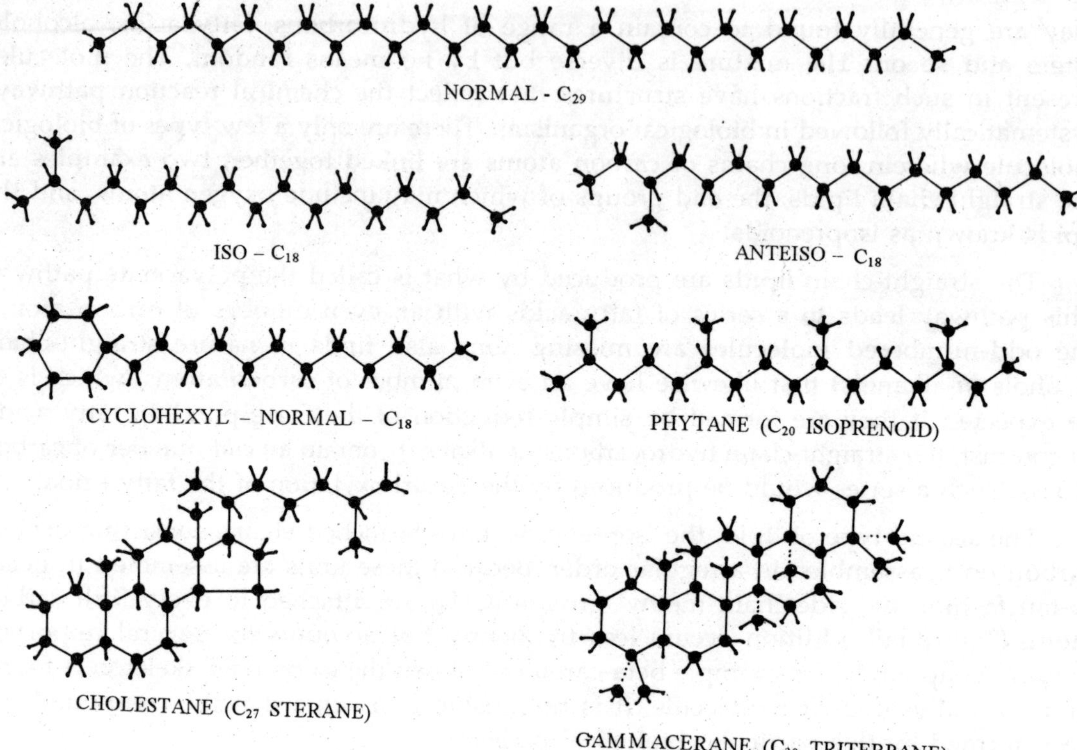

Fig. 6.5: *Alkane Hydrocarbon Molecules* can take various forms: straight chains (which are actually zigzag chains), branched chains and ring structures. Those depicted here have been found in crude oils and shales. The molecules shown in color are so closely related to well-known biological molecules that they are particularly useful in be-speaking the existence of ancient life. The broken lines indicate side chains that are directed into the page.

environment that might be found where ancient seas were evaporating. Morris Kates of the National Research Council of Canada has shown that a phytyl-containing lipid (diphytyl phospholipid) is common to bacteria with the highest salt requirement but not to the other bacteria examined so far.

This last example brings out the point that in spite of the overall oneness of present-day biochemistry, organisms do differ in the compounds they make. They also synthesize the same compounds in different proportions. These differences are making it possible to classify living species on a chemotaxonomic, or chemical, basis rather than on a morphological, or shape, basis. Eventually it may be possible to extend chemical classification to ancient organisms, creating a discipline that could be called paleochemo-taxonomy.

Our study of chemical fossils began in 1961, when we decided to probe the sedimentary rocks of the Precambrian period in a search for the earliest signs of life. This vast period of time, some four billion years, encompasses the beginnings of life on this planet and its early development to the stage of organisms consisting of more than one call. We hoped that out study would complement the efforts being made by a number of workers, including one of us (Calvin), to imitate in the laboratory the chemical evolution that must have preceded the appearance of life on earth. We also saw the possibility that our work could be adapted to the study of meteorites and of

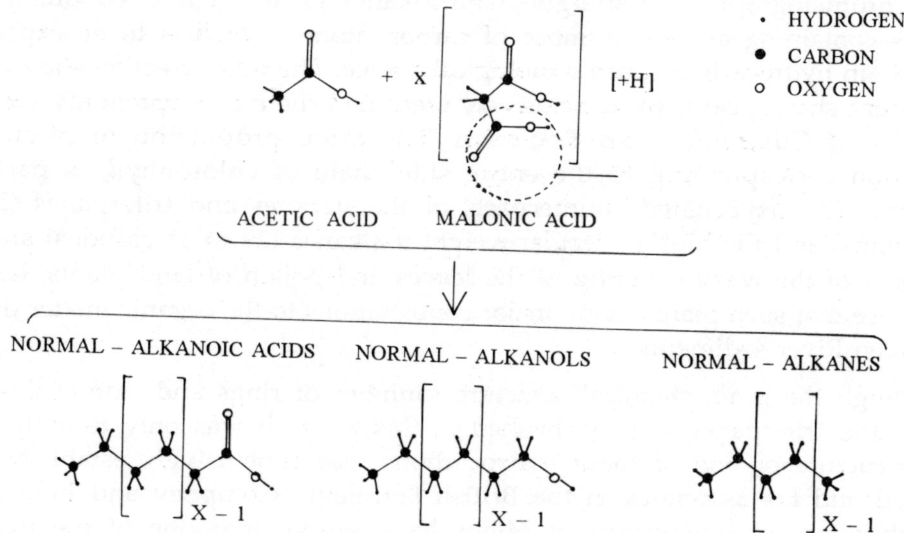

Fig. 6.6: *Straight-Chain Lipids* are created in living organisms from simple two-carbon and three-carbon compounds: acetate and malonate, shown here as their acids. The complex biological process, which involves coenzyme. *A*, is depicted schematically. The fatty acids (*n*-alkanoic acids) and fatty alcohols (*n*-alkanols) produced in this way have an even number of carbon atoms. The removal of carbon dioxide from the fatty acids, the net effect of decarboxylation, would give rise to a series of *n*-alkanes with an odd number of carbon atoms.

rocks obtained from the moon or nearby planets. Thus it even includes the possibility of uncovering exotic and alien biochemistries. The exploration of the ancient rocks of the earth provides a testing ground for the method and the concepts involved.

We chose the alkanes because one might expect them to resist fairly high temperatures and chemical attack for long period of time. Moreover, J. G. Bendoraitis of the Socony Oil Company, Warren G. Meinschein of the Esso Research Laboratory and others had already identified individual long-chain alkanes, including a range of isoprenoid types, in certain crude oils. Even more encouraging, J. J. Cummins and W.E. Robinson of the U. S. Bureau of Miens had just made a preliminary announcements of their isolation of phytane, pristine and other isoprenoids from a relatively young sedimentary rock: the Green River shale of Colorado, Utah and Wyoming. Thus the alkanes seemed to offer the biological markes we were seeking. Robinson generously provided our laboratory with samples of the Green River shale, which was deposited some 50 million years ago and constitutes the major oil-shale reserve of the U.S.

The Green River shale, which is the remains of large Eocene lakes in a rather stable environment, contains a considerable fraction (.6 per cent) of alkanes. Using the molecular-sieve technique, split the total alkane fraction into alkanes with straight chains and those with branched chains and rings and ran the resulting fractions through the gas chromatograph. The straight-chain alkanes exhibit a marked dominance of molecules containing an odd number of carbon atoms, which is to be expected for straight-chain hydrocarbons from a biological source. The other fraction shows a series of prominent sharp peaks; we conclusively identified them as isoprenoids, confirming the results of Cummings and Robinson. The large proportion of phytane, the hydrocarbon corresponding to the entire side chain of chlorophyll, is particularly noteworthy. The oxygenated counterparts of the steranes and triterpanes (27 to 30 carbon atoms) and the high-molecular-weight *n*-alkanes (29 to 31 carbons) are typical constituents of the waxy covering of the leaves and pollen of land plants, leading to the inference that such plants made major contributions to the organic matter deposited in the Green River sediments.

Although the gross chemical structure (number of rings and side chains) of the sterances and triterpanes was established in this work, it was only recently that the precise structure of one of these hydrocarbons was conclusively established. E. V. Whitehead and his associates in the British Petroleum Company and Robinson and his collaborators in the Bureau of Miens have shown that one of the triterpanes extracted from the Green River shale is identical in all respects with grammacerane. Conceivably it is produced by the reduction of a compound known as gammaceran-3-beta-ol, which was recently isolated from the common protozoon *Tetrahymena pyriformis.* Other derivatives of grammacerane are rather widely distributed in the plant kingdom.

At our laboratory in Glasgow, Sister Mary T.J. Murphy and Andrew McCormick recently identified several sterances and triterpanes and also the tetraterpane called perhydro-beta-carotene, or carotane. Presumably carotene is derived by reduction from beta-carotene, an important red pigment of plants. A similar reduction process could convert the familiar biological compound cholesterol into cholestane, one of the steranes found in the Green River shale. The mechanism and sedimentary site of such geochemical reduction processes in an important problem awaiting attack.

W.H. Bradley of the U.S. Geological Survey has sought a contemporary counterpart of the richly organic ooze that presumably gave rise to the Green River shale. So far he has located only four lakes, two in the U.S. and two in Africa, that seem to be reasonable candidates. One of them, Mud Lake in Florida, is now being studied closely. A dense belt of vegetation surrounding the lake filters out all the sand and silt that might otherwise be washed into it from the land. As a result the main source of sedimentary material is the prolific growth of microscopic algae. The lake bottom uniformly consists of a grayish-green ooze about three feet deep. The bottom of the ooze was deposited about 2,300 years ago, according to dating by the carbon-14 technique.

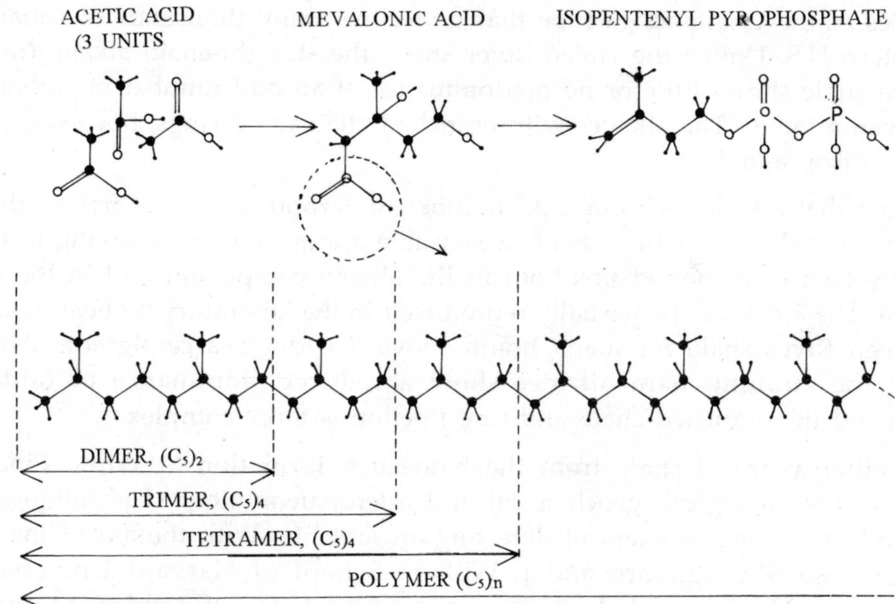

Fig. 6.7: *Branched-chain Lipids* are produced in living organisms by an enzymatically controlled process, also depicted schematically. In this process three acetate units link up to form a six-carbon compound (mevalonic acid), which subsequently loses a carbon atom and is combined with a high-energy phosphate. "Head to tail" assembly of the five-carbon subunits produces branched-chain molecules that are referred to as isoprenoid structures.

Microscopic examination of the ooze shows that it consists mainly of minute fecal pellets, made up almost exclusively of the cell walls of blue-green algae. Some pollen grains are also present. Decay is surprisingly slow in spite of the ooze's high content of bound oxygen and the temperatures characteristic of Florida. Chemical analyses in several laboratories, reported this past November at a meeting of the Geological Society of America, indicate that there is indeed considerable correspondence between the lipids of the Mud Lake ooze and those of the Green River shale. Eugene McCarthy of the University of California at Berkeley has also found beta-carotene in samples of Mud Lake ooze that are about 1,100 years old. The high oxygen content of the Mud Lake ooze seems inconsistent, however, with the dominance of oxygen-poor compounds in the Green River shale. The long-term geological mechanisms that account for the loss of oxygen may have to be sought in sediments older than those in Mud Lake.

Sediments much older than the Green River shale have now been examined by our groups in Berkeley and Glasgow, and by workers in other universities and in oil-industry laboratories. We find that the hydrocarbon fractions in these more ancient samples are usually more complex than those of the Green River shale; the gas chromatograms of the older samples tend to show a number of partially resolved peaks centered around a single maximum. One of the older sediments we have studied is the Antrim shale of Michigan. A black shale probably 350 million years old, it resembles other shales of the Chattanooga type that underline many thousands of square miles of the eastern U.S. Unlike the Green River shale, the straight-chain alkane fraction of the Antrim shale shows little or no predominance of an odd number of carbon atoms over an even number. The alkanes with branched chains and rings, however, continue to be rich in isoprenoids.

The fact that alkanes with an odd number of carbon atoms are not predominant in the Antrim shale and sediments of comparable antiquity may be owing to the slow cracking by heat of carbon chains both in the alkane component and in the kerogen component. The effect can be partially reproduced in the laboratory by heating a sample of the Green River shale for many hours above 300 degrees centigrade. After such treatment the straight-chain alkanes show a reduced dominance of odd-carbon molecules and the branched-chain-and-ring fraction is more complex.

The billion-year-old shale from the Nonesuch formation at White Pine, Mich, exemplifies how geological, geochemical and micropaleontological techniques can be brought to bear on the problem of detecting ancient life. With the aid of the electron microscope Elso S. Barghoorn and J. William Schopf of Harvard University have detected in the Nonesuch shale "disaggregared particles of condensed spheroidal organic matter." In collaboration with Meinschein the Harvard workers have also found evidence that the Nonesuch shale contains isoprenoid alkanes, steranes and porphyrins. Independently we have analyzed the Nonesuch shale and found that it contains pristine and phytane, in addition to iso-alkanes, anteiso-alkanes and cyclohexyl alkanes.

Barghoorn and S.A. Tyler have also detected microfossils in the Gunflint chert of Ontario, which is 1.9 billion years old, almost twice the age of the Nonesuch shale. They have reported that the morphology of the Gunflint microfossils "is similar to that of the existing primitive filamentous blue-green algae."

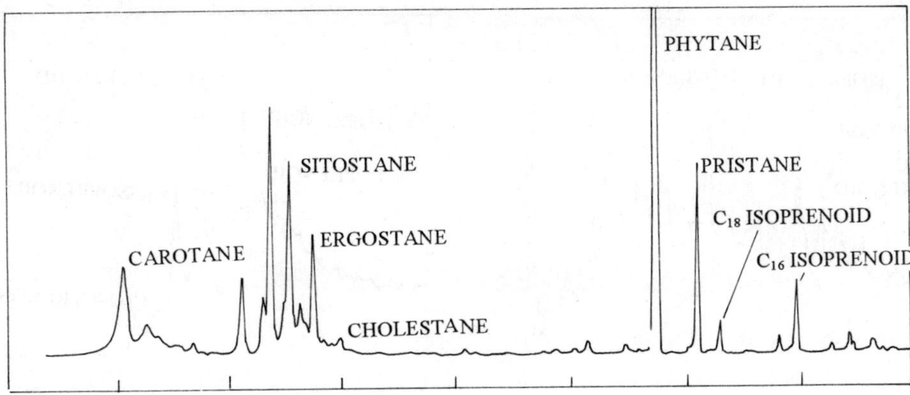

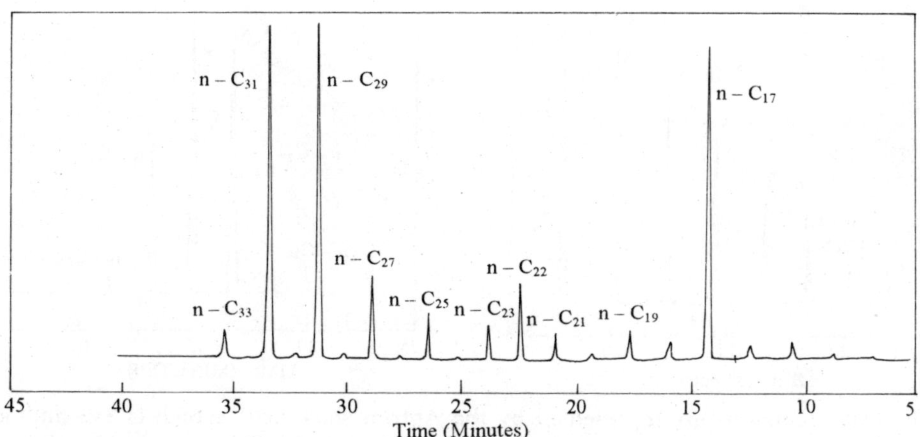

Fig. 6.8: *Hydrocarbons in Young Sediment*, the 50-million-year-old Green River shale, produced these chromatograms. Alkanes with branched chains and rings appear in the top curve, normal alkanes in the bottom curve. The alkanes in individual peaks were identified by mass spectrometry and other methods. Such alkanes as phytane and pristine and the predominance of normal alkanes with an odd number of carbon atoms affirm that the hydrocarbons are biological in origin. The biomodal distribution of the curves is also significant.

One of the oldest Precambrian sediments yet analyzed is the Soudan shale of Minnesota, which was formed about 2.7 billion years ago. Although its total hydrocarbon content is only .05 per cent, we have found that it contains a mixture of straight-chain alkanes and branched-chain-and-ring alkanes not unlike those present in the much younger Antrim shale. In the branched chain-and-ring faction we have identified pristane and phytane. Steranes and triterpanes also seem to be present, but we have not yet established their precise three-dimensional structure. Preston E. Cloud of the University of California at Los Angeles has reported that the Soudan shale

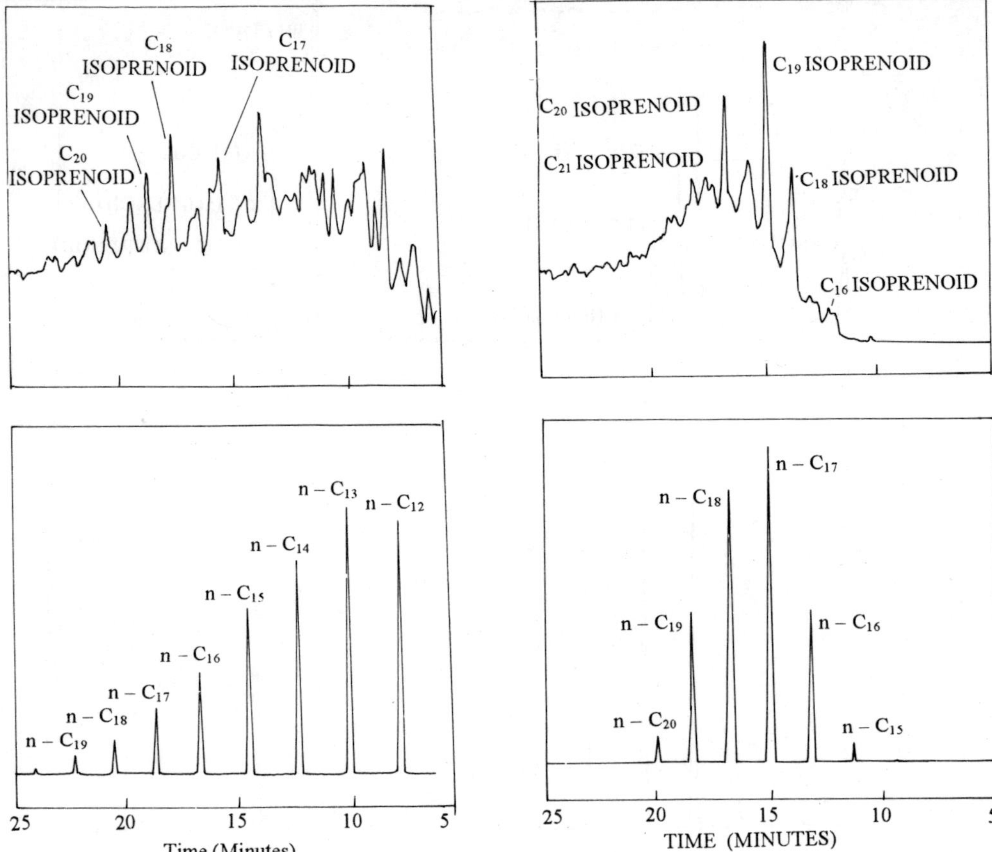

Fig. 6.9: *Older Sediments* are represented by the Antrim shale (*left*), which is 350 million years old, and by the Soudan shale (*right*), which is 2.7 billion years old. Alkanes with branched chains and rings again are shown in the top curves, normal alkanes in the bottom curves. These chromatograms lack a pronounced bimodal distribution and the normal alkanes do not show a predominance of molecules with an odd number of carbon atoms. Nevertheless, the prevalence of isoprenoids argues for a biological origin.

contains microstructures resembling bacteria or blue-green algae, but he is not satisifed that the evidence is conclusive.

A few reports are now available on the most ancient rocks yet examined: sediments from the Fig Tree system of Swaziland in Africa, some 3.1 billion years old. An appreciable fraction of the alkane component of these rocks consists of isoprenoid molecules. If one assumes that isoprenoids are chemical vestiges of chlorophyll, one is obliged to conclude that living organisms appeared on the earth only about 1.7 billion years after the earth was formed (an estimated 4.8 billion years ago).

Before reaching this conclusion, however, one would like to be sure that the isoprenoids found in ancient sediments have the precise carbon skeleton of the

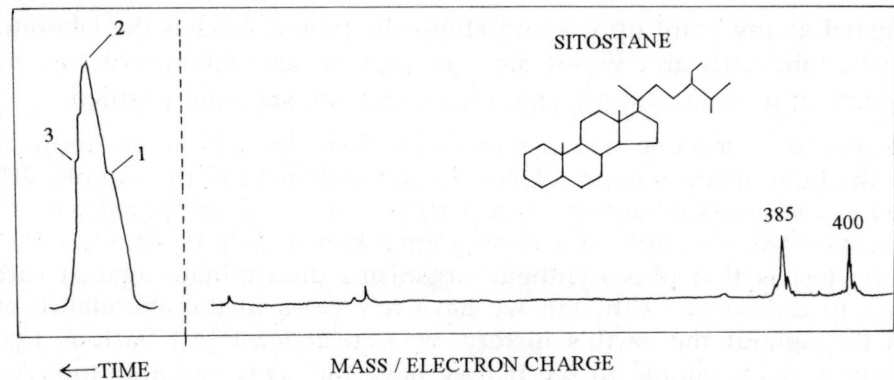

Fig. 6.10: *Identification of Sitostane* in the Green River shale was accomplished by "trapping" the alkanes that produced a major peak in the chromatogram (*colored arrow at top left on this page*) and passing them through the chromatograph-mass spectrometer. As the chromatography drew the curve at the left, three scans were made with the spectrometer. Scan 1 (*partially shown at right*) is identical with the scan produced by pure sitostane.

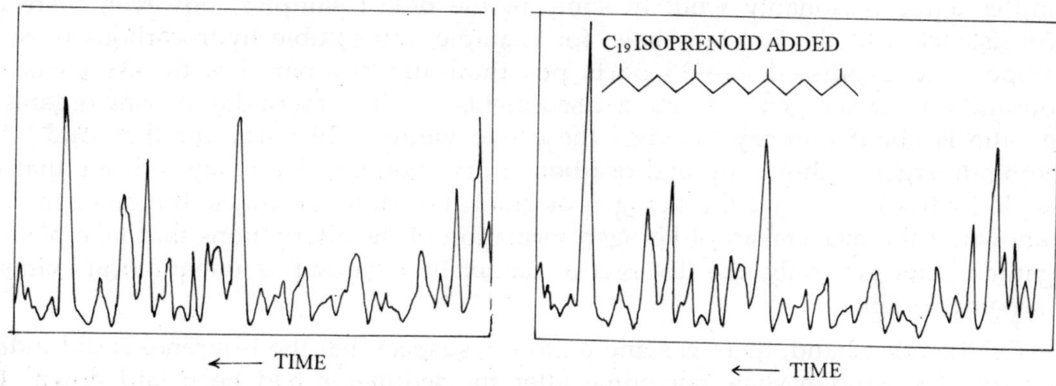

Fig. 6.11: *Nineteen Carbon Isoprenoid* was identified in the Antrim shale by using a conjection technique together with a high resolution gas chromatography. These two high-resolution curves, each taken from a much longer trace, show the change in height of a specific peak when a small amount of pure 19-carbon isoprenoid was conjected with the sample. Other peaks can be similarly identified by conjecting known alkanes.

biological molecules from which they are presumed to be derived. So far no sample of pristane or phytane—the isoprenoids that may be derived from the phytyl side chain of chlorophyll—has been shown to duplicate the precise three-dimensional structure of a pure reference sample. Vigorous efforts are being made to clinch the identification.

Assuming that one can firmly establish the presence of biologically structured isoprenoid alkanes in a sediment, further questions remain. The most serious one is: Were the hydrocarbons or their precursors deposited when the sediment was formed or did they seep in later? This question is not easily answered. A sample can be contaminated at any point up to—and after—the time it reaches the laboratory bench. Fossil fuels, lubricants and waxes are omnipresent, and laboratory solvents contain tiny amounts of pristane and phytane unless they are specially purified.

One way to determine whether or not rock hydrocarbons are indigenous is to measure the ratio of the isotopes carbon 13 and carbon 12 in the sample. (The ratio is expressed as the excess of carbon 13 in parts per thousand compared with the isotope ratio in a standard: a sample of a fossil animal known as a belemnite.) The principle behind the test is that photosynthetic organisms discriminate against carbon 13 in preference to carbon 12. Although we have few clues to the abundance of the two isotopes throughout the earth's history, we can at least test various hydrocarbon fractions in a given sample to see if they have the same isotope ratio. As a simple assumption, one would expect to find the same ratio in the soluble organic fraction as in the insoluble kerogen fraction, which could not have seeped into the rock as kerogen.

Philip II. Abelson and Thomas, C. Hoering of the Carnegie Institution of Washington have made such measurements on sediments of various geological ages and have found that the isotope ratios for soluble and insoluble fractions in most samples agree reasonably well. In some of the oldest samples, however, there are inconsistencies. In the Soudan shale, for example, the soluble hydrocarbons have an isotope ratio expressed as—25 parts per thousand compared with—34 parts per thousand for the kerogen. (In younger sediments and in present-day marine organisms the ratio is about midway between these two values:—29 parts per thousand.) The isotope divergence shown by hydrocarbons in the Soudan shale may indicate that the soluble hydrocarbons and the kerogen originated at different times. But since nothing is known of the mechanism of kerogen formation of the alternations that take place in organic matter generally, the divergence cannot be regarded as unequivocal evidence of separate origin.

On the other hand, there is some reason to suspect that the isoprenoids did indeed seep into the Soudan shale sometime after the sediments had been laid down. The Soudan formation shows evidence of having been subjected to temperatures as high as 400 degree C. The isoprenoid hydrocarbons pristane and phytane would not survive such conditions for very long. But since the exact date, extent and duration of the heating of the Soudan shale are not known, one can only speculate about whether the isoprenoids were indigenous and survived or seeped in later. In any event, they could not have seeped in much later because the sediment became compacted and relatively impervious within a few tens of millions of years.

A still more fundamental issue is whether or not isoprenoid molecules and others whose architecture follows that of known biological substances could have been formed by nonbiological processes. We and others are studying the kinds and concentrations of hydrocarbons produced by both biological and nonbiological sources. Isoprene itself, the hydrocarbon whose polymer constitutes natural rubber, is easily prepared in the laboratory, but no one has been able to demonstrate that isoprenoids can be formed nonbiologically under geologically plausible conditions. Using a computer approach,

Margaret O. Dayhoff of the National Biomedical Research Foundation and Edward Anders of the University of Chicago and their colleagues have concluded that under certain restricted conditions isoprene should be one of the products of their hypothetical reactions. But this remains to be demonstrated in the laboratory.

It is well known, of course, that complex mixtures of straight-chain, branched-chain and even ring hydrocarbons can readily be synthesized in the laboratory from simple starting materials. For example, the Fischer-Tropsch process, used by the Germans as a source of synthetic fuel in World War II, produces a mixture of saturated hydrocarbons from carbon monoxide and water. The reaction requires a catalyst (usually nickel, cobalt or iron), a pressure of about 100 atmospheres and a temperature of from 200 to 350 degree C. The hydrocarbons formed by this process, and several others that have been studied, generally show a smooth distribution of saturated hydrocarbons. Many of them have straight chains but lack the special characteristics (such as the predominance of chains with an old number of carbons) found in the similar hydrocarbons present in many sediments. Isoprenoid alkanes, if they are formed at all, cannot be detected.

Paul C. Marx of the Aerospace Corporation has made the ingenious suggestion that isoprenoids may be produced by the hydrogenation of graphite. In the layered structure of graphite the carbon atoms are held in hexagonal arrays by carbon-carbon bonds. Marx has pointed out that if the bonds were broken in certain ways during hydrogeneration, an isoprenoid structure might result. Again a laboratory demonstration is needed to support the hypothesis. What seems certain, however, is that nonbiological syntheses are extremely unlikely to produce those specific isoprenoid patterns found in the products of living cells.

Another dimension is added to this discussion by the proposal, made from time to time by geologists, that certain hydrocarbon deposits are nonbiological in origin. Two alleged examples of such a deposit are a mineral oil found enclosed in a quartz mineral at the Abbott mercury mine in California and a bitumen-like material called thucolite found in an ancient nonsedimentary rock in Ontario. Samples of both materials have been analyzed in our laboratory at Berkeley. The Abbott oil contains a significant isoprenoid fraction and probably constitutes an oil extracted and brought up from somewhat older sediments of normal biological origin. The thucolite consists chiefly of carbon from which only a tiny hydrocarbon fraction can be extracted. Our analysis shows, however, that the fraction contains trace amounts of pristane and phytane. Recognizing the hazards of contamination, we are repeating the analysis, but on the basis of our preliminary findings we suspect that the thucolite sample represents an oil of biological origin that has been almost completely carbonized. We are aware, of course, that one runs the risk of invoking circular arguments in such discussions. Do isoprenoids demonstrate biological origin (as we and others are suggesting) or does the presence of isoprenoids in such unlikely substances indicate that they were formed nonbiologically? The debate may not be quickly settled.

There is little doubt, in any case, that organic compounds of considerable variety and complexity must have accumulated on the primitive earth during the prolonged

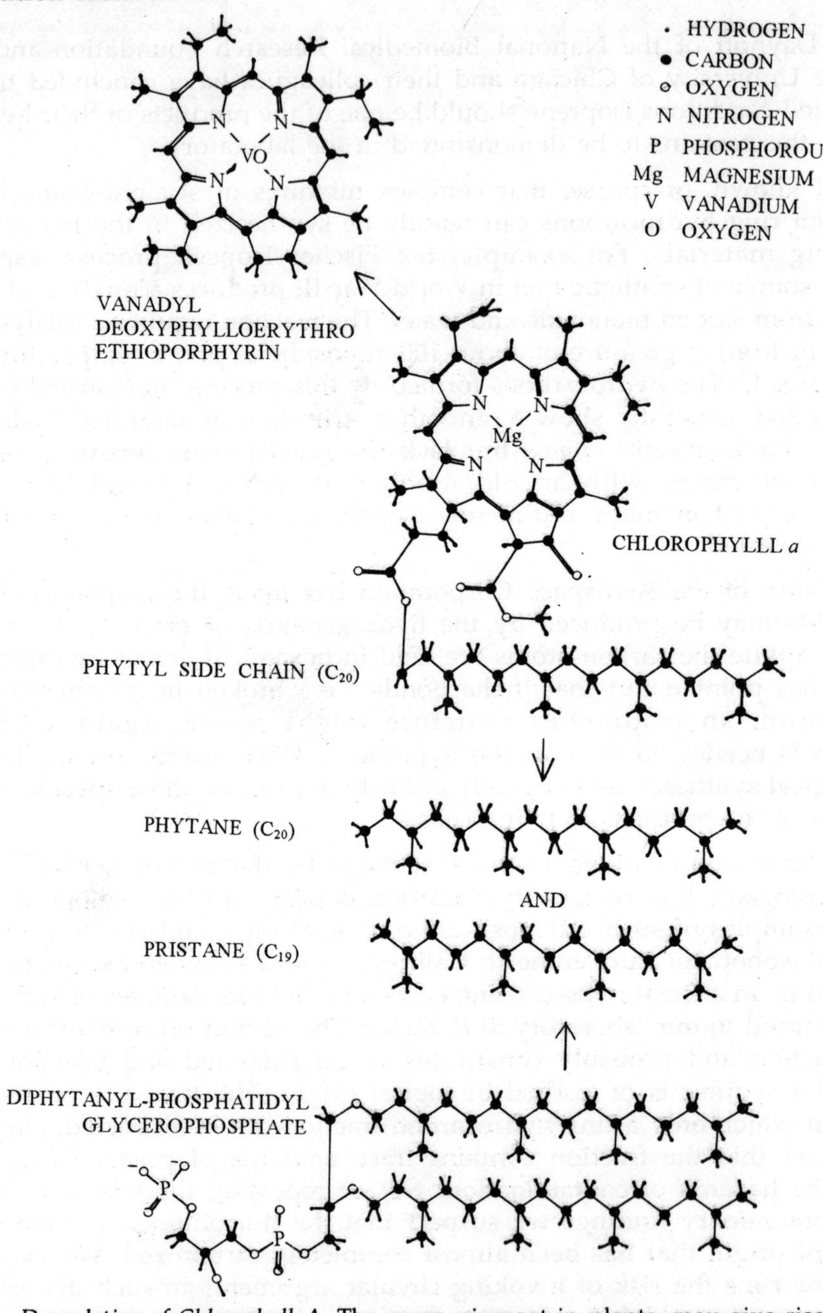

Fig. 6.12: *Degradation of Chlorophyll A.* The green pigment in plants, may give rise to two kinds of isoprenoid molecules, phytane and pristane, that have been identified in many ancient sediments. It also seems likely that phytane and pristane can be derived from the isoprenoid side chains of a phosphate-containing lipid (*bottom structure*) that is a major constituent of salt-loving bacteria. The porphyrin ring of chlorophyll *a* is the probable source of vanadyl porphyrin (*upper left*) that is widely found in crude oils and shales.

period of nonbiological chemical development—the period of chemical evolution. With the appearance of the first living organisms biological evolution took command and presumably the "food stock" of nonbiological compounds was rapidly altered. If the changeover was abrupt on a geological time scale, one would expect to find evidence of it in the chemical composition of sediments whose age happens to bracket the period of transition. Such a discontinuity would make an intensely exciting find for organic geochemistry. The transition from chemical to biological evolution must have occurred earlier than three billion years ago. As yet, however, no criteria have been established for distinguishing between the two types of evolutionary process.

We suggest that an important distinction should exist between the kinds of molecules formed by the two processes. In the period of chemical evolution autocatalysis must have been one of the dominant mechanisms for creating large molecules. An autocatalytic system is one in which a particular substance promotes the formation of more of itself. In biological evolution, on the other hand, two different molecular

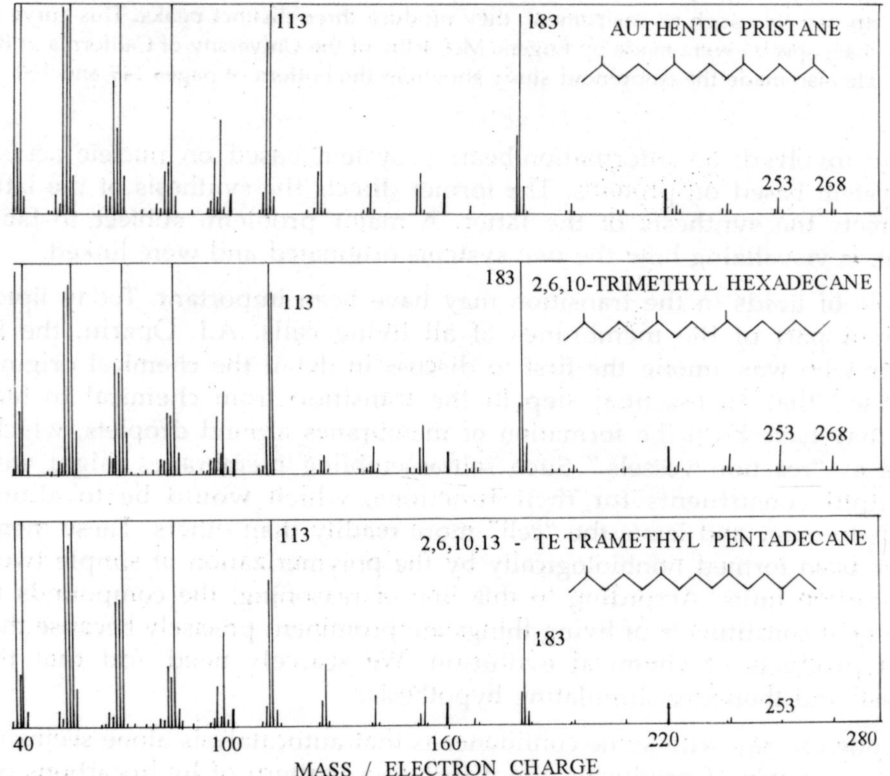

Fig. 6.13: *Similarly of Mass Spectra* makes it difficult to distinguish the 19-carbon isoprenoid pristane from two of its many isomers (molecules with the same number of carbon and hydrogen atoms). The three records shown here are replotted from the actual tracings produced by pure compounds. When the sample contains impurities, as is normally the case, the difficulty of identifying authentic pristane by mass spectrometry is even greater.

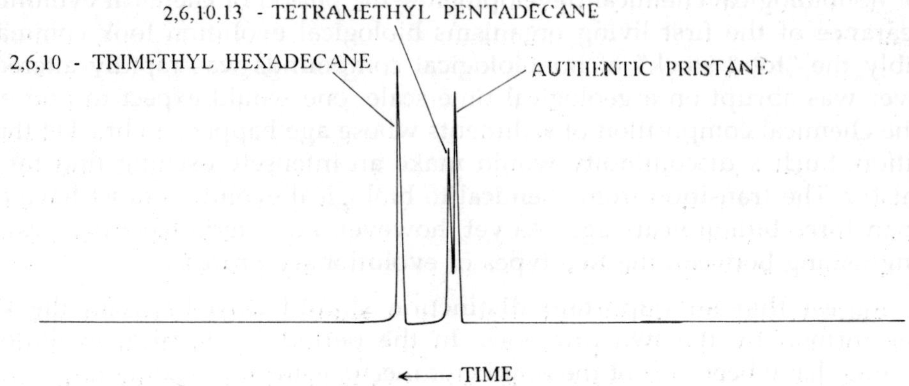

Fig. 6.14: *Identification of Pristane* can be done more successfully with the aid of a high resolution gas chromatograph. When pure pristane and the isomers shown in the illustration above are fed into such an instrument, they produce three distinct peaks. This curve and the mass spectra were made by Eugene McCarthy of the University of California at Berkeley. He also made the isoprenoid study shown at the bottom of pages 149 and 150.

systems are involved: an information-bearing system based on nucleic acids and a catalytic system based on proteins. The former directs the synthesis of the latter. The former directs the synthesis of the latter. A major problem, subject to laboratory experiment, is visualizing how the two systems originated and were linked.

The role of lipids in the transition may have been important. Today lipids form an important part of the membranes of all living cells. A.I. Oparin, the Russian investigator who was among the first to discuss in detail the chemical origin of life, has suggested that an essential step in the transition from chemical to biological evolution may have been the formation of membranes around droplets, which could then serve as "reaction vessels." Such self-assembling membranes might well have required lipid constituents for their functions, which would be to allow some compounds to enter and leave the "cell" more readily than others. These membranes might have been formed nonbiologically by the polymerization of simple two-carbon and three-carbon units. According to this line of reasoning, the compounds that are now prominent constituents of living things are prominent precisely because they were prominent products of chemical evolution. We scarcely need add that this is a controversial and therefore stimulating hypothesis.

What one can say with some confidence is that autocatalysis alone seems unlikely to have been capable of producing the distribution pattern of hydrocarbons observed in ancient Precambrian rocks, even when some allowance is made for subsequent reactions over the course of geologic time. That it could have produced compounds of the observed type is undoubtedly possible but it seems to us that the observed pattern could not have arisen without the operation of those molecular systems we now recognize as the basis of living things. Eventually it should be possible to find in the geological record certain molecular fossils that will mark the boundary between chemical and biological evolution.

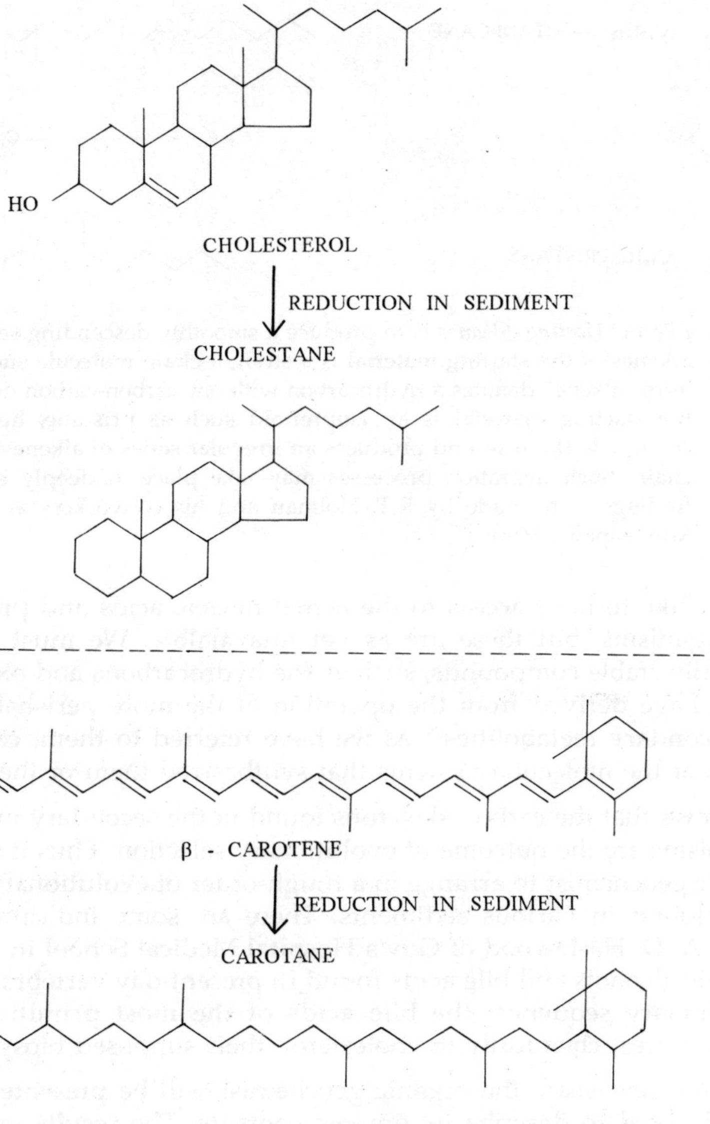

HO

CHOLESTEROL

REDUCTION IN SEDIMENT

CHOLESTANE

β - CAROTENE

REDUCTION IN SEDIMENT

CAROTANE

Fig. 6.15: *Two Alkanes in Green River Shale,* cholestane and carotane, probably have been derived from two well-known biological substance; cholesterol and beta-carotene. The former is closely related to the steroid hormones; the latter is a red pigment widely distributed in plants. These two natural substances can be converted to their alkane form by reduction: a process that adds hydrogen at the site of double bonds and removes oxygen.

Another and more immediate goal for the organic geochemist is to attempt to trade on the molecular level the direction of biological evolution. For such a study

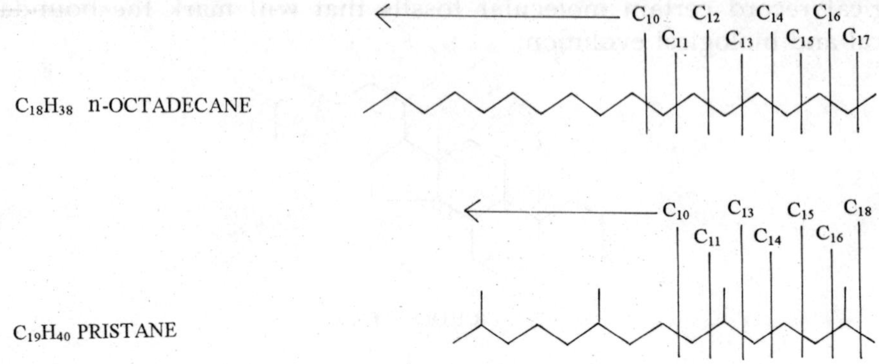

$C_{18}H_{38}$ n-OCTADECANE

C_{10} C_{11} C_{12} C_{13} C_{14} C_{15} C_{16} C_{17}

$C_{19}H_{40}$ PRISTANE

C_{10} C_{11} C_{13} C_{14} C_{15} C_{16} C_{18}

Fig. 6.16: *Effect of Heating Alkanes* is to produce a smoothly descending series of products (normal alkenes) if the starting material is a straight-chain molecule such as *n*-octadecane. (The term "alkene" denotes a hydrocarbon with one carbon-carbon double bond.) If, however, the starting material is an isoprenoid such as pristane, heating it is 600 degrees centigrade for .6 second produces an irregular series of alkenes because of the branched chain. Such degration processes may take place in deeply buried sediments. These findings were made by R.T. Holman and his co-workers at the Hormel Institute in Minneapolis, Minn.

one would like to have access to the actual nucleic acids and proteins synthesized by ancient organisms, but these are as yet unavaialble. We must therefore turn to the geochemially stable compounds, such as the hydrocarbons and oxygenated compounds that must have derived from the operation of the more perishable molecular system. These "secondary metabolities," as we have referred to them, can be regarded as the signatures of the molecular systems that synthesized them or their close relatives.

It follows that the carbon skeletons found in the secondary metabolities of present-day organisms are the outcome of evolutionary selection. Thus it should be possible for the organic geochemist to arrange in a rough-order of evolutionary sequence the carbon skeletons found in various sediments. There are some indications that this may be feasile, G. A. D. Haslewood of Guy's Hospital Medical School in London has proposed that the bile alcohols and bile acids found in present-day vertebrates can be arranged in an evolutionary sequence: the bile acids of the most primitive organisms contain molecules nearest chemically to cholesterol, their supposed biosynthetic precursor.

Within a few years the organic geochemist will be presented with a piece of the moon and asked to describe its organic contents. The results of this analysis will be awaited with immense curiosity. Will we find that the moon is a barren rock or will we discover traces of organic compounds—some perhaps resembling the complex carbon skeletons we had thought could be produced only by living systems? During the 1970's and 1980's we can expect to receive reports from robot sampling and analystical instruments landed on Mars, Venus and perhaps Jupiter. Whatever the results and their possible indications of alien forms of life, we shall be very eager to learn what carbon compounds are present elsewhere in the solar system.

Chapter—7

The Oldest Fossils

How did life originate on the earth? It was not so long ago that attempts to answer this question defined more areas of uncertainty than of agreement. Today a fossil record that was virtually unknown before the 1950's has been found to bear witness to three of the key events in the earliest stages of organic evolution. The fossils come from widely separated parts of the earth. All are preserved in unusual rocks of the Precambrian, the first and by far the longest interval in geologic history. The earliest of the fossils are more than three billion years old.

All that is known or conjectured about the terrestrial origin of life suggests that the first appearance of living organisms was preceded by the gradual development of a complex chemical environment. This environment is usually pictured as a kind of primordial broth, filled with such "organic" molecules as amino acids, sugars and other biologically important substances that came into existence through nonorganic processes. Millions of years must have been required for the accumulation, elaboration and differentiation of the broth. That period may be called the time of chemical evolution, a concept that owes much to the work of a leading student of abiotic synthesis, Cyril A. Ponnamperuma of the Ames Research Center of the National Aeronautics and Space Administration. As a preliminary stage in the total process of organic evolution, chemical evolution of course reaches its climax when life-less organic molecules are assembled by chance into a living organism. This first form of life is what the Russian biochemist A. I. Oparin calls a "protobiont."

Judging from the various forms of life we know today, the first protobionts were probably microscopic in size and single-celled in structure, perhaps resembling the modern coccoid, or spherical, bacteria. Rather than imagining some specific organism, however, let us consider this first form of life more abstractly and give it a name that describes how it lives rather than how it looks. Call it a heterotroph, which is to say an organism that cannot manufacture all its own nutrients but must feed on organic molecules in the broth that surrounds it. (This assumes that the organism is immersed in an aqueous medium or at least rests on a wet surface; a supply of water is essential to protoplasmic life). It seems reasonable to suppose that the organism was a heterotroph; to have at this stage a full-fledged autotroph, an organism that can manufacture its organic nutrients out of inorganic substances, would be asking too much of chance.

We can however, demand that heterotrophs rather promptly give rise to autotrophs. Otherwise, as Preston Cloud of the University of California at Santa Barbara puts it, once the heterotrophs had "gobbled up all the goodies" within reach they would die off, making another accident of biosynthesis necessary in order to get things going again.

One way or another, then, autotrophs must evolve from an original heterotrophic population. This event in the development of life marks the time when the organic mutrients of the primordial broth are depleted and photosynthesis—the most plausible form of organic self—nutrition-must have been invented. Such an assumption is not based solely on logical considerations of thermodynamics and physiology. It is also supported by geological evidence that early organisms depended on photosynthesis to sustain themselves. Among the most ancient formations in the geological record are some showing signs that small amounts of free oxygen—the gaseous by-product of photosynthesis—were present early in the history of the earth. This chemical evidence coincides with indications in the fossil record that some of the most primitive forms of terrestrial life, organisms that resembled modern bacteria and blue-green algae, were then increasing in numbers and diversity.

Many modern bacteria are photosynthetic, although they do not produce free oxygen; all modern blue-green algae are photosynthetic and all produce free oxygen. The evidence suggesting that the first autotrophs were organisms of the same primitive kind is significant for an additional reason. Bacteria and blue-green algae are alone among living things in the simplicity of their cells. They have neither a membrane-enclosed nucleus nor such specialized cellular organelles as mitochondria. Their genetic material is diffused throughout the cell, and they are incapable of either mitosis (body-cell division) or meiosis (germ-cell division). Both kinds of cell division require that the genetic material be organized into chromosomes.

Bacteria and blue-green algae are thus fundamentally different from other organisms: all other plants, all animals and the many forms that are neither entirely plant nor entirely animal. These other organisms have cells with nuclei and specialized organelles or specialized intracellular structures; they are called eukaryotic (truly nucleated), whereas bacteria and blue-green algae are prokaryotic (prenuclear). It would be surprising if the autotrophs on the lowest rungs of the evolutionary ladder were anything but prokaryotic.

The prokaryotic cell is notable for still another reason. Any organism whose genetic material is diffused throughout the cell and whose reproduction does not involve a recombination of parental genes is genetically conservative. In such an organism random mutations, instead of being preserved when they benefit the organism, tend to be damped out in a few generations. The blue-green algae are an outstanding example of such genetic conservatism. Many living species are almost indistinguishable in structure from species that flourished a billion or more years ago.

Against this background of fact and conjecture, how many of the events in the early stages of organic evolution can be labeled as being outstanding? There appear to be three such events, each of them a kind of threshold-crossing. The first, obviously the *sine qua non,* is successful biosynthesis: a crossing of the threshold separating the initial period of abiotic chemical evolution from the subsequent organic period. Perhaps someday fossil evidence of the earth's earliest organisms, the heterotrophs that crossed this first threshold, will be discovered. In the meantime the remains of their successors, the early photosynthetic autotrophs, provide the necessary proof that the first threshold had been passed.

The next threshold can be characterized as the threshold of diversification. As evolution progresses the first photosynthetic organism should not be limited to a few similar forms. Instead they should develop differences in shape and structure indicative of roles in a variety of ecological niches.

The third threshold divides prokaryotic organisms from all others. A world populated solely by bacteria and blue-green algae is conceivable; indeed, that is apparently the way it was. Viewed from our present vantage such a world seems poor in possibilities for further evolution. The extraordinary variety of plant and animal life that has arisen on the earth over the past 600 million years is due entirely to the invention of the eukaryotic cell, with its potential for genetic diversity.

Thanks to a lucky accident of fossil preservation, we now have evidence that each of these thresholds was successfully crossed during the vast span of Precambrian times. The Precambrian represents nearly four billion of the earth's first 4.5 billion years. It began with the formation of the earth and ended some 600 million years ago with the dawn of the Paleozoic era. Nothing is known of the first billion years or so; the world's oldest-known rocks, found in Africa, are not much more than three billion years old.

Precambrian rocks are found not only in Africa but also on every other continent. The best-known areas are the Canadian Shield in North America and the Fenno-Scandian Shield in Europe, but more than a third of Australia is also a Precambrian shield and there are sizable Precambrian areas in South America and Asia. Some Precambrian rocks are of igneous origin and some are of sedimentary origin. Most of the sediments have been heavily metamorphosed: changed in form and chemical composition by heat and pressure. In these metamorphic rocks not only fossils but also the faintest traces of organic matter have been obliterated.

A few Precambrian sediments have escaped substantial alteration. Extensive deposits of black shale, black chert and other stratified sediments are scattered through the major shield areas in virtually unmetamorphosed condition. These carbon-rich rocks-for example the formations in the Lake Superior region of North America in the Transvaal of South Africa and in parts of western Australia—look even to the experienced eye very much like certain sediments of the Carboniferous epoch that are a mere 300 million years old.

The oldest-known group of Precambrian sediments is located in the border region between the Republic of South Africa and Swaziland. The formation is called the Swaziland Sequence; its stratified rocks are thousands of feet thick. One series of strata in the middle of the sequence, known as the Fig Tree formation, is well exposed in the Barberton Mountain Land, a gold-mining district near the town of Barberton in the eastern Transvaal. The Fig Tree rocks consist of black, gray and greenish cherts, interbedded with jaspers, ironstones, slates, shales and graywackes. In places the chert beds are 400 feet thick. The chert is usually fractured, but it is cemented together and the fractures are filled with quartz; parts of the formation show little evidence of metamorphism. The Fig Tree cherts contains traces of organic matter and a few microfossils.

The Barberton Mountain Land has long been an active mining area, and its geology has been studied in considerable detail. The age of the graywackes and shales has been determined independently by several laboratories using a method based on the decary of radioactive strontium and rubidium. This particular radioactive clock started to run 3.1 billion years ago, but there is good reason to believe the sediments were deposited somewhat earlier. Recently rocks lying close to the base of the Swaziland Sequence were shown to be 3.36 billion years old. In light of these age determinations it seems probable that the age of the Fig Tree cherts is in excess of 3.2 billion years.

In 1965 I collected cherts from several localities in the Fig tree series, and samples were prepared for examination in my laboratory at Harvard University that summer.

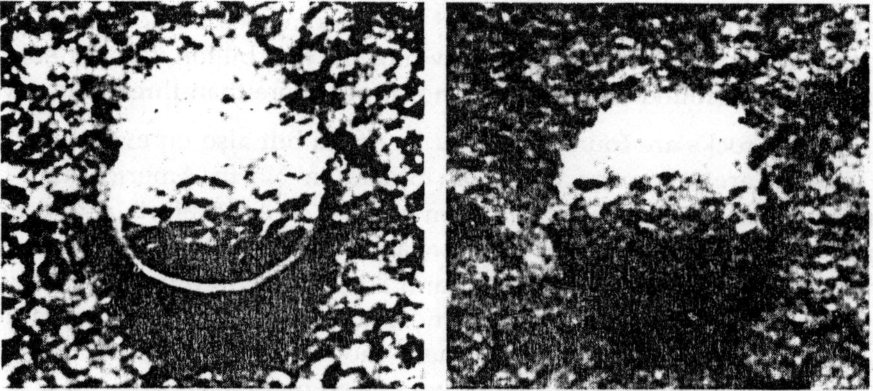

Fig. 7.1: *Two Cross Sections of Eobacterium are seen in electron micrographs of metal-shadowed carbon replicas. In the fossil at the left the outer and inner layers of the cell wall are visible. The wall, .015 micron thick, resembles the wall of living bacteria of the bacillus type.*

Two techniques were employed: thin sections of chert were cut for examination by reffected and transmitted white light under the microscope, and carbon replicas of etched and polished chert surfaces were made. The carbon replicas were then "shadowed" with metal and examined by transmission electron microscopy. J. William Schoph joined me in the study of the specimens.

When Schopf and I examined the thin sections under the light microscope, we could see that the rock matrix contained abundant laminations of dark-colored and virtually opaque organic matter. The laminations were irregular but were usually aligned parallel to the strata of the chert, suggesting that they had originally been formed as part of an aqueous sedimentary deposit. No distortion was evident where the organic matter crossed the boundaries of the individual grains of chalcedony comprising the chert which suggested to us that the process of deposition had emplaced the organic material within a silica-rich matrix before the silica was crystallized into chert. There was no evidence whatever that the silica was of secondary origin.

It was difficult to discern distinct bodies within the layers of organic matter under the light microscope. Our first success in isolating a Fig Tree organism was achieved with the carbon-replica technique; the electron microscope revealed a number of rod-shaped structures, preserved both in profile and in cross section. The rods are very small. They range in length from slightly under .5 micron to a little less than .7 micron, and in diameter from about .2 micron to a little more than .3 micron. In cross section the cell wall is sometimes seen to consist of an inner and an outer layer, with a total wall thickness of .015 micron. This is comparable to the cell wall of many modern bacteria in both structure and dimension.

Electron microscopy also revealed the presence of organic material in the form of irregular, threadlike filaments lacking discernible structural detail. The threads are clearly native to the chert and not the products of contamination. They are as much as

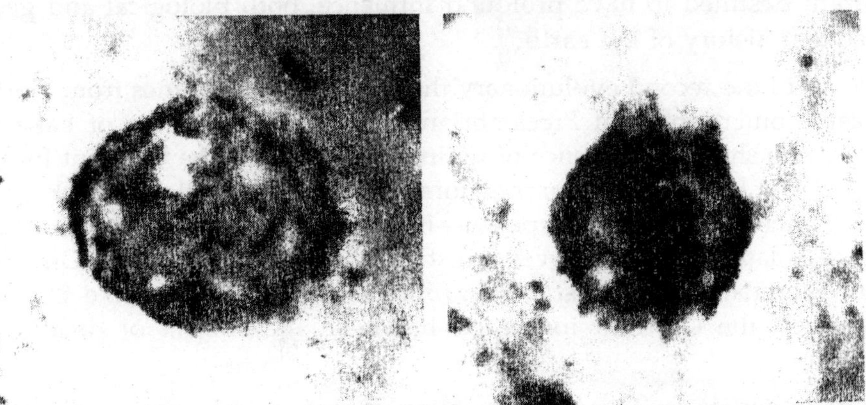

Fig. 7.2: *Alga-like spheres, seen in photomicrographs of thin sections of rock, are the other fossil organisms that are found in the Fig Tree formation. The diameter of the spheres is usually less than 20 microns. The organism is named Archaeophaeroides barbertonesis.*

nine microns long and resemble decomposed plant material. Although these filaments are almost certainly of biological origin they cannot be identified with any known type of organism. It has been suggested, probably as a result of wishful thinking, that they may be polymerized strands of abiotic organic matter from the primordial broth.

Schopf and I later succeeded in resolving larger microfossils in thin sections of Fig Tree chert under the light microscope. These fossils are spheroidal; measurements of 28 particularly well-defined specimens show that the majority are between 17 and 20 microns in diameter. Some have a darkened interior, as if cytoplasm within the spheroid had coalesced and become "coalified." Just as the rod-shaped organisms revealed by the electron microscope resemble certain modern bacteria, so the spheroids are not unlike some modern blue-green algae of the cocoid group. They may even be among the evolutionary precursors of such algae.

We have named the rods *Eobacterium isolatum,* a new genus and species. The generic name (*eo*—is the Greek root for "dawn") points to the great antiquity of the organism and to its bacterium-like appearance; the specific name defines its noncolonical, single-cell habit of growth. The spheroids we have named *Archaeosphaeroides barbertonensis,* also a new genus and species. Again the generic name refers to the organism's great age and its appearance; the specific name identifies its place of discovery. The existence of these two organisms, successful inhabitants of an aquatic environment more than three billion years ago, is evidence that the first evolutionary threshold—the transition from chemical evolution to organic evolution—had been safely crossed at some even earlier date. We now know that at least two living organisms appeared well before the first third of earth history had passed. If we accept the evidence that the alga-like Fig Tree organisms were photosynthetic, an important geochemical event must also have occurred. With the onset of photosynthesis free oxygen would have begun to appear among the other constituents of the environment. The appearance of free oxygen was an event destined to have profound influence, both biological and geological, on the subsequent history of the earth.

Evidence of the second evolutionary threshold-crossing comes from North America. A remarkable outcropping of Precambrian rocks along the shore of Lake Superior in western Ontario shows a sequence of sediments known as the Gunflint Iron formation. The rocks at the base of the Gunflint formation include beds of black chert three to nine inches thick. The beds are exposed—in some place more or less continuously and in others as isolated outcrops—over a distance of some 115 miles in Ontario, from the vicinity of Schreiber on the east to Gunflint Lake on the west. Like the much earlier Fig Tree series, the Gunflint formation has been the subject of detailed geological investigation.

Granite underlies the Gunflint formation unconformably, that is, the basement rock had been eroded before the Gunflint sediments were deposited on its surface. Radioactive clocks have yielded two independent dates for the granite. A concentrate of its biotite contents indicates an age, in terms of the argon-potassium ratio, of 2.5±.75 billion years. The age of a whole-rock sample, in terms of its rubidium-strontium ratio, is 2.36±.70 billion years. The age of the granite thus provides a "floor" of maximum age under the Gunflint formation. The Gunflint cherts cannot be any older than these dates.

Micas separated from rocks in the upper levels of the Gunflint formation, collected near Thunder Bay, indicate an age in terms of the argon-potassium ration of 1.60±.05 billion years. For technical reasons this figure is only 80 per cent of the true age; it must therefore be adjusted to 1.90±.20 billion years. The micas thus provide a "ceiling" of minimum age for the cherts that lie at the base of the Gunflint formation. It seems reasonable to set the age of the cherts at approximately two billion years, which makes them a billion years or so younger than the Fig Tree cherts.

The only rocks in the Gunflint formation that contain microfossils are the cherts. Like the Fig Tree cherts, they are evidently the product of deposition in an aqueous environment that was rich in silica. Most of the Gunflint fossils are three-dimensional, and many show exquisite anatomical details. It has been argued that the structure of these fossils has been preserved by the infiltration of silica from the surrounding sediments. In my opinion the organisms were preserved without distortion by being deposited in a siliceous solution that later crystallized into chert, much as a modern biological specimen is preserved by being embedded in plastic. The soft structure of the organism owes its preservation to the almost complete incompressibility of the silica matrix. This is as unusual as it is fortunate. In most instances of fossil preservation the matrix is composed of relatively plastic sediments that are much compressed during consolidation, with the result that any preservation of soft tissues, let alone preservation in three-dimensional form, is a rarity.

How were the Gunflint cherts deposited? The picture in Ontario is rather clearer than it is in the Transvaal. The Gunflint cherts were apparently precipitated and consolidated around an underlying complex of basement rocks, consisting of greenstone boulders and finer conglomerates, that seems to have been continuously submerged at the time. "Domes" of algae, which are visible to the unaided eye in samples of Gunflint rock, grew on the surface of the boulders. Algal "pillars" grew perpendicularly to the domes; their fossil remains consist of alternate layers of coarsely crystalline quartz and fine-grained black chert.

In the 1950's I was privileged to be associated with Stanley A. Tyler of the University of Wisconsin in collecting and analyzing specimens of Gunflint chert from the Schreiber area. Only a few preliminary studies of these and other Gunflint specimens were published before Tyler died, although be then we both knew that his work had added to the fossil record an entire new group of very ancient, primitive photosynthetic organisms whose existence had not been suspected. Indeed, even though many years of study have now been devoted to the Gunflint organisms, the formation continues to yield new finds. Eight genera of primitive Gunflint plants, comprising 12

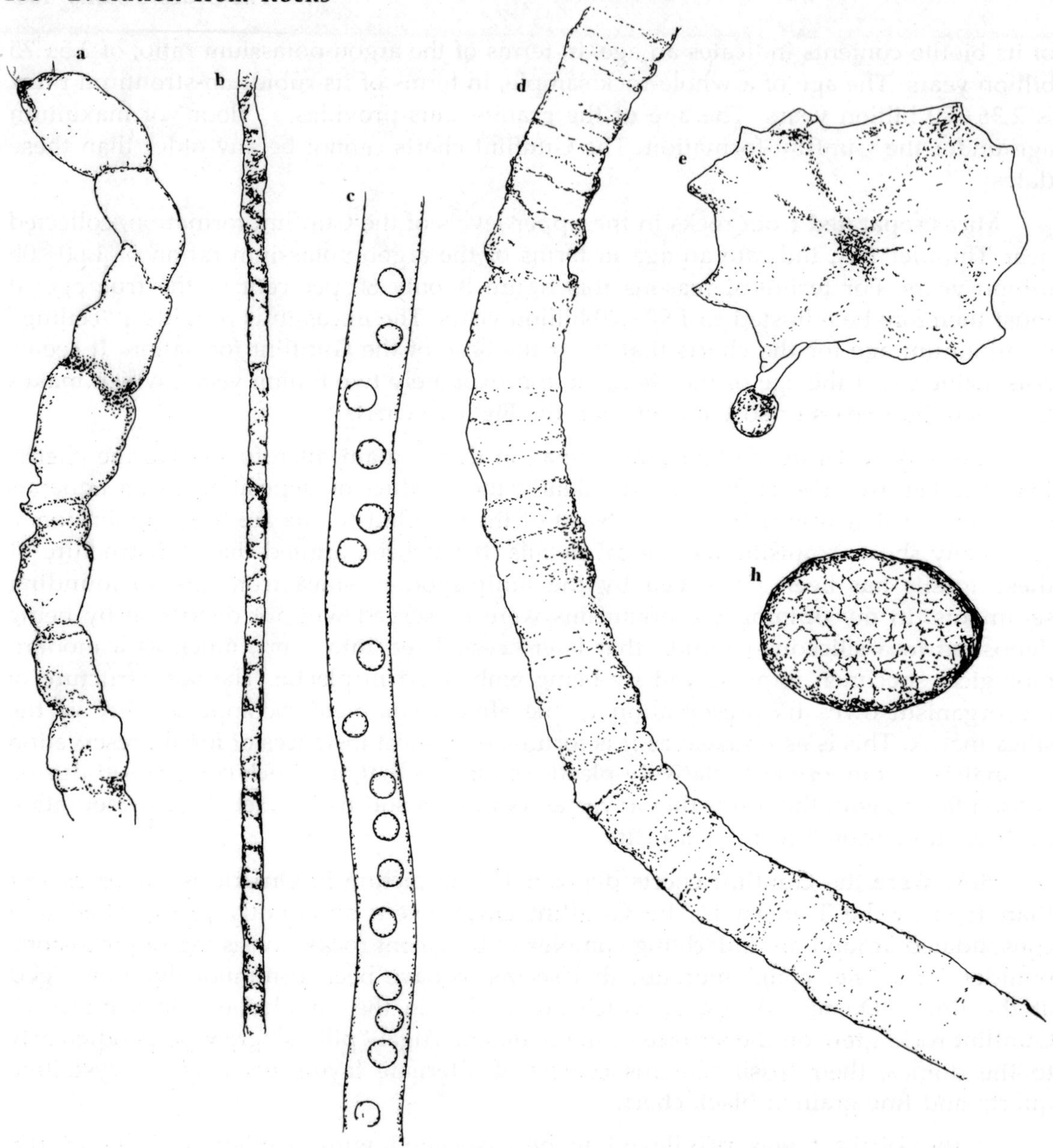

Fig. 7.3: *Diversity of Form,* evident among the plants fossilized in the Gunflint cherts, is indicative of evolutionary progress during the billion years that separate Fig Tree from Gunflint times. These composite drawing are based on a study of several specimens of each species and include three of the most unusual ones found. All are shown at the same scale: 2,500 times actual size. Filamentous organisms include two of the genus *Gunflintia:* G. *grandis* (a) and *G. minuta* (b). The other two are *Entosphaeroides amplus* (c) and *Animikiea septata* (d). The hydra-like *kakabekia umbellata* (e) has a modern counterpart, if not a descendant, in a newfound soil organism. The globular Eosphaera tyleri (f)

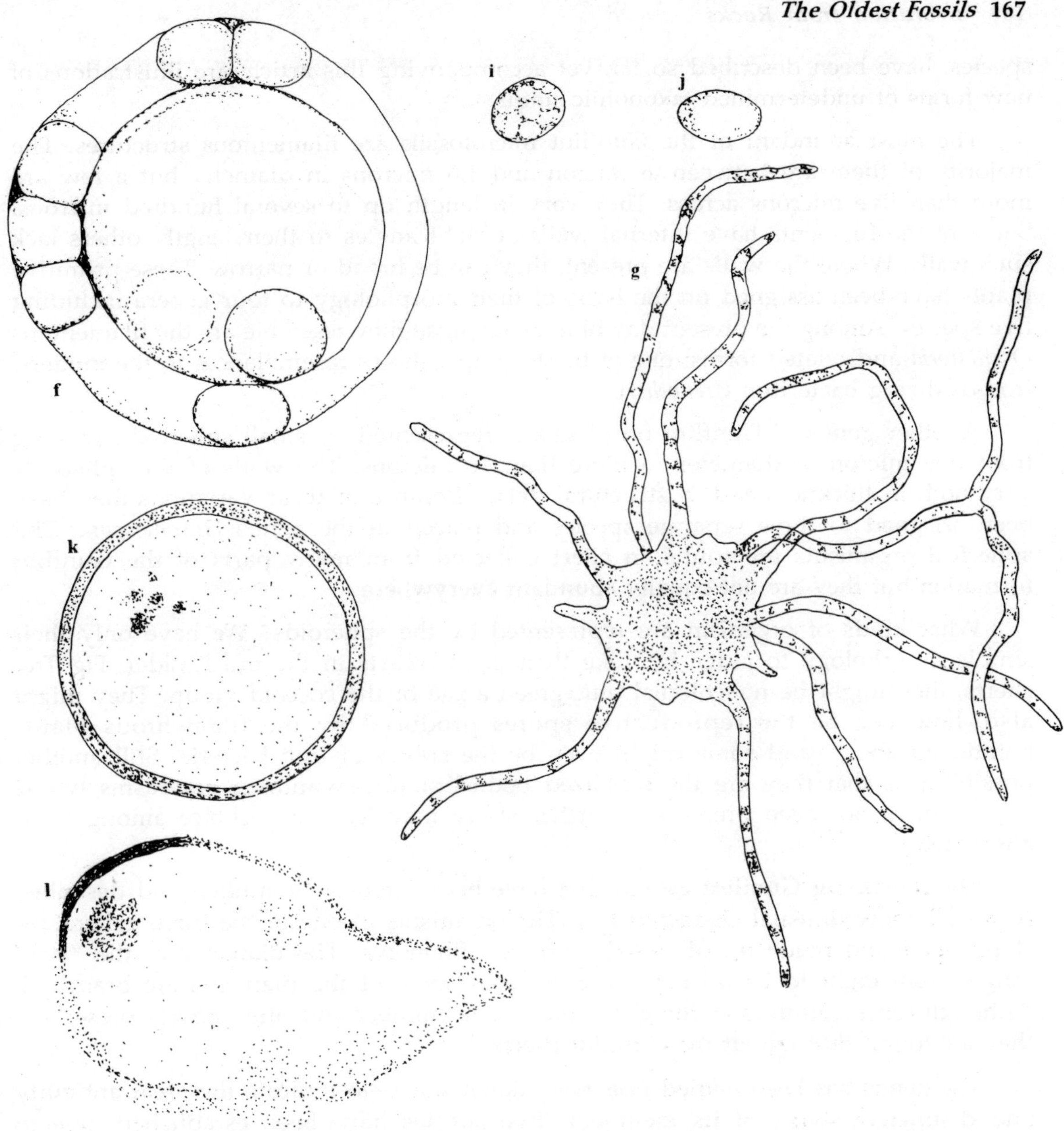

evidently failed to survive. Precambrian times. *Eoastrion bifurcatum* (g), with its array of filaments, is one of two such species. Three spherical organisms that differ chiefly in surface markings are assigned to the single genus *Huroniospora;* they are *H. microreticulata* (h) *H. macroreticulata* (i) and *H. psilata* (j). A one-celled organism that has not yet been classified (*k and l*) has internal features suggestive of a nucleus; the second specimen may have been fossilized as the process of cell division began. If these are eukaryotic cells, a key evolutionary event took place far earlier than is now thought.

species, have been described so far, yet accompanying this article are illustrations of new forms of undetermined taxonomic status.

The most abundant of the Gunflint microfossils are filamentous structures. The majority of them are between .6 micron and 1.6 microns in diameter but a few are more than five microns across. They vary in length up to several hundred microns. Some of the filaments have internal walls at right angles to their length; others lack such walls. Where the walls are present, they can be broad or narrow. These primitive plants have been assigned on the basis of their morphology to four genera including five species. Among the present day blue-green algae they resemble are the filamentous *Oscillatoria* and related forms; one of the four also shows resemblances to the modern iron-oxidizing bacterium *Crenothrix*.

Another genus of Gunflint organisms is represented by small spheroids, ranging from one micron in diameter to more than 16 microns. The walls of the spheroids vary both in thickness and in structural detail. Because of these variations they have been assigned to three separate species and placed in the genus *Huroniospora*. The spherical organisms are found in chert collected from many parts of the Gunflint formation but they are not equally abundant everywhere.

What kinds of organism are represented by the spheroids? We have only their simple morphology to judge by. Like their counterparts in the much older. Fig Tree cherts, they might be noncolonial blue-green algae of the coccoid group. They might also, however, be the reproductive spores produced by the filamentous plants mentioned above, and some might even be the spores of iron bacteria. Still another possibility is that they are the fossilized bodies of free-swimming organisms whose flagella have not been preserved. Further study may lead to a choice among these alternatives.

The remaining Gunflint genera that have been intensively studied and described so far all show unusual characteristics. The organisms in one of the three genera are star-shaped and made up of radially arrayed filaments. The diameter of the "star" ranges from eight to 25 microns; in rare cases some of the filaments are branched. Although representatives of the genus are few in number and often poorly preserved, they are found throughout the Gunflint cherts.

The genus has been named *Eoastrion* "dawn star"—to indicate the great antiquity and distinctive shape of its members. Two species have been established, one to accommodate the fossils with non-branched filaments and the other the fossils with branched filaments. There are no clear analogies between the fossils and modern organisms, although in some respects *Eoastrion* resembles the curious iron-and manganese-oxidizing organism *metallogenium personatum*.

A most peculiar organism comprises the next of these genera. Its fossils are abundant in the cherts exposed near Kakabeka Falls, some 20 miles west of Thunder Bay. It consists of a bulb with a narrow stalk, surmounted by a structure that in some cases resembles an umbrella. The relative size of the three parts varies widely from specimen to specimen; the size of the bulb and the stalk together or of the umbrella alone ranges between 10 and 30 microns.

This odd plant has been assigned to the new genus *Kakabekia* and the species *umbellata*. Living organisms that superficially resemble it in form are certain multicellular polyps. All modern polyps, however, are many times larger.

The story of *Kakabekia* has had some interesting and continuing overtones. Late in 1964, quite independently (in fact, unaware) of paleontological investigations of the Gunflint fossils, sanford M. Siegel discovered a strange new form while he was studying soil microorganisms that can survive extreme atmospheric conditions. The organism defied identification or assignment to any known taxonomic category. Siegel set it aside as being an enigma, although he kept his preparations, drawings and photomicrographs. A few months later a description of *Kakabekia* was published, and Siegel immediately noted a striking resemblance between his soil organism and some of the Gunflint specimens. Siegel's organism is very slow growing, contains no chlorophyll and apparently has no nucleus; it may therefore be representative of a new group of prokaryotic microorganisms. It was first found in ammonia-rich soil collected at Harlech Castle in Wales, and it has since been recognized in soils from Alaska and Ice-land and recently in soil from the slopes of the volcano Haleakala in Hawaii. Whether or not it is related to the two-billion-year-old *Kakabekia* is questionable, but the existence of two such bizarre forms is at least a remarkable evolutionary coincidence.

The eighth genus of Gunflint organisms comes from a single area near the easternmost outcropping of the chert, in the vicinity of Schreiber Beach. The organism consists of two concentric spheres; its outside diameter ranges from 28 to 32 microns. In most of the fossils the inner sphere is kept from contact with the outer one by as many as a dozen "spacers" in the form of small flattened spheroids. I have assigned this distinctive organism to the new genus *Eosphaera* ("dawn sphere") and the species *tyleri* (in honor of Tyler). No analogous living organism is known, nor has an organism resembling *Eosphaera* been found in any other Precambrian rocks. *Eosphaera* may be somewhat fancifully regarded as a "mistake" in evolution that did not survive the middle Precambrian.

Anyone who wanted to question that the organisms preserved in the Fig Tree cherts three billion years ago were photosynthetic could defend the negative position quite eloquently. Where the billion-years-younger Gunflint organisms are concerned, however, the affirmative evidence seems overwhelming. For one thing, the chemical analysis of organic material from the Gunflint cherts in several laboratories reveals the presence of the hydrocarbons pristane and phytane: two "chemical fossils" that can most reasonably be regarded as being breakdown products of chlorophyII. A second datum is the striking resemblance between the filamentous fossils that are the most abundant Gunflint forms and modern photosynthetic blue-green algae. A third is the small domes and pillars in the Gunflint chert that resemble the structures formed in shallow water today by dome building photosynthetic algae.

One may add the evidence of the relative abundance in Gunflint organic matter of the two nonradioactive carbon isotopes cabon 12 and carbon 13. The carbon in the carbon dioxide of the earth's atmosphere normally consists of about 99 per cent carbon 12 and 1 per cent carbon 13. In the process of photosynthesis, however, plants tend to fix-slightly more carbon 12 than carbon 13, so that plant tissues are even poorer in the heavier isotope. Measurements of the ration of the two isotopes in Gunflint organic material were made by Thomas C. Hoering of the Geophysical Laboratory of the Carnegie Institution of Washington. The results indicate that the Gunflint material too is poor in carbon 13, and to a degree that is almost identical with the depletion in modern algae and other photosynthetic plants. The billion-years-older Fig Tree organic material shows a carbon-12 to carbon-13 ratio that is about the same as the ratio in the Gunflint material. This result lends credence to the view that the alga-like Fig Tree organisms probably were photosynthetic too.

It seems reasonable to conclude that even if oxygen was only a trivial component of the environment is Fig Tree times, the Gunflint organisms represent the intermediate agency that brought about the oxygen-rich environment at the end of the Precambrian. This is a development of major importance. It is not, however, the only development or even the most important development of Gunflint times. The variety of form represented by the eight genera of Gunflint plants demonstrates that terrestrial life had crossed the second evolutionary threshold—the threshold of diversification—no less than two billion years ago.

The crossing of the third threshold is clearly documented in a series of late Precambrian formations, consisting primarily of limestones, sandstones and dolomites, found along the northern rim of the Amadeus basin in the Northern Territory of Australia. One member of the series is the Bitter Springs formation; a ridge in the Ross River area, about 40 miles from Alica Springs, consists of strata from its lower and middle levels. The exposed formations include isolated beds of dense black chert and laminated rocks that are associated with fossil structures built by colonial algae.

No absolute age is known for the Bitter Springs formation. Its top strata, however, lie some 4,000 feet below the lowest of the rocks that in the Ross River area are the boundary between formations of Precambrian age and those of the succeeding Cambrian period. The generally accepted date for the beginning of the Cambrian period is 600 million years, ago, so that the lower position of the Bitter Springs cherts makes them considerably older. The Bitter Springs formation also underlies Precambrian sediments that are known to be some 820 million years old on the basis of their rubidium-strontium ratio. It is reasonable to assume that the Bitter Springs cherts are roughly a billion years old. This makes them only half as old as the Gunflint cherts and less than a third as old as the Fig Tree cherts.

A preliminary analysis of the abundant microfossils in the Bitter Springs cherts indicates that at least four general groups of lower plants lived in the shallow seas or

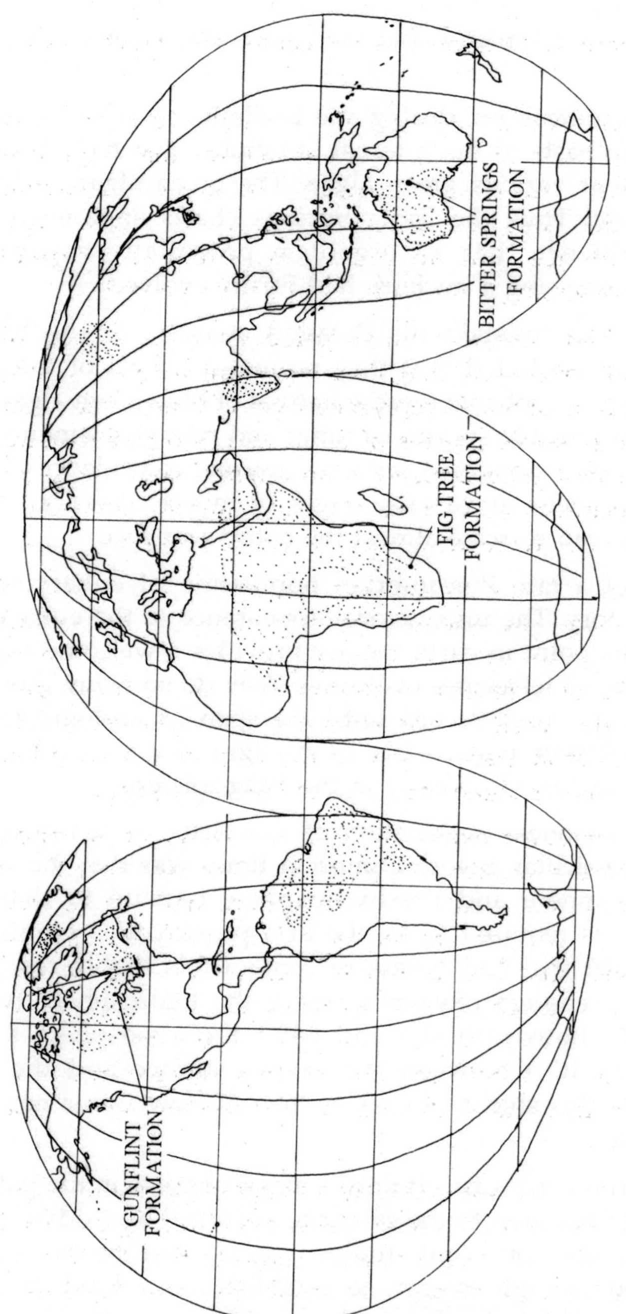

Fig. 7.4: *Precambrian Rocks* occupy the old, low-lying "core" areas of continents throughout the world. Many, formed from accumulated sediments, are so altered by heat and pressure that any fossils have been obliterated. Some relatively unaltered exposures of Precambrian sediments, however, are rich in organic remains. Three such fossil-bearing Precambrian formations are shown on this map.

embayments that covered this part of central Australia in late Precambrian times. As one would anticipate, the plants included filamentous blue-green algae akin to such living forms as *Oscillatoria* and *Nostoc.* Some of these filaments are more than 75 microns long; they are thickest (about 1.4 microns) at the center and taper toward the ends to less than one micron.

Our most exciting identifications during the preliminary analysis concerned the other three groups. On the basis of the internal structures that have been preserved, all three appear to represent various green algae. The green algae, unlike the blue-green algae, are eukaryotic! Thus the Bitter Springs cherts apparently contain the earliest-known fossil evidence that an organism potentially capable of sexual reproduction, or at least possessing a nucleus, had finally evolved.

By the time Schopf had finished his detailed analysis of the Bitter Springs specimens in 1968, he had concluded that they represent a total of three bacterium-like species, some 20 certain or probable representatives of blue-green algae, two certain genera of green algae, two possible species of fungi and two problematical forms. For one of the green algae, *Glenobotrydion aenigmatis,* an unusual coincidence of fossilization has preserved several specimens at various stages of mitotic division. By arranging the specimens in order one can recreate almost the entire sequence.

With the discovery of a late Precambrian population of eukaryotic plants we approach the end of our story. The unequivocal appearance of the eukaryotic level of cellular organization at this point in earth history provides a welcome explanation for one of the principal puzzles of terrestrial evolution: Why do so many groups of higher organisms not appear in the fossil record until the span of geologic time is seven-eight past? It is a good question, particularly in the light of what we know about the multicellular animals and higher algae early in the Paleozoic era.

Until recently the commonest explanation of the rarity of advanced organisms, multicellular animals in particular, before Cambrian times was that the supply of free oxygen in both the atmosphere and the hydrosphere had up to that point been extremely limited. Time was required to let the first prokaryotic organisms adjust to existence in an environment that had begun to contain this highly reactive element. Time was also needed for enough oxygen to leave the hydrosphere (where it was being produced by algal photosynthesis) and enter the atmosphere to establish a protective shield of ozone (O_3) between the earth's surface and the sun's harsh ultraviolet radiation. Until this shield existed neither shallow water nor bare ground would have been habitable.

An oxygen-poor environment was certainly a major obstacle in the path of oxygen-dependent heterotrophs. In addition to the evidence provided by oxidized Precambrian formations, however, a number of recent studies indicate that the earth's atmosphere had actually accumulated enough oxygen to establish some kind of ozone shield considerably before the end of the Precambrian. The appearance of eukaryotic organisms late in the Precambrian, as indicated by the green algae of the Bitter Springs formation, provides a better explanation for the failure of higher organisms to appear until an even later time. The fundamental key to evolutionary progress is genetic variability. Sexual reproduction, which involves the recombination of heritable

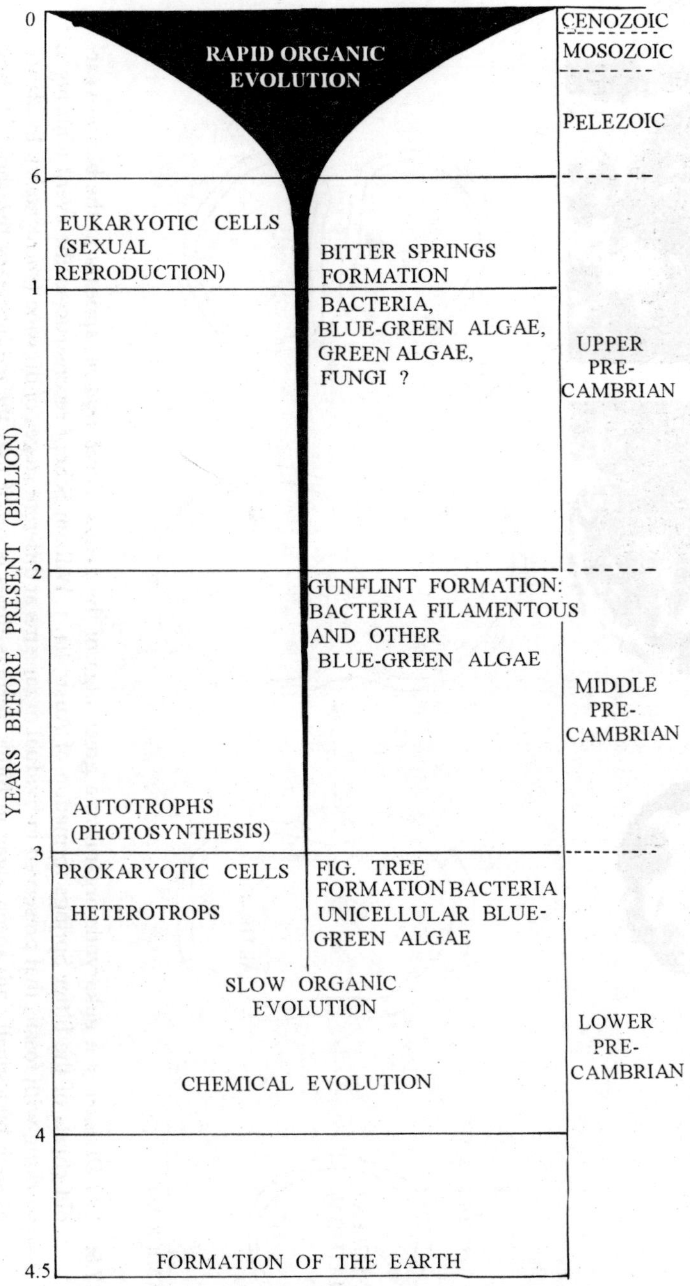

Fig. 7.5: *Organic Evolution* is presented in terms of successively briefer episodes of biological advance. The Precambrian interval, the earliest any by far the longest, began when the earth was formed 4.5 billion years ago and ended 600 million years ago with the beginning of the Paleozoic era. The increasing abundance of species with the passage of time is indicated in color. Once organisms with eukaryotic, or truly nucleated, cells evolved, hastening evolutionary progress in late Precambrian times, the number of species multiplied explosively.

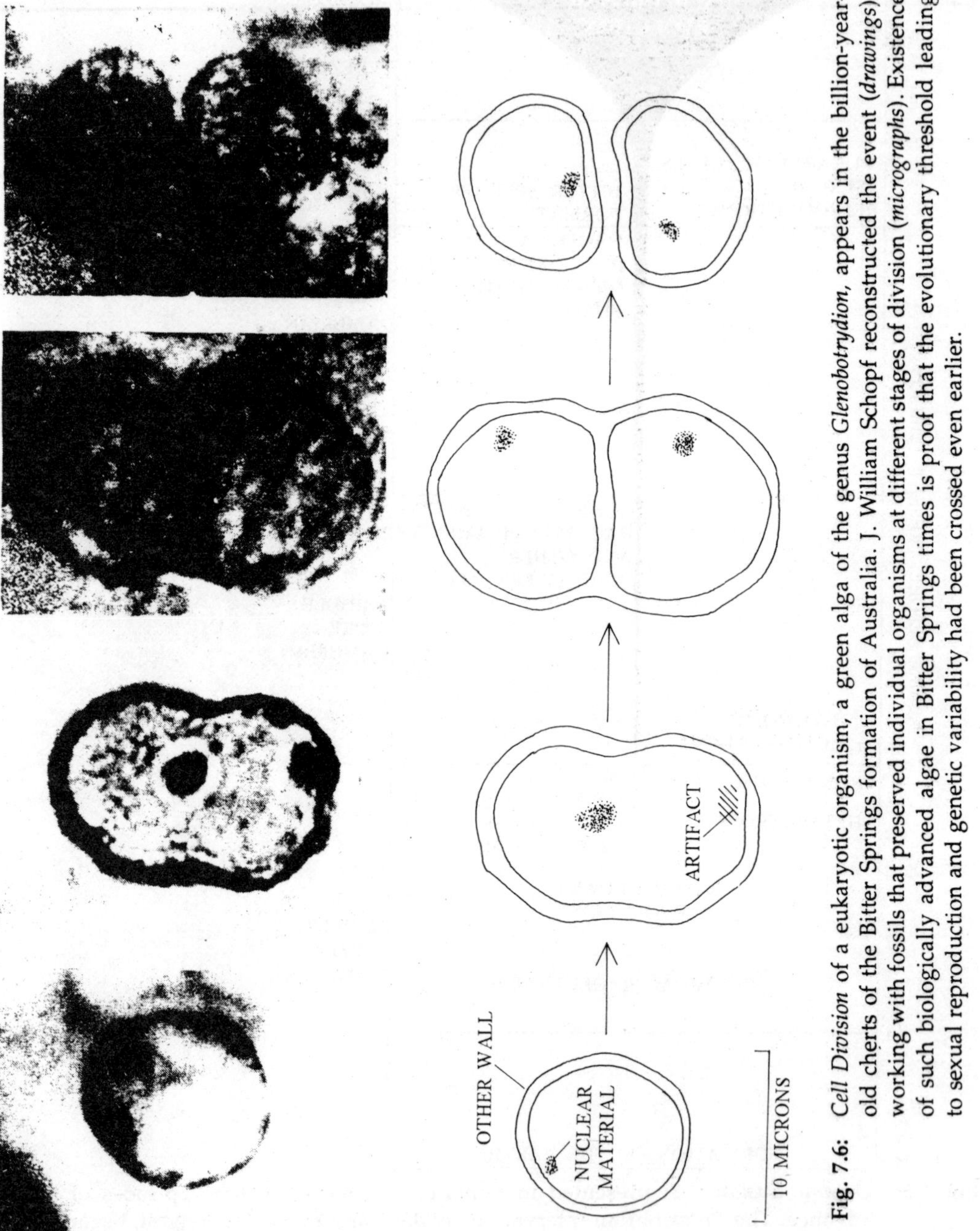

Fig. 7.6: *Cell Division* of a eukaryotic organism, a green alga of the genus *Glenobotrydion*, appears in the billion-year-old cherts of the Bitter Springs formation of Australia. J. William Schopf reconstructed the event (*drawings*), working with fossils that preserved individual organisms at different stages of division (*micrographs*). Existence of such biologically advanced algae in Bitter Springs times is proof that the evolutionary threshold leading to sexual reproduction and genetic variability had been crossed even earlier.

characteristics, is the highway to genetic variability and all its consequences, including the increased complexity of form and function at all levels of organization, that are thereafter apparent in the course of evolution.

This last of three thresholds, which separates the primitive world of cells without a nucleus from a world where sexual reproduction is possible, must have been crossed a long time before the formation of the Bitter Springs cherts. Only half a billion years or so later the Paleozoic seas were swarming with highly differentiated aquatic plants and animals, evolved from primitive forebears that had managed to cross all three precambrian thresholds. Half a billion years does not seem to be evolutionary "room" enough to account for such epic progress. Moreover, there is evidence to suggest that developments in this direction may have begun in Gunflint times.

The Fig Tree, Gunflint and Bitter Springs cherts are not the only source of Precambrian fossils, nor have Tyler, Schop and been the only investigators of the Precambrian fossil record. Indeed, work in the field has gone on for nearly a century. The emphasis has changed in recent years from an earlier concern with the curious "boundary" that has been traditionally accepted as separating the Precambrian from the beginning of the Paleozoic. It is a field that requires all levels of observation from the macroscopic down to the electron microscopic. Workers in the field agree that the search for other stages and thresholds in the evolution of life must be focused on fine structures wherever these have been fortuitously preserved. The Precambrian fossil record is still meager, but significant gaps in the record are steadily being filled. Indeed, the traditional concept of a clearly delimited Precambrian boundary may soon disappear from reconstructions of the history of terrestrial life.

Two final comments may be of interest to those who, like myself, are stimulated by the search for first causes no matter how often the search goes unrewarded. Some biologists now suggest that the organelles characteristic of the eukaryotic cell may once have been independent organisms that somehow came to live symbiotically inside larger host cells. It is not clear whether or not one of the host cell's responses to the presence of such a guest was a regrouping into a nucleus of the genetic material formerly scattered throughout the host's cytoplasm. If symbiosis was indeed the first step toward evolution of th eukaryotic cell, it may be that certain of the Gunflint organisms currently being studied show initial steps in this direction formation of various bacterial and algal organisms into components of the eukaryotic cell is given by Lynn Margulis of Boston University in a recent book. Mrs. Margulis' argument rests on biological grounds; the increasingly detailed Precambrian fossil record supports her thesis.

Astronomers and physicists have found evidence in recent years that molecules such as the hydroxyl radical (OH), carbon monoxide (CO), ammonia (NH_3), hydrogen cyanide (HCN) and formaldehyde (HCHO) are formed in the "empty" reaches of space. So far the ultimate in interstellar chemical complexity is represented by organic material that some investigators maintain they find in a peculiar class of meteorites known as carbonaceous chondrites. Until recently no carbonaceous chondrite had been recovered under circumstances that completely ruled out the possibility of accidental contamination by terrestrial organic material. As a result nagging questions about the chondrites have remained, particularly with regard to such biologically important molecules as amino acids (which are a *sine qua non* of terrestrial life). Recently, however, using stringent laboratory procedures to analyze carefully documented samples of the Murchison chondrite, Ponnamperuma and his associates at the Ames Research Center (in collaboration with Carleton B. Moore of Arizone State University and Ian R. Kaplan of the University of California at Los Angeles) seem to have proved the existence of extraterrestrial amino acids. Not only is the quantity of amino acids in the Murchison meteorite surprisingly high but also certain of them are unknown in terrestrial organisms and hence cannot be contaminants from the soil.

This finding opens up a new world of chemical evolution: a world of random synthetic processes not on the earth but in space, including the extraterrestrial formation of bodies (of which the carbonaceous chondrites seem to be fragments) that are rich in organic materials. The finding brings us back to the discussion at the beginning of this article. The chemical evolution of organic matter, the prelude to biogenesis on the earth, seems to have occurred elsewhere in the solar system or outside it. For knowledge of the earliest stages of organic evolution the biologist and paleontologist must rely on the chemist and astrophysicist.

Chapter—8

Geological Distribution

To understand the complete evolutionary changes in the organisms (plants and animals, one has to study the organisms of the distant past which are now extinct. The extinct forms of life, which are also the connecting links between the most primitive forms and recent forms, are preserved in the rocks as *fossils*. The study of these fossils constitutes the subject-matter of *Palaentology*.

A number of estimates have been proposed from time to time about the age of earth. For examples, *Archbishop Ussher* of Ireland in 1645 declared that the creation of earth took place 26th October 4004 B.C., at 9 o'clock in the morning. This was the adopted version of the *Bible* for a long time. *James Hutton* believed that geological time was unimaginably long. *Lord Kelvin* (1897), from a study of the flow of heat from the interior to the surface of earth, estimated the age of earth from 20-40 million years.

The most accurate method of determining the age of a rock is by radioactive decay studies. Certain radioactive elements like uranium and thorium, found in certain rocks, decay into lead. The rate of this radioactive disintegration is constant and in unaffected by external physical conditions like temperature, pressure, etc. The formula may be written as $U^{238}=Pb^{206}+8He^{4}+heat$. It has been determined that on gram of uranium gives on decay 1/7,600,000 gram of lead in one year. The lead: uranium ratio in a rock would thus give the age of the rock. The maximum age of the earth has thus been estimated to be around 5500 m.y. (million years). Other methods of dating, *e.g.*, potassium—argon method, radiocarbon method, etc. are also known today.

GEOLOGICAL TIME SCALE

By studying the fossils, present in the rocks, the geologists have divided the earth's past history into *Era* which are major divisions. The eras are divided into *periods*. The periods are subdivided into *Epoches*. These eras, periods and epoches are arranged on the time scale in an order of their age, and this arrangement is called, *'geological time scale'*. The first geological time scale was developed by *Giovanni Avduina*, an Italian Scientist, in 1760.

ERAS

Based on the nature of the forms of life preserved as fossils in the rock, the earth's past history is divided into six eras.

I. Azoic
II. Archaeozoic or Eozoic (Primal life or dawn of life)
III. Proterozoic (Early life)
IV. Palaeozoic (Ancient life)
V. Mesozoic (Mediaeval life)
VI. Cenozoic (Recent life)

PERIODS

These are the fundamental units of the standard geologic time scale. There are 12 periods:

I. Periods of Palaeozoic Era

There are 7 periods in Palaeozoic era:

1. Cambrian
2. Ordovician
3. Silurian
4. Devonian
5. Mississipian
6. Pennsylvanian
7. Permian.

II. Periods of Mesozoic Era

There are three periods:

8. Triassic
9. Jurassic
10. Cretaceous.

III. Periods of Cenozoic Era

There are two periods:

11. Tertiary
12. Quaternary.

EPOCHES

There are seven epoches. The epoches of cenozoic's period are:

I. Quaternary Period

There are two epoches:

1. Recent
2. Pleistocene.

II. Tertiary Period

It is subdivided into five epoches:

3. Pliocene
4. Miocene
5. Oligocene
6. Eocene
7. Palaeocene.

I. AZOIC ERA

This was the time, in which life was not present on the earth. During this time, earth was formed, cooled and underwent many changes. These changes made favourable conditions for the appearance of living organisms. The rock of this time are igneous type without any fossil.

II. ARCHAEOZOIC ERA

The manner in which life originated is not known. It is probable that sometime in early Archaeozoic Eras the earliest living matter that could reproduce itself got generated under very favourable conditions of temperature, humidity, light, etc., by the combination of certain chemicals. This first living matter may have been very similar to protoplasm. Filterable viruses which represent an intermediate stage between living and non-living entities, may have been the first to appear on the surface of the globe in the early Archaeozoic Period. Unicellular Protozoa and Protophyta may have been also present.

According to *Stirton* (1959), the oldest known fossils are blue-green algae, fungi, and probably a flagellate described from Ontario, Canada. The age of the rock in which these algae occur has been estimated to be about 2000 m.y. (million years) by radiometric methods.

PROTEROZOIC ERA

It may be called the '*Age of Primitive Invertebrates*'. Life was in the form of algae, precipitations, siliceous sponge spicules, remains of radiolarians, impressions of jelly fishes, a simple brachiopod (*Lingullela montana*) as well as deposits of graphite. These first organisms were quite possibly soft bodied. Reducing bacteria must also have been predominant during this era because the chief iron-ore deposits of the world are found in the Proterozoic era.

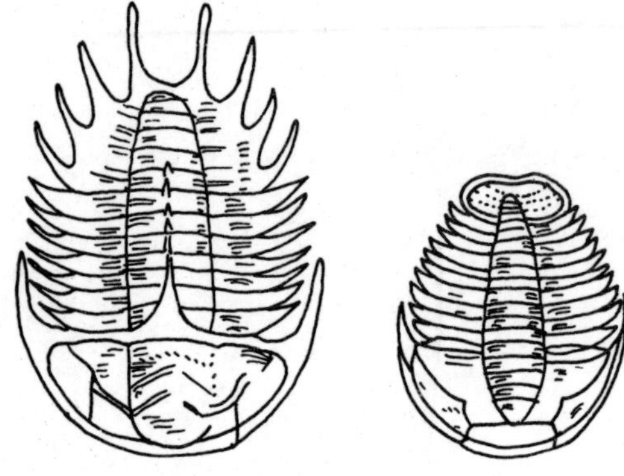

Fig. 8.1:

The fossil record from the rocks is much less probably due to absence of hard parts in the organisms to be preserved as fossils. The fossils which were formed must have been destroyed due to effect of pressure and high temperatures.

This era is estimated to have lasted 3.5 billion years.

III. PALAEOZOIC ERA

This fossil record in this era are very extensive. Almost all the major invertebrate phyla are represented even in the early Palaeozoic era. The fossils of vertebrates also present late in the era because the early chordates were soft-bodied and could not be preserved as fossils.

Palaeozoic era is divided into following periods:

1. Cambrian Period

The name Cambrian was proposed by *A. Sedgwick* in 1835. It is taken from *Cambria* the Latin name of City of Wales.

This period started by the melting of glaciers, slow rise of sea level and warm climate.

Almost all the main invertebrate phyla had existed during this period as also the main phyla of plants. Trilobite were the most dominant forms of life, the important genera being *Olenellus, Redlichia* in the early Cambrian times; *Paradoxides* in the middle Cambrian and *Crepicephalus* in late Cambrian. Small primitive brachiopods and gastropods with simple shells are found in early Cambrian rocks. Some foraminifers and radiolarians have also been found. A few graptolites are recorded in the Cambrian rocks. Echinoderms have been recorded from Cambrian rocks of Australia while some bryozoans from Russia and Canada.

The first land plants also appear in this period.

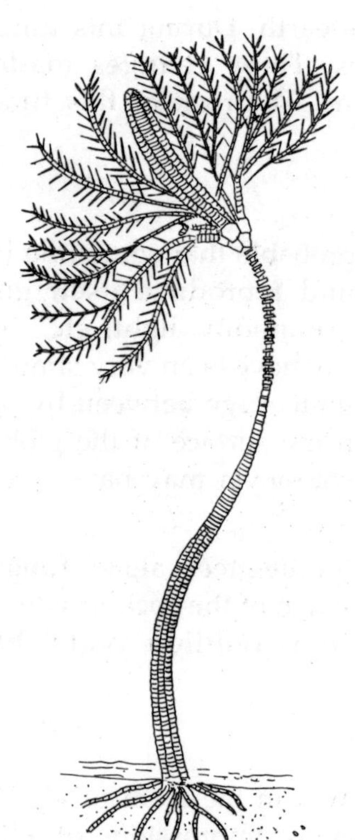

Fig. 8.2: Crinoids or sealilies which look like flowers growing in the ocean.

2. Ordovician Period

The name Ordovician was proposed by *Charles*

Lapworth in 1879, on the basis of the ordovices, an ancient tribe, that formerly inhibited the City of Wales.

During this period, conditions became more favourable as the climate was uniformly warm with glacial activities in certain regions.

This was the *"Age of Giant Mollusca"*. Graptolites, which started in late Cambrian, show their maximum development during Ordovician. They in fact serve as index fossils in Ordovician rocks. Trilobites were still abundant and reached the climax of their evolution. First ostracods appeared on the scene and brachiopods developed harder shells. Cephalopods, both straight-shelled and coiled type, were abundant. Echinoderms appeared in large numbers and show diversification, the starfish being very common.

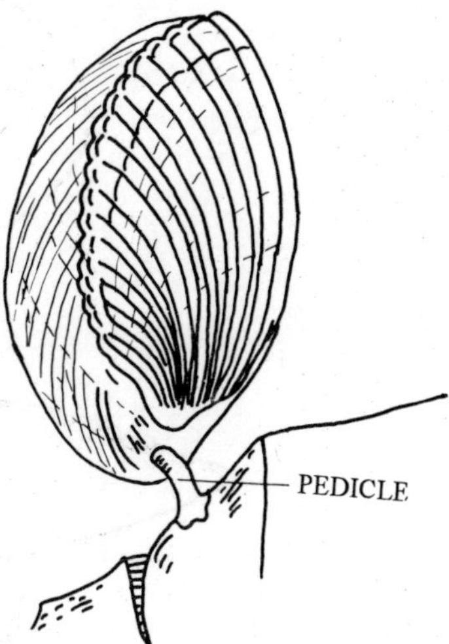

Fig. 8.3: Brachiopods attached to rocks by their pedicles.

Ordovician is also characterised by first appearance of conodonts which are very important marine environment indicators. Land plants are meagre. First appearance of fish-like animals Ostracoderms is also recorded in Ordovician rocks. This Period probably marks the first appearance of vertebrates.

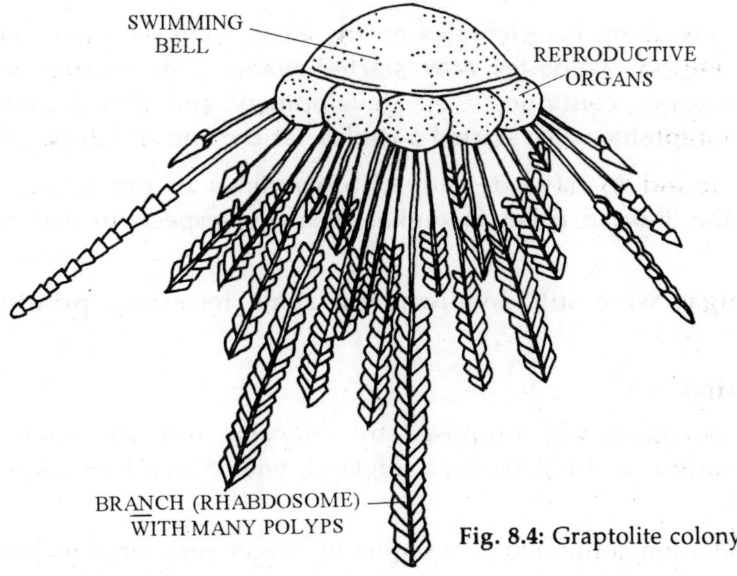

Fig. 8.4: Graptolite colony.

3. Silurian Period

The term Silurian was proposed by *Murchison* in 1835 to designate the rocks present

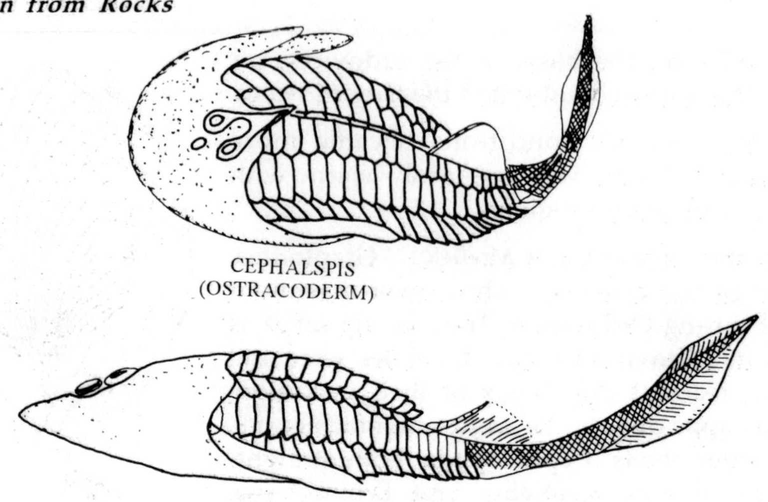

Fig. 8.5: An ostracoderm, *Cephalaspis*.

on the borders of Wales and England. This area was inhibited by Silures, an ancient tribe, inhibited Wales.

The climatic conditions were mild and some areas were arid. Land elevation at places caused formation of deep continental seas. Rise of *Caledonian Mountains* in British Isles, Scandinavia and N. Greenland and in N. Africa and East Central Asia.

The first air-breathing invertebrates on the earth were scorpions which appeared in the Silurian Period. Trilobites now started waning. Bryozoans were extremely abundant. Brachiopods continued their development and first *Spirifer* appeared in Silurian Period. Graptolites also started a decline at the end of Silurian Period.

The fish is found to have developed lower jaws in the Silurian Period. The organisation of the Silurian fishes is similar in many respects to that of the fishes of today.

Although algae were still dominant but some terrestrial primitive ferns also evolved.

4. Devonian Period

The name Devonian was proposed by *Sedgwick* and *Murchison* in 1839 after Devonshire, a country of South-Western England, where rock formation of this period were first found.

The land elevation continued. Consequently larger seas were further divided into smaller seas. Due to fall of temperature, climate remained cold any dry. This period was marked by great volcanic activities and the formation of coal, oil and gas.

During the period, brachiopod developed wonderfully. Corals and bryozans were also abundant.

The Devonian Period is called the *`Age of Fishes'* because of their great number. Plate-skinned fishes, sharks and bony fishes were common. The Devonian ancestors of modern lung fish, Choanichthyes, were able to survive dry period by burrowing in the moist beds of streams or lakes. The lobed-fin fish, crossopterygians were characterized by the presence of lobed rounded bases of the paired fins. These were on the direct line of evolution from fish to terrestrial vertebrates. They became extinct at the closer of Palaeozoic except *Latimeria* and *Melania*.

From the late Devonian Rocks of Pennsylvania, the foot-print named, *Thinopus antiquas* gave the evidence of first amphibian life.

Tall terrestrial Psilophytes, club-mosses and horse-tails occur in the rock of Scotland. First wingless insect also appeared in this period.

5. Mississippian Period

The name `Mississippian' was given by *Alexander Winchell* in 1869 to designate the area in eastern Mississipi basin where rock formation of this period were found.

The climatic condition were hot and damp which caused formation of wide-spread swamps.

During this period, sharks were the dominant fishes in the seas and oceans. They were of many kinds but consisted principally of the more ancient shell-feeding types which were subsequently almost wholly blotted out.

This period is called *Age of Crinoids* because they were most abundant and diversified. Amphibian spread and the insects evolved wings during Mississippian Period.

No evolutionary change in terrestrial plants in comparison to those of Devonian Period is observed in this period.

6. Pennsylvanian Period

Term `Pennsylvanian' was proposed by *H.S. Williams* in 1819 which refers to the state Pennsylvania.

The climatic conditions were disturbed so that the forests burried under the swamps and turned, in due course of time, into present-day coal mines. For this reason, Mississippian and Pennsylvanian Periods are often combined together under a head, *Carboniferous Period (i.e.,* coal bearing).

Amongst animals, insects developed abundently both in size and diversity. Cockroaches reached such a large number that sometimes this period is called the *`Age of Cockroaches'*. Dragon flies were also abundant. Among vertebrates, amphibians showed maximum diversity and specialization during this period. True sharks and sea-likes were common. Towards the close of this period appeared the first animal, *Tuditanus punctulatus,* showing the reptilian affinities, from Stegocephalians.

Among plants, ferns dominated and bryophytes also appeared.

7. Permian Period

The term `Permian' was given by *Murchison* in 1841, because fossil record of the period were found in Russian province of Perm.

The cold and dry condition caused widespread glaciation. Extensive land elevation caused the formation of high mountains. Consequently, shallow seas were emptied and their bottom became exposed as salty deserts. Melting of glaciers gave rise to rivers upon the land.

Great diversification in reptiles took place and first mammal like reptiles appeared in the Permian times. Many important groups of marine invertebrates now dwindled. Permian marks the complete extinction of Trilobites which were very important and dominant during the early Palaeozoic. Due to the cold climates, the insect developed larval adaptations seen in the metamorphosis of present day insects.

Among terrestrial animals, spiders, scorpions, centipedes, snails etc. were present.

Among plants swampy trees declined and were replaced by tall gymnosperms. The true conifers and cycads became most abundant and trees somewhat similar to date palms appeared.

V. MESOZOIC ERA

In contrast to the dominance of marine animals throughout Palaeozoic Era, the Mesozoic Era is characterized by the dominance of land animals, the most spectacular of which were giant dinosaurs which ruled the earth for thousands of years. That is why this era is known as *'Age of Reptiles'*. This Era is divided into three Periods.

1. Triassic Period

This name was give by *F. Von Alberti.*

This period was characterized by hot and dry climate and land elevation which helped in expression of deserts and continents.

Snails, bivalved molluscs, insects and sea-urchins diversified and spread. This period is characterized by the diversification of reptiles. Dinosaurs spread upon land; Ichthyosaurs (fish-like) in water; lizard-like Plesiosaurs in water; and bird like Pterosaus in air. The egg-laying mammals also made their appearance.

During this period horstails, clubmosses and seed fern declined, Gymnosperms progressed fast.

Fig. 8.6: Brontosaurus.

2. Jurassic Period

The name was given by *A. Von Humboldt* in 1799. This Period has been named after the Jura Mountains of Switzerland and France. The climates during this Period were mild throughout the world.

Consequently both invertebrate and vertebrate fauna thrived. Among the invertebrates, belemnites and ammonites were most abundant; the latter, in fact, served as a basis for division of Jurassic rocks into eighty *stages* and thirty-three *zones* by Oppel.

Jurassic is the age of giant dinosaurs like Ichthyosaurs which were more than 50 feet long, and whose skull along measured more than 8 feet. The first bird *Archaeopteryx,* which is the connecting link between reptiles and birds has been found in the Solnhofen limestone of Jurassic Period. Similarly the first mammals are recorded in Jurassic Period.

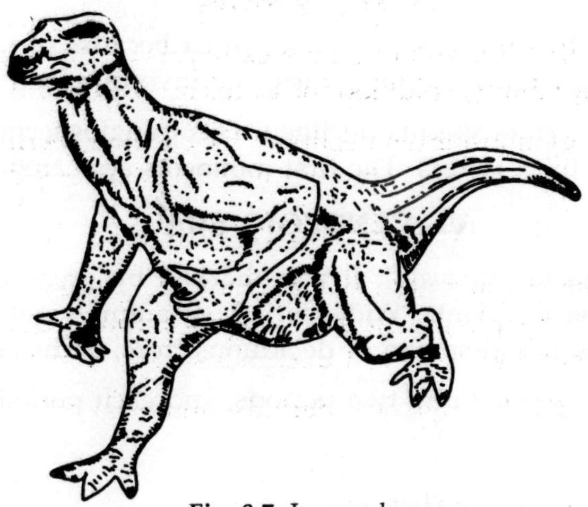

Fig. 8.7: Iguanodon.

Among plants gymnosperms continued spreading. Advanced seed ferns gave rise to the first Angiosperms which were dicotyledonous.

3. Crataceous Period

The term was given by *J.J. d'omalus* d'Halloy in 1882 derived from Latin word *Creta*=chalk.

The climate was warm and somewhat uniform over the most earth surface, but became cooler later on. The seas spread widely over the continents in the earliest past of this period. The Larmide Orogeny changed the surface relief of the earth which affected the climates. The changed climate accelerated the evolution of mammals, birds, flowering plants etc. while caused complete extinction of certain groups of animals like the dinosaurs, ammonites and rudistids.

In the sea, sharks and bony fishes flourished. Snakes first appeared in this period. New varieties of birds appeared. Mammals were still small and less numerous. Marsupials diverged from the archaic forms. Placental mammals appeared.

Fig. 8.8: Stegosaurus.

During this period the dinosaurs became extinct because of their heavy body and small brain, lack of adaptability, coldbloodness, herbivorous food etc.

Among plants, the gymnosperms declined. Dicot angiosperms flourished and tall oak, maple etc. formed thick forests. The first monocots appeared.

VI. CENOZOIC ERA

It is known as *"Age of mammals"*. It was marked by great adaptive radiation in birds, insects and flowering plants. Wide variety of animals got food and protection as a result of development and spread of deciduous trees, herbs and grasses.

The cenozoic era is divided into two periods, and each period into epochs.

1. Tertiary Period

This period includes five apoches.

(i) Palaeocene Epoch

In the beginning the climate was fairly warm but gradually it became colder.

After complete extinction of dinosaurs, nine new orders of terrestrial mammals evolved. Among these were the first primitive carnivores, lemurs, tarsiers etc. Modern groups of birds evolved. Plants also underwent a corresponding evolutionary change.

(ii) Eocene Epoch

Climate remained less hot. These was a widespread erosion of mountains.

All the modern orders of mammals were present. The first horse *Eohippus* or Hyracotherium evolved also first camels which were originally of the size of a dog or even smaller evolved during this period. The horse had progressed to the *Epihippus* stage by this time. Pigs, rats, monkeys, whales and seals appeared.

Angiosperms spread and evolved, giving rise to several modern types. Heavy forest and limited grass land characterized in this period.

(iii) Oligocene Epoch

Climate remained hot. The level of land decreased.

Most of the families of mammals were recognizable by this time. Larger brontotheres reached their climax in early Oligocene Period and suddenly died out before middle Oligocene-Period. The Oligocene camel *Poebrotherium* was of the size of a small sheep. The true flesh-eaters like dogs, cats, sabertoothed cats and other small animals, as well as rodents and insect-eating animals became most abundant. The horse had reached the *Miohippus* stage. It was of the size of a sheep with three toes in each foot. Tortoises of medium-size were abundant. Hawks and vultures hovered in the sky and the 'American pigs' on the ground. Near the end of the Oligocene Period the first New World monkeys appeared.

Among plants tropical and subtropical forests, flowering plants and monocots became widespread.

(iv) Miocene Epoch

In Miocene epoch, the Alps reached their hights and Himalayas were pushed up. So the climate remained somewhat cold. New mountain ranges such as Cascade, Sierra etc., formed upon America continent.

This Period saw the rise and rapid evolution of grazing mammals, long slender-limbed camels, *Oxydactylus,* bear dogs, small foxes, American pigs etc. were particularly abundant. The horse had reached *Parahippus* stage, a major step in evolution towards modern horse. A free interchange of old and new world terrestrial mammals took place among which cats, dogs, horses and mastodonts were especially common. In Late Miocene Period, the horse had reached *Merychippus* stage of evolution in North America. In Eurasia, the Miocene Epoch is marked by the appearance of hyenas, deer, giraffes, and true antelopes.

Terrestrial plants were more progressive.

(v) Pliocene Epoch

This period started with subsidence of land in many places and ended with the beginning of one or two Ice Ages. The climate became still cooler.

The mammals became more specialized and reached their peak of evolution. Camels, horses, rhinoceroses, etc. were abundant during this period. Camels had grown larger in size, and the horse had evolved into several genera. *Pliohippus,* .which ultimately gave rise to modern horse, evolved during this period from its Miocene ancestors. *Hipparion* fauna was typical of Pliocene Period in Eurasia and has distributed from Western Europe and North Africa to South-western Asia, India, Burma and China. Characters found in the fossil jaws and teeth of *Dryopithecus,* found in Siwalik rocks of India, showed affinities with chimpanzee, gorilla and modern man. The lack of a land bridge between North and South America in early Pliocene Period inhibited the free migration and dispersal.

Thick forests declined due to cold. Tail and woody plants were replaced by soft monocot plants.

2. Quaternary Period

This period is divided into two epoches:

(i) Pleistocene Epoch

There are evidences in four *Ice Ages* alternating with warmer intervals. Widespread glaciation during this epoch lowered the sea level. Land connection between European continent and Britain, and Siberia and Alaska were established. Due to these land connections dispersal of horses, camels, deer, cats, saber, dogs, rodents etc., took place.

The modern horse, *Equus* originated.

The first evidence of a primitive man *Homo heidelbergensis,* recognised by a massive mandible, was found in the middle Pleistocene rocks of Heidelberg, Germany. The Peking man, *Pithecanthropus pekinensis,* lived in the mildy cooler climates in parts of China and adjoining areas. The ape man of Java, *Pithecanthropus erectus,* found among Pleistocene fauna of Java, attests to the evidence of land bridge between Java an mainland of Asia. The entire Pleisotocene Period shows dominance of man.

(ii) Recent Epoch

The climate has again become hot. Supremacy of man as the wisest and dominant species has been established. With increasing mental development, gradually he got control over water, air and land. He is trying to get control over plants other than earth. Thus we can say that this epoch is the *'Age of man'*.

Table: 8.1
Geologic Time-table

Era	*Period*	*Epoch*	*Duration in million of years*	*Time from beginning of period to present (millions of years)*	*Geologic conditions*	*Plant Life*	*Animal Life*
Cenozoic (Age of Mammals)	Quaternary	Recent	0.025	0.025	End of last ice age; climate of warmer.	Decline of woody plants; rise of herbaceous ones.	Age of man
		Pleistocene	1	1	Repeated glaciation; 4 ice ages.	Great extinction of species species.	Extinction of great mammals; first human social life.
	Tertiary	Pliocene	11	12	Continued rise of mountains of western North America; volcanic activity.	Decline of forests; spread of grasslands; flowering plants, monocotyledons developed.	Man evolving, elephants, horses, camels almost like modern species.
		Miocene	16	28	Sierra and Cascade mountains formed; volcanic activity in north-weste U.S., climate cooler		Mammals at height of evolution; first manlike apes.
		Oligocene	11	39	Lands lower; climate warmer.	Maximum spread of fore- forests; rise of monocotyledons, flowering plants.	Archaic mammals extinct; rise of anthropoids; forerunners of most living genera of mammals.
		Eocene	19	58	Mountains eroded; no continental seas; climate warmer.		Placental mammals diversified and specialized hoofed mammals and carnivores established.
		Paleocene	17	75			Spread of archaic mammals.
Mesozoic Age of (Reptiles)		Cretaceous	60	135	Andes, Alps, Himalayas, Rockies formed late; earlier inland seas and swamps; chalk, shale deposited.	First monocotyledons, first oak and maple for ests, gymnosperms declined.	Dinosaurs reached peak, became extinct, toothed birds became extinct, first modern birds, archaic mammals common.
		Jurassic	30	165	Continents fairly high, shallow seas over some of Europe and western U.S.	Increase of dicotyledons, cycads and conifers comm- fers common	First toothed birds, dinosaurs larger and specialized, insectivorous marsupials.

Table 8.1 (Contd.)

Era	*Period*	*Epoch*	*Duration in million of years*	*Time from beginning of period to present (millions of years)*	*Geologic conditions*	*Plant Life*	*Animal Life*
		Triassic	40	205	Continents exposed widespread desert conditions, many land deposits.	Gymnosperms dominant, declining toward end, extinction of seed ferns.	First dinosaurs, pterosaurs are egglaying mammals, extinction of primitive amphibians.
Paleozoic (age of Ancient (Life)		Permian	25	230	Continents rose, Appalachians formed, increasing glaciation and aridity.	Decline of lycopods and horsetails.	Many ancient animals died out, mammal-like reptiles, modern insects arose.
		Pennsylvanian (Carboniferous)	25	255	Lands at first low, great coal swamps	Great forests of seed ferns and gymnosperms).	First reptiles, insects common, spread of ancient amphibians.
		Missis-sippin (Carboniferous)	25	280	Climate warm and humid at first, cooler later as land rose.	Lycopods and horsetails dominant gymnosperms increasingly widespread.	Sea lillies at hight spread of ancient sharks.
		Devonian	45	325	Smaller inland seas, land higher, more arid, glaciation.	First forests; land plants well established first gymnosperms.	First amphibians; lungfishes, sharks abundant.
		Silurian	35	360	Extensive continental seas, lowlands, increasingly arid, glaciation.	First definite evidence of land plants; algae dominant.	Marine arachnids dominant, first (wingless) insects, rise of fishes.
		Ordovician	65	425	Great submergence of land, warm climate even in Arctic.	Land plants probably first appeared; marine algae abundant.	First fishes, probably freshwater, corals, trilobites abundant, diversified molluscs.
		Cambrian	80	505	Lands, low, climate mild, earliest rocks with abundant fossils.	Marine algae.	Trilobites, brachiopods, most modern phyla established.
Archeozoic			???	???	Great volcanic activity, some sedimentary deposition, extensive erosion.	No recognizable fossils, indirect evidence of living things from deposits of organic material in rock.	

Chapter—9

Zoogeographical Distribution

Zoogeography is the science which deals with the distribution of all the animals over world. First serious attempted to map out the geographical regions was made by *Sclater* (1857) who based his conclusions upon the distribution of birds. These regions are: 1. Palacartic region, 2. Ethiopian region, 3. Indian region, 4. Australian region, 5. Neotropical region, 6. Neartic region. These are, however, very apparent objections to utilizing vagrant and barrier defying creatures like birds for this purpose. Accordingly, *Mkurray* in 1866 and more in detail *Alfred Russel Wallace* in 1876, divided the surface of earth into zoological regions based chiefly on the distribution of mammals.

Many other regional classifications have been made, based on either discontinuous distribution of one particular class or upon the distribution of temperature variations and similar climatic factors. None of these is satisfactory. Therefore, *Wallace* combined Sclater's system. *Blanford* (1980), *Lydekk* (1896), *Heilprin* (1887), *Gadow* (1913), *Schmidt* (1954) etc. also made their contribution for the study of zoogrophy.

Karl P. Schmidt (1954) divided the earth three zoogeographical realms:

1. Arctogaean,
2. Neogaean, and
3. Notogaean.

Arctogaean includes two regions the (*a*) *Holarctic* with sub-region: *Arctic, Nearctic, Caribbean* and *Palearctic.*

(*b*) *Paleotropical* with sub-regions: *Oriental, Ethiopian* and *Malagasy.*

Neogaean includes a single sub-region the *Neotropical*. *Notogaean* consists of 2 regions: the *Australian* having the Austalian and Papuan sub-regions and the *Oceanian,* which is formed of the New Zealandian, oceanic and Antarctic sub-regions.

Wallace's classification is generally considered to be accepted system. The Indian region of *Sclater* is termed as Oriental region in the classification of *Wallace.*

Thus, six zoogeographical regious have been distinguished as follow:

Palaerctic Region

It includes Europe, temperate Asia, North Africa and Arabia.

Neartic Region

It includes whole of North America and Greenland.

Ethiopian Region

It includes whole of Africa, Arabia, South of the tropic of Cancer, and Madagascar.

Neotropical Region

It includes whole of South and Central America and West Indies.

Australian Region

It includes Australia, New Zealand, New Guinea and neighbouring islands.

Oriental Region

It includes India, Sri Lanka, Indochina and Malaya.

The exact boundaries of these regions are very difficult to recognize. The fauna of these regions are described only with reference to the vertebrates, although they form only 3-4% of animal kingdom.

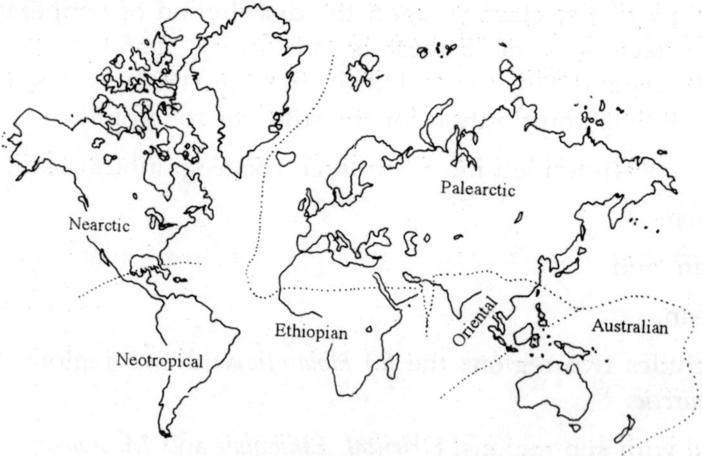

Fig. 9.1: The biogeographical regions of world.

PALAEARCTIC REGION

Extent

This, the largest of the six regions, is estimated to cover an approximate area of 14,00,000 sq. miles. It includes the whole of Europe, Iceland, the Azores, Madeira,

Canary Island, Cape Verde Islands, all that portion of Africa and Arabia which lies to the north of the tropic of Cancer, Asia Minor, Persia, Afghanistan and Baluchistan and the whole of Asia north of a line which runs up to valley of the Indus, along the great Himalayan range, thence eastwards to the Nanling Mountains, south of the Yang-tse-kiang and out to sea just south of the Japanese Islands.

Climate

The climate is temperate. This region includes wet forest lands as well as dry open steppe lands. It also includes coniferous forests and a fringe of Tundra.

In the western portion there is a great in land seas which have on equalising effect on the climate. The Gulf stream materially helps to raise the temperature of Western Europe. The eastern region is much cooler. The extreme of cold are felt in North-Eastern Siberia. However, the highest summer temperature of the area is experienced in North-Western India, Arabia and Afghanistan.

The Northern part of this region is mostly low and flat, Scandinavian and the Ural mountains are the only elevated lands. In south a series of mountain ranges run across the region from west to east, including in Europe the Alps, Balkans, Caucasus etc.; in Asia—the Tian Shan, Altai, Himalayan ranges etc.

Zoological Characteristic

The fauna of this region as a whole is very similar to the Nearctic region, that is why *Heilprin* proposed a great region, *Holarctic Realm* for both Palearctic and Nearctic regions.

The region possesses representatives of 135 families of terrestrial vertebrates, namely 33 of mammals, 68 of birds 24 of reptiles and 10 of amphibia. None of these, however, is peculiar to this region but 9 are common and confined to the Palearctic and Nearctic Regions. The nine families are the Talpidae (Moles) Ochotonidae (Picas) and Castoridae (Beavers) among mammals, the Regulidae (Gold-crests), Colymbidae (Divers) and Tetraomidae among birds; the Proteidae Protous), Salamanders and *Amphiuma* among Amphibia. The fresh water fishes also support the union of the two regions, for five families namely the perches, stickle backs, Pikes, Sturgeons and Polyodontidae are distinctly Holarctic.

The Palaearctic region is subdivided into four sub-regions: (1) European sub-region, (2) Mediterranean sub-region, (3) Siberian sub-region, and (4) Manchurian sub-region.

1. *European sub-region* comprises Northern and Central Europe. In this sub-region only 85 families of terrestrial vertebrates are represented and of these the reptiles and amphibia number only 6 each. Only one form of mammals are peculiar namely *Myogale* but such animals as the wolf, Hedgehog, shrew, mole and Darmouse are very characteristic as are also the was tails, Pipits, Tits, Thrushes and many other birds.

2. *The Mediterranean sub-region* includes the rest of Europe, all the African and Arabian portion, Asia Mion, Persia, Afghanistan and Baluchistan. This is the richest region possessing representatives of 120 families of terrestrial vertebrates. The Fallow-Deer, Elephant Shrew, Civet, Ichneumon, Haena, Hyrax and Porcupine are all characteristic mammals.
3. *Siberian sub-region* embraces all Northern Asia, southwards to the Himalayas. It possess representatives of 94 families of the terrestrial vertebrates. Four genera of mammals *i.e.,* the Yak, Musk, deer and a mole are almost confined to this sub-region but also ranged into the Oriental region. One of the most important members of the Siberian fauna is the freshwater seal, *Phoca ibirica* which inhabits the great fresh water lake—Baikal.
4. *The Manchurian sub-region* includes China, Mongolia, Manchurian and Corea together with the whole of Japan. It has a rich and varied fauna with representatives of 102 families of terrestrial vertebrates. At least a dozen genera of mammals including Tibetan langur, Great Panda etc. are peculiar.

NEARCTIC REGION

Extent

It includes the whole of North America and extends south as for as the middle of Mexico. It includes Greenland in the east and Aleutian islands in the west.

Climate

It has a great range of temperature. The chief physical features of the region are: (1) Large lakes and inland sea in the north eastern portion and the important ranges of high mountains in the west. (2) In the east are smaller ranges, constituting the so called Appalachian high land. (3) In the centre of this great continent is a vast extent of plain which in the north is frozen and hence barren, between latitudes 50° and 60° covered with forest, and in the south dry treeless desert.

Zoological Characteristics

The number of families of terrestrial vertebrates represented in the Nearctic Region is 120 *viz.* 26 of Mammals, 59 of Birds, 21 of Reptiles and 140 of Amphibians. The Nearctic Region possesses five peculiar families; *viz.* The Haplodontidae and Antilocapridae among Mammals; the Chamaeidae among Birds; the Aniellidae among Reptiles and Sirenidae among Amphibians.

The Nearctic differs from the palaearctic region, in the possession of several characteristic Mammals, such as Opossums (Didelphyidae) and Racoon etc., many birds such as the Blue—jays, and Turkey-buzzards etc., Reptiles such as Rattle snakes and Iguanas; Amphibia including *Axolotl, Nectures, Siren* and other large Urodeles.

There are, several peculiar genera of importance of which Scalops (web-footed Moles), Toxidae (American Badger) *Haplocerus* (Rocky mountain Goat) and Musk may be taken as examples.

Wallace has divided Nearctic Region into four sub-regions:

(1) California sub-region.

(2) Rocky Mountain sub-region.

(3) Alleghany sub-region.

(4) Canadian sub-region.

1. *California Sub-region* embraces a narrow strip of country between Sierra Nevada and the Pacific. In the North it includes island of Vancouver and the southern part of British Columbia and in the South it extends upto the head of the Gulf of California. It has 80 families of terrestrial vertebrates and of which 21 are mammals, 49 of birds, 8 of reptiles and 8 of amphibian. Three families are peculiar to this region, namely Haplodontidae, Chamaeidoe and Aniellidae. Vampires and free tailed Bats are characteristic of this region.

2. *Rocky Mountain Sub-region*. It lies immediately to the east of California and includes the whole of the dry and elevated area covered by the mountains. It is the richest portion Nearctic region. It has 107 families of terrestrial vertebrates out of which 25 are mammals, 55 of birds, 18 of reptiles and 9 of amphibian. Although there are no peculiar families but are several characteristic genera like Prong-buck (*Antilocapra*), Rocky mountain goat (*Haplocerus*), Prairie dog (*Cynomys*), American Bison and poisonous lizard (*Heloderma*).

3. *Alleghany sub-region* comprises the United States east of the Rocky mountain sub-region and South of the Great Lakes and includes Novo Scotia. Some 99 families of terrestrial vertebrates including, mammals 8, birds 53 reptiles 16 and amphibia, 12 are represented in this sub-region. The peculiar animals are: opossum, star nosed moles, vampire bats etc. Turkeys and Sirenids (muded) are the other two families.

4. *Canadian-sub region*. All the remaining portion of North America and Greenland constitute the great Canadian sub-region. In this sub-region are to be found the representative of only 75 families including 20 of mammals, 44 of birds, 3 of birds, 3 or reptiles and 8 of amphibia. Thus it is the poorest region. But there are characteristic families including Deer, Reindeer, Elk, Bison, Sheep, Gluttons, Lemmings, the polar deer etc.

ETHIOPIAN REGION

Extent

The Ethiopian Region consists of the whole of Africa and Arabia, south of the tropic of Cancer, together with Madagascar and the small adjacent is lands. *Darlington* (1957) has not included Madagascar but has considered it as separate region on account of its distinctive fauna.

Climate

This region is mainly tropical. It is a large block of rain forests and isolated mountains. The eastern part has wide grassy plains. Its southern part is warm temperate with mixed vegetations.

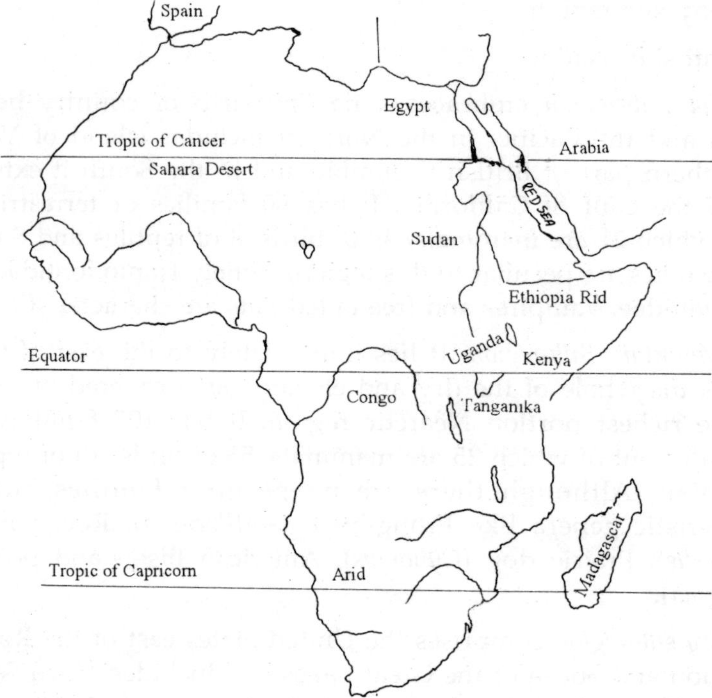

Fig. 9.2: The Ethiopian region with adjacent regions (Madagascar, Spain and Arabia).

Zoological Characteristics

The fauna of Ethiopean Region is rich, varied and well-marked. There are about 161 families of terrestrial vertebrates. Out of these 30 are peculiar.

Among mammals 12 families are peculiar which are: Chiromyidae (Aye Aye), Chrysochloridae (Golden moles), Centetidae (Tenress), Potamogalidae (Potamogale), Protelidae (Earth wolf), Battryergidae (African mole rats), Lophimyidae (created rats), Pedetidae (African Jumping Hares), Anomaluridae (African flying Squerrels), Giraffidae (Giraffes), Hippopotamidae (Hippopotamus) and Orycteropodidae (Aard-Varks).

Among birds 13 families are peculiar; Promeropidae (Promerops), Acerocharidae (Helmet Birds), Vangidae (Vangidal), Philepittidae, Musophagidae (Plankton eators), Coliidae (Colies), Serprentariidae (Secretary birds) Scopidae (Hammer-head bird) Mesoenatidae, Numididae (Guinea fowl) and Balaenicipitidae (whale head).

Among reptiles following families are peculiar: Rhachiodontidae (Egg-eating snakes), Gerrhosauridae, Zonuridae (Girdled lizard) and Uroplatidae.

Among amphibian family Dectylethridae (Clawed foads) is peculiar.

In addition to these families, a large number of important genera are confined to this region, some of them are:

Macroscelidae (Elephant—shrews), Hyaenidae (Hyaena) Elephantidae (Elephants) Rhinocerotidae (Rhinoceroses), Procavidae (Hyraces) Equidae (Horse etc.) Manidae (Pangolins); and among Birds the Ploceidae, Nectariniidae (Sunbirds) Zosteropidae (White Eyes); Indicatoridae (Honey Guides), Capitonidae (Barbets), Bucerotidae (Hornbills) and Struthionidae (Osriches).

The Ethiopean Region is divided into four sub-regions:

1. South African sub-region.
2. East African sub-region.
3. West African sub-region.
4. Malagasy sub-region.

1. *South African sub-region* comprises the southern portion of the continent from the Cape northwards. It is represented by 133 families of terrestrial vertebrates of which only one *i.e.,* Prome ropidae is peculiar. The Golden Moles, Elephant shrews, Earth Wolf, African Mole Rats, African Jumping Hares and Aard Varks are all characteristic mammals. Among birds—Secretary birds, ostriches and in reptiles Egg-eating snakes are characteristic to this part.

2. *The East African sub-region* include the rest of the tropical Africa and tropical Arabia. It has 145 families of terrestrial vertebrates of which only two namely Crested-Rats among mammals and Whale-head among birds are peculiar. Certain other faniiliar mammals, though not absolutely confined but are highly characteristic, are Giraffes, Zebras and the Rhinoceroses.

3. *The West African sub-region* includes most of the African forest region from the river Gambia eastwards to beyond Lake Chan and Southwards to embrace the water shed of the Congo. This area has no peculiar families although 134 occur in this subregion. The most characteristic mammals are the Gorilla, Chimpanzee, the Monkeys, Potamogal, the African flying squirrel and Okapi.

4. *Malagasy sub-region.* It includes Madagascar, Mauritius, Secychelles and neighbouring islands. The fauna of this sub-region is one of most interesting in the whole world. 86 families of terrestrial vertebrates found here, no fewer than 8 are absolutely confined to it. These includes two of mammals Chiromyidae (Aya Aye) and Centetidae (Trenecs); 5 or birds, *i.e.,* Aerocharidae (Helmet Bird), Vangidae, Philepittidae, Mesoenatidae and Leptosomatidae, and one of reptiles, *i.e.,* Uroplatidae.

The most characteristic feature is that all the 35 known species of sub-family Lemurinae of Family—Lemuridae are confined to this sub-region. Of birds 55 families are represented 15 families of reptiles occur here, presence of 2 genera of 9 guanidae is of special interest. Since this is an essentially New world group. Chamaelions are very characteristic of Madagascar because they are more abundant than in any other part of the world.

NEOTROPICAL REGION

Extent

It embraces South America, most of Mexico and West Indies. It is joined to the Nearctic Region by Central American isthmus and separated from other regions by sea.

Climate

Neotropical region is mainly tropical but southern South America continues into south temperate zone. From the west to the east runs the river Amazon with its hundreds of square miles of evergreen forests. In the west the long range of the Andes has high mountain forests, plateau land and gentle slopes.

Zoological Characteristics

155 families of terrestrial vertebrates found here and not less than 39 families are absolutely confined to it.

Among mammals 10 families are peculiar: Cebidae (American Monkeys), Callitrichidae (Marmosets), Solenodontidae (Solenodonts), Dasyproctidae (Agouties), Chinchillidae (Chinchillas), Caviidae (Cavies), Dinomydidae (Dinomys), Bradypodidae (Sloths), Calnolestidae (Selvas) and Myrmecophagidae (Ant-eaters).

Among birds 23 families are peculiar *viz.* Coerebidae (Honey Creepers), Phytotomidae (Plant Cutters), Pipridae (Manakins), Oxyrhamphidae, Dendrocolaptidae, Conopophagidae, Formicarridae, Pteroptochidae, Galbuldidae, Bucconidae, Rhamphastidae Tinamidae, Steatornithidae, Momotidae, Todidae, Palamediedae, Psophiidae, Aramidae, Cariamidae, Eurypygidae, Thinocorythidae, Opisthoconidae and Rheidae.

Among reptiles two families are peculiar *viz.* Xenosauridae and Dermatemydidae.

Among amphibians, 4 families are peculiar; Dendrobratidae, Dendrophryniscidae. Hemiphractidae and Amphignathodontidae.

Neotropical Region is divided into four sub-regions:

1. Chillian sub-region.
2. Brazalian sub-region.
3. Mexican sub-region.
4. Antilean sub-region.

1. *Chillian sub-region.* It includes western coast of South America and embraces the summits of Andes of Peru and Bolivia. This region does not contain any peculiar family but characteristic forms are chinchillas, Llamas, Rhears etc.
2. *Brazilian sub-region* includes all the rest of South America terminating Northwards at the Isthmus of Panama. This is the richest of the Neotropical sub-regions, containing representatives of 133 families of which 27 are of mammals, 71 birds, 23 reptiles and 12 amphibia. This sub-region is the great home of rearly all the arboreal vertebrates of South American such as American monkeys, Vampire bats, American porcupines, sloths and opossums.

There are in addition many other animals as Spiny-mice, Cavies, American Tapirs etc. found in the forest. Among reptiles and amphibians, the tree snakes, coral snakes, iguanas, alligators, Solid-chested tree frogs, typical tree frogs etc. are common.

3. *Mexican sub-region* constitutes all the Neotropical country North of the Isthmus of Panama. It possesses representatives of 127 families of terrestrial vertebrates which include 24 of mammals, 67 of birds, 26 of reptiles and 10 of amphibia.
4. *The Artillian sub-region* includes all the West Indies except Tobago and Trinidad. Since the area is wholly made up of islands, most of which are small, it is hardly surprising of find that the number of families of terrestrial vertebrates is much less. Some 76 families, including 7 mammals, 47 birds, 16 reptiles and 6 amphibia are represented. There is a remarkable absence of mammals in this region as in other insular sub-region for there are no Primates, Carnivores, Ungulates or Edentata. The only rodents are the spiny mice. The birds of the following important families are quite absent *i.e.*, Plant cutters Manakins, American creepers, Ant thrushes etc.

The negative characteristics of this region are also very remarkable. Except in Central America and West Indies, these are no insectivores, civets, oxen, sheep and antelopes.

AUSTRALIAN REGION

Extent

The Australian Region includes the whole of Australia New Zealand, Newguinea, the Molluca and other neighbouring islands, and practically the whole of islands in the Pacific ocean. Its boundary on the west is a line drawn between the islands of Bali and Lombok (Wallace's line), thence to the east of Celebes or the Philippine Islands. The line then due eastwards along the tropic of Cancer to include the Sandwich Islands, thence curves round to the south of New Zealand and Auckland Islands, Tasmania and Australia.

Climate

The northern part of the Region, north Australia and New Guinea, lies with in the tropics with high summer temperature and much of the area is covered by rain forest. The interior of the Australia continent is also hot, but dry, while further south the climate becomes mainly temperate.

Zoological Characteristics

There are 134 families of terrestrial vertebrates. About 30 families are peculiar.

Among mammals 8 families are peculiar *viz.*, Macropodiae (Kangaroos), Phalangeridae (Phalangers), Phascolomydae (Wombats), Peramelidae (Bandicoots), Notoryctidae (Marsupial), Dasyuridae (Dasyures), Echidnidae (*Echidna*) and Ornithorhynchidae (*Ornithorhynchus*).

Among birds 17 families are peculiar namely: Paradiscidae (Birds of Paradise), ptilonorthynchidae (Bower Birds), Meliphagidae (Honey Eaters), Drepanididae (Drepanis), Artichornithidae (Scrub Birds), Xenicidae Menuridae (Nestor Parrots), Loriidae (Loreis), Cyclopsittacidae Stringopidae (Owl-Parrots), Phinochaetidae (Kagu), Gonudiae (Crowned Pigeons), Didunculidae(Tooth-billed Pigeon), Apterygidae (Kiwis), Dromaeidae (Emus), and Casuariidae (Cassowaries).

Among reptiles 3 families are peculiar: Pygopodidae (Scalefooted Lizards), Hatteriidae (Turtra), and Carettchelydidae (Fly-river turtle).

Among amphibians 2 families are peculiar. Ceratobatrachidae and Genyophrynidae.

The Australian Region is also characterized by the absence of certain important groups. Thus, only few mammals are there except marsupials and monotremes. The other orders being represented by some bats and small rodents. Apes and monkeys, insectivores, carnivores, ungulates and edentates are entirely absent. Among birds the finches, buntings and wood peckers are absent.

The Australian Region has been divided into four sub-regions:

1. Australo-Malayan sub-region.
2. Australian sub-region.
3. Polynesian sub-region.
4. New Zealand sub-region.

1. *Austro-Malayan sub-region* comprises all the islands of the Malya Archipelago not included in the Oriental Region together with New Guniea, the Moluccas and the Solomon Islands. In this sub-region, there are 113 families of terrestrial vertebrates of which 4 are peculiar. These are crowned-pigeon and Fly-River Turtle, both confined to New Guinea and two little known Amphibian families Ceratobatrachidae and Genyophrynidae from the Solomon Islands and Sudest Island respectively. Since most of this area

consists of small islands, a large number of peculiar genera and species exist. In New Guinea, three marsupial genera are peculiar. The Babirussa is another remarkable and peculiar member of the Austro-Malayan fauna occurring in Celebes, Buru and the Sulu Islands. Among birds, Birds of Paradise, Bower Birds, Honey-Eaters, White eyes, Frog mouth, Lories and Megapodes attain their highest development in this area; while Honey-peckers, Cuckoo-Shrikes, Fly-catchers, Pitatas and Fruit-pigeons are very numerously represented. Lastly in New Guinea and the neighbouring islands, the amphibian families Narrow-mouthed Toads and Tree frogs are characteristic.

2. *Australian Sub-region* consists of the whole of Australia and Tasmania. In the sub-region 98 families of terrestrial vertebrates are represented, namely 15 mammals, 67 birds, 13 of reptiles and 3 of amphibians. Of these about half a dozen, which include wombats, marsupial mole. Duck-Bills, Scrub-Birds, Lyre-birds and Emus are confined to this region. This area is notable as being the great home of marsupials, for out of 41 known genera, 34 are represented and 24 confined absolutely to it. The avifauna too is highly peculiar; though only 3 families are confined to this subregion yet the proportion and peculiar species is larger than in any other sub-region in any part of the world.

The characteristic but not peculiar animals are Kangaroos, bandicoots, Thylacine, Bower-birds, Honey-eaters, Creepers, Swallow-shrikes, Cockatoos, Bustard-Quails, Cobras. Scale-footed lizards, Varanus, Side-necked Tortoise. The Cobras form about 2/3 of the all snakes found in Australia and all of them are poisonous so that the absence of Vipers and Rattle snakes is more than compensated. Lastly, the entire absence of all tailed amphibians is noteworthy.

3. *Polynesian sub-region*. The rest of the Islands as far north as the Tropic of Cancer and including the Sandwich Islands are embraced in the Plynesian sub-region. This is made up entirely of small islands, and as such the absence of certain forms is to be regarded as more characteristic. The families of terrestrial vertebrates is only 53, of these 3 are of bats and 37 of birds, whose occurrence in these remote islands is probably due to their superior powers of dispersal. Of reptiles, 9 families occur and of amphibians only 2. The other two families of mammals include the Muridae (Mice) and the Cervidae (Deer).

The families peculiar to the Plynesian are the Drepanididae confined to Sandwich Islands; Kagu only found in New Caledonia and tooth-billed pigeons confined to Samoa.

4. *New Zealand sub-region*. In includes New Zealand, Norfolk island, Auckland, Campbell and Macquaire islands. These are 34 families of terrestrial vertebrates. Among mammals there are 3 families, family of Noctiliomidae (free-tailed bats) Vespertilinoidae (typical bats) and Muradiae (Mice rats).

Among birds 5 families namely the Xeicides, Nestorids (Nestor parrots), Stringópids (Owl-parrots), Apterygids (Kiwis) and Hatterids are confined to this sub-region.

Among reptiles 13 families are peculiar. Snakes are absent. The amphibian is only represented by one frog, *Liopelma*.

The peculiar animal known as Tuatra (*Sphenodon punctatus*) belonging to family Rhynchocephalidae (class—Reptilia). It is commonly known as *living fossil*.

ORIENTAL REGION

Extent

This Region includes those portions of continental Asia which are not comprised in the Palaearctic and Ethiopian Region together with the Malaya Archipelago as for east including Bali, Borneo, the Philippine islands and Formosa.

Climate

The Indian sub-region is in its northern portion, chiefly composed of plain and desert, more particularly in the watersheds of great rivers Indus and Ganges. Its fauna as a whole shows a desert affinity to that of the Ethiopian Region; while the desert in the north-west is debatable ground, and may be regarded as a transitional tract between the Oriental and Palaearctic Regions. The southern portion of India is more luxurient than the northern one, and is largely covered with tropical forest, with a series of elevated tracts culminating in Western and Eastern Ghats. Ceylon, the Indo-Chinese sub-region and most of the Malayan islands are almost entirely covered with tropical forests of the most luxurient character and posses varied and extremely rich fauna.

Fig. 9.3: The Oriental region (solid lines) and adjacent land (broken lines).

Zoological Characteristics

The terrestrial vertebrates are represented with 153 families of which 10 are peculiar.

Among mammals 4 families are represented by Hyalobatidae (Gibbons), Tarsiidae (Tarsiers), Galeopithecidae (Flying lemures) and Tupaiidae (Tree-shrews).

Birds are with only one family—*i.e.,* Eurylaemidae (Broad Bills).

Among reptiles 5 families are peculiar they are Elachistodontidae, Uropeltidae (Shield tails), Lanthanotidae, Gavialidae (*Gavialis*) and Platysternide (Big Head Tortoise).

In addition to the above, several well known species and genera are quite characteristic. Among Mammals the Orangutan (*Simia satyrus*) the Macaque monkey (*Macacus*), the Tiger (*Felis tigris*) and Indian elephant (*Elphas maximus*), the Malayan Tapir (*Tapirus indicus*), and three out of the five known species of Rhinoceros are nearly or quite confined to the Region, while families Tragulidae (Chevrotains) and Manidae (Pangolin) are very characteristic. Although only one family of Birds is given above as peculiar, yet many other have their monopolies in the Oriental Region. They are Starlings drongos, Orioles, Honey Peckers, Bulbuls, Pittas and many others. Reptiles and Amphibians are very well-represented in this region for besides the five peculiar families above enumerated, are following the characteristic Illysiidae (Cylinder Snakes). Xenopeltidae, Acrochordidae (Wart Snakes) and pitvipers etc.

Wallace divided Oriental region into four sub-regions:

(1) Indian sub-region

(2) Ceylonese sub-region

(3) Indo-Chinese sub-region

(4) Indo-Malayan sub-region

1. *Indian sub-region* consists of Central and Northern India, from the river Indus and the foot of the Himalayan southwards to Goa and the River Krishna, the line of demarcation taking a southward bend nearly as far as Mysore. It is inhabited by 123 families of terrestrial vertebrates of which only one is peculiar *i.e.,* Elachistodontidae which include a single species of Colubrine Snake. There is no other peculiar feature. Indeed by Scalter and others, the first two of the Wallace's sub-regions which together include the whole of Indian Peninsula are regarded as one.

2. *Ceylonese sub region.* It includes the remaining portion of Indian Peninsula, not included in the Indian sub-region and the island of Ceylon. The main peculiarities of this area are the exclusive possession of the curious family of snakes known as Shield-Tails and the presence of Loris which does not occur in Northern India, though recorded from Eastern Burma. The genere Platacanthomys (Spiny-Rat) is also characteristic, only occuring elsewhere in Cochin China. On the whole, 122 families of terrestrial vertebrates are represented and are identical to the fauna of Indian sub-region.

3. *Indo-Chinese sub region.* It includes China, South to the Palaeartic region, Burma, Thailand and the islands of Hainan, Formosa and Andamans. It contains as many as 138 families.

Among mammals three genera *i.e.,* Takin, Hapalomys and Panda are confined to this sub-region. It also contains Gibbons, Flying lemurs, Java Rhinoceros, Malyan Tapir etc.

Among birds, Wrens are common. One family of tortoise *i.e.,* Platysternidae is peculiar. Among amphibians Salamander and Disc-tongued frogs occurs.

4. *Indo-Malayan sub-region.* Includes the Malayan Peninsula and all those islands of the Malaya Archipelago which fall with Oriental Region. In this region occur 132 families of terrestrial vertebrates of which one, the Lanthanotidae is peculiar and another *i.e.,* Tarsidae is practically so. In addition, several genera of mammals are peculiar. Among these may be mentioned Orang-Utan, Proboscis-monkey, Heringale, Cynogale, Malayan Bedger and some small Rodents. The green Tupaia (Tree-shrews) is well-represented and characteristic while the Gibbons, Flying Lemurs, two species of Rhinoceros, the Malayan Tapirs and the Broad-bill are peculiarly Oriental forms which this region shows with the Indo-Chinese. The typical Australian Cockatoos are represented by a single species in the Philippines while the Megapodes another essentially Australian group occur in the Philippines, Borneo and the Nicobar Islands.

ZOOGEOGRAPHY OF INDIAN SUB-REGION

Blanford (1901) in *Phil. Trans. Royal Soc. London* discussed the geographical distribution of animals on the earth in British India. In his discussion he included India Pakistan, Kashmir, Gilgit, Ladakh, Nepal, Bhutan, Sikkim, others Cis-Himalayan States, Assam, countries between Assam and Burma such as Garo, Khasi and Naga Hills, Manipur, Burma with Karenni, Tenasserim and the Mergui Archipelago, Andaman and Nicobar Islands. *B. Prasad* (1942) added Laccadive and Malolive islands.

Sub-regions of India

Owing to the heterogeny nature of the fanuas of the different parts further divisions of Indian sub-region is not an easy task. Its divisions has been attempted by a number of scientists. A brief history of such attempts is as follows:

Jorden (1862) divided it on the basis of birds; *Gunther* (1864) on the basis of reptiles; *Blanford* (1876) on molluscs; *Wallace* (1876) considered the distribution of all the groups of animals. *Blanford* (1888) on mammals, *M. Smith* (1931) based his findings of reptiles and amphibians; *B. Prasad* (1921) sub-divided Indian sub-region into 5 division *Mahendra* (1942) divided. Indian sub-region into 10 sub-division on the basis of distribution of reptiles and amphibians. These are:

1. The arid and semi-arid province of North India.
2. The Western Himalayas.
3. Southern Burmese province.
4. Trans-Gangatic province.

5. Gangatic plain and adjacent part as far as South as 20° latitude.
6. South India below 20° latitude excluding Trav encore.
7. Travencore province.
8. Ceylon.
9. Andaman Islands.
10 Nocobar Islands.

1. *The arid and semi-arid province of North India*. It consists of W. Pakistan, Punjab, Western Rajasthan and Cutch. The fauna is allied to that of the countries to the west of present W. Pakistan and also almost all general and species are characteristic.

Among lizards are *Teratoscincus, Stenodactylus, Alsophylax, Agamura, Pristrus, Eremias* etc. Among shakes are *Leptotyphlops Contia, Lytorhynchus* etc., are present. Many species are endemic. Very few turtles occur but none are endemic or peculiar to it.

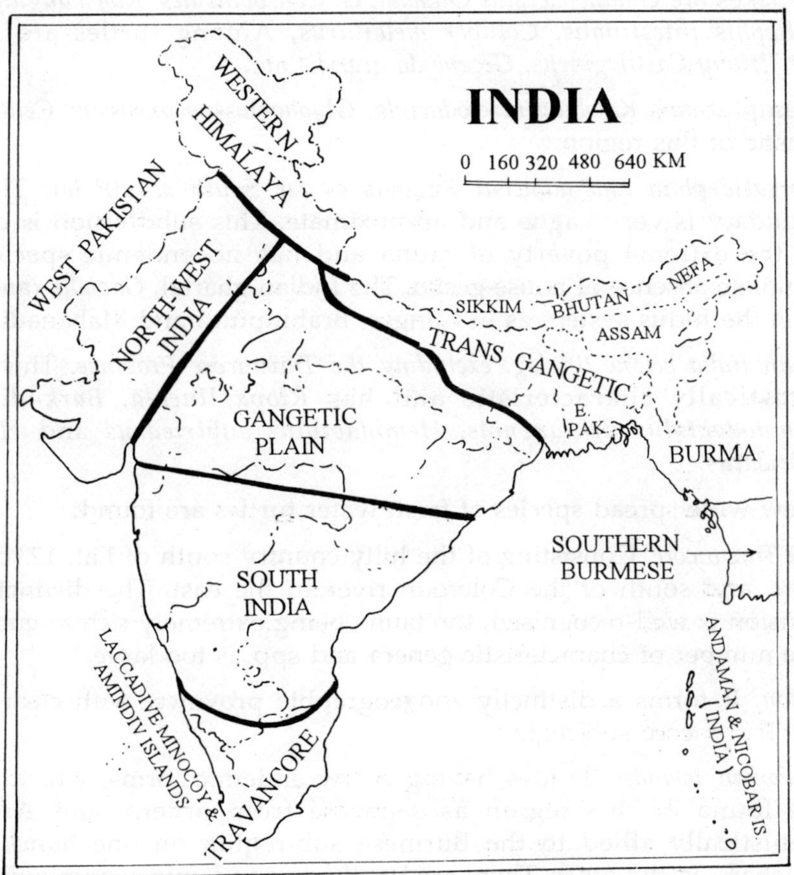

Fig. 9.4: The Indian sub-region.

Amphibians are rare the only genera found being the cosmopolitan *Rana* and *Bufo*.

2. *Western Himalayas*. It embraces Kashmir, Simla, Kumaun and Garhwal Districts, adjacent mountainous regions of Western Tibet, Alpine and Punjab etc. This zone is poor in amphibians and reptiles. These are certain endemic species *e.g. Gymnodactilus fasciolatas, G. lawderonus, Japalura major, J. Kumaonesis* and *Phrynocephalus theobaldi* and *Leiolopisma ladacense*.

3. *The Trans-gangetic province*. It consists of the northern part of Bihar and undivided Bengal, the whole of Assam, Burma north of 20° latitude Nepal, Himalayas to the east of Nepal, Sikkim and Bhutan. The herpetological fauna is both rich and characteristic. Amongst the endemic genera are *Ptyctoaemus, Mictopholis, Oriocalotes, Stolizteaia, Tylotriton* etc.

4. *Southern Burmese Province* comprises southern Burmese province including Burma approximately south of 20° latitude. It includes certain characteristic snake species and some endemic amphibians.

Among snakes are *Gymnodactylus Oldham, G. Consobinoides, Riopa anguina, Bungarus flaviceps, Doliophis intestinalis, Coluber melanurus*. Among turtles are *Platysternus megacephalum, Triony Cartilagineus, Geoemyda grandis* etc.

Among amphibians *Kalobula macrodactyla, Glyphoglossus molossus, Cellulla gnttulata* are characteristic of this region.

5. *Gangatic plain and adjacent Regions as far South as 20° lat*. The southern boundary is very vague and approximate. This subdivision is characterised by the extreme poverty of fauna and has no endemic species. There is complete absence of house gecko. The Indian gharial, *Gavialis* ranges as much as in the Indus system as in Ganges, Brahmputra and Mahanadi.

6. *South India below 20° lag. excluding the Travancore Province*. This province is funistically characteristic and has *Riopa lineala, Barkudia insularis, Gomnodactylus dekkanenois, Hemidactylus subtriedrus* and *Hemidactylus reticulatus*.

Only a few wide spread species of fresh water turtles are found.

7. *The Travancore*. Consisting of the hilly country south of Lat. 12° or 13° on the west, and south of the Coleroon river in the east. The distinctness of this division is well-recognised, the fauna being extremely rich in endemic forms. The number of characteristic genera and spp. is too large.

8. *Cylon*. It forms a distinctly zoogeographic province, with clear affinities to the Travancore sub-region.

9. *Andaman Islands*. Besides having a few endemic forms, which differentiate the fauna of this region as separate from others, and Andaman are faunistically allied to the Burmese sub-region on one hand, and to the Nicobars on the other. There are hardly any endemic genera owing evidently to the fauna being imported from the continent in comparatively recent times.

The endemic species are *Gymnodactylus rubidus, Phelsuma andamanesis, Calotes andamanisis, Typhlope andamansis, Boiga andamanesis* etc. There are no amphibians (except 2 spp. of *Rana*) and no fresh water turtles.

10. *Nicobar Islands.* It fauna has affinities with Andaman Islands. It has fauna like *Calotes mystaceus Mabuya andamanesis, Gonicephalus suberistatus.* The endemic species are few in number like *Mabuya rugifera, Glotes jubatuo Leiolopisma macrotis* and some other.

WALLACE'S LINE

The Wallace's line is the supposed boundary between the Oriental and Australian faunas. The faunas of the Oriental and Australian Regions are extra-ordinarily different. What happens where they meet in the Indo-Australian Archipelago has long fascinated Zoogeographers. As first drawn by *Wallace* (1860), the line ran between Bali and Lombom, between Borneo and Celebes and between the Philippines and Sangi and Taland islands. Later authors debated the position of the line especially the northern part of it. *Huxley,* for example, who named Wallace's line, thought it should split the Philippine, putting Palawan in the Orient and the rest of the Philippine in the Australian region. Still later even the existence of the line was debated.

Later zoogeographers doubted the validity of Wallace's line. Majority of the scientists thought it to be imaginary, some regarded that such a sharp boundary could not be drawn between any two faunal regions. Now Wallace's line is not longer accepted as the boundary between the two regions, as there is great difference between the faunas of Bali and Lombok, that are separated by only about 32 km. The rich fauna of Borneo and comparatively poor faunas of Celebes and the Philippines cannot be denied. The difference is regarded due to impoverished young fauna in one area and an older richer fauna in the other.

Weber's Line

In 1904 Weber draw another line between the Oriental and Australian regions. It demar cates fauna to the west as predominentally Oriental and to the east Australian. However, one problem its that if a line is thus drawn to separate 50 per cent Oriental and 50 per cent Australian elements for one group of animals, it may not be the same for another group. Majority of the zoogeographers do not recognize Weber's line and prefer a broad belt as a separate zone. This is due to the difference in mobility of different kinds of animals. A further proof of these differences is that the flora of New Guinea is mainly Oriental while its fauna is strictly Australian.

Dickenson and others (1928) gave the term `Wallacea' to separate eastern boundary of the Oriental region at or near Wallace's line and western boundary of the Australian Region just west of New Guinea. In that the intervening area may be regarded as a separate subtraction-transition area. *Darlington* (1957) also recognized that this is an appropriate and useful term. Wallacea bounds the Oriental and Australian regions.

Chapter—10

Geographical Distribution

Life occurs in almost all the habitats ranging from high mountains to the deepest sea bottom. One would find a series of contrasting conditions which of necessity profoundly affect the organism.

The distribution of animals and plants may be of following two types:

Distribution in Space

It is of two types:

(*i*) *Geographical distribution*. It deals with horizontal or surficial distribution of animals on land and in water in different continents and on islands. The study of geographical distribution is called *Zoogeography.*

(ii) *Bathymetric distribution*. It deals with the vertical or altitudinal distribution of animals on land and in water. The bathymetric distribution is further divided into:

(a) *Geobiotic* or *Terrestrial*

(b) *Limnobiotic* or *fresh water inhibiting*

(c) *Holobiotic* or *marine water inhibiting.*

Distribution in Time

It is *geological distribution* or the *durational distribution* of animals from the very beginning of the life on the earth upto recent time.

Pattern of Distribution

The distribution of life over the surface of earth is world wide. There is no such place where the living organisms are not found. In spite of that the distribution of animals and plants is not uniform for example, a particular species of animals is present in one area but entirely absent from other. Three patterns have been observed regarding distribution of animals.

1. *Cosmopolitan Distribution*
2. *Bipolar Distribution*
3. *Discontinuous Distribution.*

Cosmopolitan Distribution

The animals which are found all over the world are said to have cosmopolitan distribution. *Mytilus, Artemia* (Shrimp), bats, rats, hawks, cockroaches etc., are world wide in distribution. These animals have a wide range of adaptations to a wide variety of environmental conditions.

There are two types of cosmopolitan animals (a) *Eurytopic,*when they have wide range and uniform distribution and (b) *Steno-topic,*when they have restricted range of distribution.

Bipolar Distribution

In bipolar distribution the animals are only confined to the polar regions i.e., only in the waters of arctica and antarctica, A few examples are *Lampta, Grammaria, Sardina* etc. (Phylum Coelenterata); *Limacina* (Mollusca); *Oncorhynchus,*(Pacific Salmon) and *Lamma cornubica* (shark).

Discontinuous Distribution

Some of the present day animals which are closely related are found distributed in widely separated areas. Some examples will illustrate the discontinuous distribution. Marsupials are found in Australia, and South America. In Australia there are several kinds of marsupials like Kangaroos, Wombets, Koalas and Bandicoots. Only the Opossum is found in the eastern states of America and a few marsupials in the South America.

Another example is that of Tigers. The Tigers are found only in India and neighboring regions like Nepal. Lions are only found in India and Africa. Elephants are found in the forests of India, Nepal and Africa. Two species of Alligators are found one in South-Eastern United States and the other in China. The lung-fishes (Dipnoi) are represented by only three living genera, *Neocera-todus* in Australia, *Protopterus* in Africa and *Lepidosiren* in South America.

Another example of discontinuous distribution is the Camel. They are present in parts of Africa, Asia and South America. The Llama found in Andes. Different species of Tapirs are found in Malaya, and South America and Islands of Jawa and Sumatra.

Many more examples are there for the discontinuous distribution. It can be explained if we understand the geological data and animal evolution.

From the Palaeozoic to the Mesozoic era, there were only two major land masses Gondwana and Laurasia and these were in contact at times. Gondwana Centered around the South Pole while Laurasia overlapped the equator and extended well into the northern hemisphere. During the Cretaceous, these masses fragmented to form the present continents and these have since drifted apart very slowly toward their present positions. Gondwana gave rise to Africa, South America, Australia, Antarctica, Arabia and India. Laurasia broke up into North America and Eurasia.

A species originates in a definte more or less restricted area and then tends to spread in ali directions occupying suitable habitates. When an impassable barrier is encountered, the migration of the species stops unless it is removed, by geologic or climatic changes. Sometimes *bridges* are established and the animals are able to cross the barriers. Three type of bridges have been recognised (1) A *corridor bridge* is a broadly continuous connection as exists between Europe and Asia (2) A *filter bridge* is more temporary in duration as is the Beringstrait which connects Asia with North America (3) A *Sweepstakes bridge* is the one which depends upon accidental transportation on floating ice, log of wood clinging mud of the birds claw or even on board the ships.

It is in connection with this type of history of the earth and animals that the problem of widely discontinuous distribution must be understood. Marsupials, Ratitae birds and lung fishes had a world wide distribution to begin with. Later on by subsedence of land and continental drift, they became isolated. Where the competition with superior animals was severe, they have been eliminated, in order places they survive till the present day. The barrier of deep sea between Bali and Lombock has existed since Palaeozoic times and it has never been bridged. This is the reason for the distinctness of fauna in the two islands.

DISPERSAL

All the animals are capable of multiplying in large numbers. If all the progenies of a particular species in allowed to live in the same area, the result will be shortage of food, and the struggle for existence. Moreover, there is a population pressure due to which the individuals migrate to expand to boundaries of their range or, sometimes, are perished. The geographical range of the species is increased due to dispersal. *Darlington* (1957) defined dispersal as, "The total geographical movement of groups of animals and the movement of individuals (eggs, youngs and adults) especially their outward scattering."

Means of Dispersal

There are several agencies which help in the dispersal of animals. These are:

By wind

Some animals are dispersed either by the power of flight or by being carried passively by means of wind. Several small insects are blown by air to a few hundred metres from their place of birth. Sometimes, by air, they are also taken away to higher levels. Air-dispersal is found in insects, birds and bats.

By hurricanes

Small living animals are transported by cyclonic storms of great velocity about 120 km. an hour. Some American birds have reached in this way the coasts of Britain and France.

By water

Some aquatic animals and larve migrate from one place to another with water currents. Rafts of trees or floating debris also carry the animals, such as many species of arthropods to pass down large rivers and sometimes are seen far at sea. Small arctic animals move on to the ice.

By land

Active movements of the species are plausible by land. For finding new pastures, active movements of animals are well-known. Land-bridges are used as passages of migration.

By human agency

Several animals reach new localities by accidental or international transportation.

By hosts

By this means some parasitic and commensal animals are transported to various new localities.

Migration

It is also an important, factor for the dispersal of animals. There are two types of migration (*i*) permanent and (*ii*) seasonal migration.

Thus the distribution of animals into different regions is only possible due to the dispersal. But there are a large number of barriers to these dispersals.

BARRIERS TO DISPERSAL

Barriers may be defined as the factors which hinder in the normal distribution of animals. Following types of barriers are there:

1. Typographical barriers.
2. Climatic barriers.
3. Vegetative barriers.
4. Large bodies of water as barriers.
5. Impurity and lack of salinity of sea water.

Topographical Barriers

High and extensive mountain ranges act as barriers in the distribution of many terrestrial animals. They are effective if they are parallel with the equator as in Asia and Europe. Here we find a remarkable difference between the species occupying the northern and those occupying southern slopes. This is true of the Great Himalayan Range which is covered by snow. On the south of it, there are hot, moist plains of

India having tropical fauna which resembles largely with that of Africa. In the north, the climatic conditions are changed so that the animals of this region resemble largely with that of Europe.

In the New World, the mountain ranges run North to South and their influence upon the distribution is very less. *Mayr* found that there are distinct altitudinal races including lowland, midmountain, and alpine races.

Climatic Barriers

Climate seems to limit the range of several animals. But it is believed that, in many cases, it is not the climate itself so much as the change of vegetation consequent on climate which produces the effect.

Degree of heat is very important in limiting animal distribution. However, the distribution of tiger and elephant shows that this does not very much alter their distributions. For instance, tiger has its normal home in the hot districts of India and the Indian archipelago but is also found in colder climates such as in the elevated regions of the Caucasus and the Altai chain and in the Himalayas perhaps on snow and also in the cold plains of Manchuria. Elephants likewise do not suffer from cold, provided a sufficient amount of water is available.

The influence of temperature is much more marked in the cold-blooded animals. Thus the amphibians and reptiles are tropical and temperate in their distribution, rapidly diminishing in numbers towards the poles.

Lack of moisture also controls the distribution and the effect is more marked in the extreme cases where it produces the desert conditions where only forms adapted to withstand drought can survive, for others it forms insuperable barrier. The most notable desert barrier is Sahara, due to presence of this desert the fauna of North Africa and S. Africa are totally distinct. Similarly, the Kalahari desert is responsible for the distinct fauna of Central Africa from that of cape of Good Hope.

The increase of moisture may produce swampy conditions which may make the area impassable for creatures not adapted to them. Increased humidity also have marked effect in the vegetation and also in the spread of insect-life.

Vegetation Barriers

The influence of vegetation on dispersal is both direct as well as indirect. Its direct effect is on arboreal animals, which can not cross the regions where forests do not occur. In the same way, the larger terrestrial animals can not penetrate through the dense forest of tropics. For example, several species of elephants i.e., *Elephas* and *Mastodon* of N. America did not succeed in crossing the southern region of Mexican plateau to the South.

The vegetation also effects the distribution indirectly as it forms the important source of food. For example the primates live in tropical forests as they feed on nuts, fruits and blossoms. The animals with short-crowned browsing teeth would needtrees and shrubbery for the food while long-crowned grazing teeth need extensive pasturage of harsh-grasses. many insects and catterpillars feed upon certain plants and if these plants are absent, those insects and catterpillar will not exist. Similarly some plants are dependent upon certain insects for pollination.

Large Bodies of Water

Larger bodies of water such as giant river systems and oceans act as most effective barriers in the dispersal of terrestrial vertebrates like amphibians, reptiles, mammals and fightless birds. Fresh water fishes like carps, catfishes etc., are unable to migrate through the large bodies of salt water.

To the modern amphibia salt water constitutes a most effective barrier as the common salt is poisonous to them, even 1 per cent solution preventing the development of their larvae. Amphibia are, therefore, almost completely absent from oceanic islands. Moreover, the tailed amphibians such as sirens, newts, salamanders etc. are restricted to northern hemisphere for the southern land masses, Australia, Africa and S. America are isolated by oceans and are incapable of crossing through small migratory roots. The anurans have a wider range of distribution while the burrowing caecilians are confined to S. America and Africa.

For crocodiles and marine turtles, the seas do not act as barrier. Giant tortoises are confined to certain oceanic islands e.g., Galapagos islands of the western Indian ocean and totally absent on the mainland of S. America, Africa and Eurasia.

The serpents, though many of which are good swimmers, but are incapable of passing large bodies of salt water with the exception of some sea snakes. Adult lizards are also incapable of passing an oceanic barrier, through the eggs in some cases could be transported to the islands.

With the birds, it is only the flightless forms such as the ostrich, rhea, apteryx etc. which are incapable of oceanic migration for even small birds are carried far on favouring winds.

Among mammals, except for whales and seals, bodies of water of a greater expanse than 20 to 50 miles are impassable when not frozen.

Similarly land masses form barriers to the spread of sea-life.

Impurity and Lack of Salinity of Sea-water

Impurity and salinity of water also act as barriers. Sea-water forms a barrier for sponges, branchiopods, cephalopods, asteroids and echinoids.

Moody (1962) described two types of barriers:

(*i*) Physical barriers include bodies of water, dry land, high mountain ranges, deserts, open plains, forests as well as climatic factors.

(*ii*) Biological barriers include absence of food suitable to species in question, presence of competitors for the same food supply or nesting sites, or presence of predator animals.

However, a species is limited in distribution by the sum-total of external influences. The range of equilibrium is finally governed by *Liebig's Law of Minimum*, which is the most valuable attribute. Animals having adaptability can extend their ranges to regions, which offer conditions of life differing from those in the centre of dispersal. Lack of adaptability prevents dispersal.

Chapter—11

Bathymetric Distribution

The Bathymetric distribution concerns itself with the vertical range of organisms in space, i.e., from the highest Alpine peak to the abyssal depth of the sea. One would find a series of contrasting conditions which of necessity profoundly effect the organism. Of these the primary conditions are:

(1) It determines the method of breathing. With a few exceptions, if the animals are terrestrial and breath in air they can not live in the water and those who are aquatic animals and breathe in water cannot survive outside of water.

(2) The presence or absence of light not only modifies the animal directly but indirectly through the effect upon the food supply for assimilative plants, which form the ultimate nourishment of all animate nature, cannot exist where light is wholly absent.

(3) The third primary condition is the presence or absence of a substratum, without which the organisms must be self supporting, either buoyant or able to swim. This condition therefore determines the bionomic group to which the animal belongs, whether plankton or nekton on the one hand, or benthos on the other.

Like fresh water ponds and lakes marine environment has been classified into two zones by *Hedgepath* (1957). (1) *Pelagic Zone.* It includes the entire water mass lying above the ocean floor. Thus the animals which live here are called *Pelagic*. They may be *plankton* (microscopic) or *nekton* (large swimming animals). (2) *Benthic zone.* It includes all the sea bottom, thus the animals live in or on the sea bottom are called *benthonic*.

The secondary conditions limiting bathymetric distribution are whether the water be fresh or salt and the increase of pressure with the depth.

ORGANIC REALMS

Hence from the bathymetric as well as from the geographic stand point three organic realms may be recognised. The bathymetric divisions are:

(1) Geobiotic or terrestial

(2) Limnobiotic or fresh water inhabiting (lakes and rivers)

(3) Halobiotic or marine or salt-water inhabiting (the sea).

Geobiotic or Terrestrial Realm

It extends from high tide mark along the shore of continents and islands to the summit of the highest elevation. It ranges in altitude, therefore, from lowlands to the Alpine peak and each division—low land up land, prairie, high plain or mountain range has its own fauna and flora, governed by many factors but in part at least by altitude. The so called timber line, the limit of tree growth is governed by altitude, although vary in different regions due to climate and latitude.

One additional terrestrial subrealm of Geobiotic realm is the *eryptozoic* or *subterranean caves.* The dryness and absence of light is very harmful to the plants which are the food supplier to the animals. Examples of permanent cave dwellers are *Peromyscus leucopus* (white-footed mouse) having bulging eyes, and ling tectile whiskers and ears; *Typhlotriton spelaeus* with normal eyes in larval stage but in adult the eyes degenerate; *Gronias nigrilabris* (cave cat fish) is partially blind. Internal parasites and wood boring larvae of insects also dwell in dark environment and are consequently modified.

Limnobiotic or Freshwater Realm

The terrestrial waters such as ponds, lakes and rivers contain very limited fauna due to freshness of water and continuous flowing character. This is a condition to which the great majority of invertebrates with their planktonic larvae can not adapt themselves. Certain lakes and relic seas are the only bodies of fresh water of sufficient depth to have deep sea characteristic but we do not find such profound modifications in lacustrine forms as in those which live in the deep sea bottom. For example biolumieniscence is characteristic of deep sea nocturnal animals but none are found in the deep lakes, although in each instance they might be very useful in the struggle for existence.

Halobiotic or Marine Realm

It is a very important realm, for here we find all the contrasting characteristics abundantly developed. From the ages during which the ocean has existed, it has given sufficient time to the inhabitants for evolution.

The halobiotic realm is divided into four subrealm as follows:

(a) Strand

(b) Flat sea or shallow sea

(c) Pelagic

(d) Abyssal.

Strand

The strand or tidal zone is the transitional area between the maine and terrestrial realms, for here the inhabitants are left bare twice daily by the receding ride and have to endure drying, either by means of closable shells or other devices or burrow down into the moist sand, or must be able to breathe both the air and the water. The tidal Zone is of course of very variable extent, owing to the differing height of tides and range from a width of a few feet to several and in the rare cases to many miles. The tides, when run into a constantly narrowing area, may grow to a height of 60 ft at the time.

Flat Sea

The flat sea of *Neritic province* is the ocean upto a depth of 200 meters above the continental shelf below low tide mark. The continental shelf is formed by the action of storm waves, which are continually cutting back the shore line nd depositing the debris, together with other land waste, upto the sea especially at its outer edge. This margin is 200 meters and marks the extreme limit of wave action.

This province is very important t biologically for it has both light and its attendant vegetation and a substratum where upon benthonic organisms may dwell. All of these factors, light, plant food, movement of water, warmth, and isolation, make this area a variable hot bed for evolution.

The shore fauna is certainly the most representative of all faunas.

Foraminifera with beautiful shells of lime slowly gliding on the ponds of sea-wind. Calcareous sponges like little vases and more irregularly moving sponges, hydroed zoophytes, sea-anemones and corals often like beds of flowers; unsegmented worms such as the living films which glide on the sea weeds and the nemertines or ribbon worms provided with a remarkable protrusible proboscis. The higher ringed worms or annelids like *Nereis, Aphrodite* etc. The starfishes creeping up the rocks, the brittle stars, the sea urchins and the sluggish sea cucumbers. The beautiful colonies of moss animals or Bryozoa: myriads of crustaceans such as water fleas, beach fleas, sand hoppers, shrimps, hermit crabs, shore crabs etc. Bivalves innumerable such as cockles and mussels, oysters and razor fish, many gastropods, cuttle fishes; a large representation of ascidians, the lancelets, true shark fishes like sandribs, poisonous sea snakes, numerous shore birds and an occasional mammals like otter and seal—on the whole a more representative fauna than anywhere else.

The flat sea is divisible into (a) A *pelagic zone* and (b) The *Littoral zone.* The pelagic zone is characterized by the absence of substratum while the littoral zone is the substratum of sea bed.

Pelagic Realm

The pelagic realm embraces all the open seas down to the depth to which effective sunlight penetrates. It is characterised physically by the absence of a substratum. In the upper portion there is variable temperature and frequent and violent wave action, while in its lower strata the movement of the waters and the temperature are greatly reduced.

Assimilating plant life, which forms the ultimate food supply of all animals is dependent upon the presence of red, orange, and yellow rays which virtually restricts it to the upper 200 m. of oceanic water.

Due to absence of substratum no *benthonic* form is found in this realm. Only either the *planktonic* or *nektonic* forms exist here.

4. Abyssal Realm

Beneath the limits of the continental shelf and pelagic realm, the abyssal realm is found which extends from below 200 meters. Two region abyssal can be distinguished: the region between 200-600 meters may be referred as *abyssal pelagic* (where no substratum is there so all the animals swim or float); and below 600 meters as *abyssal benthos* (the substratum is present).

Characteristics of Abyssal Realm

These are as follow:

(i) *Absence of Light*. Light may exist in the upper transitional strata but it lacks the rays which are essential for assimilating plants, hence non exists. The animals, therefore, are all carnivorous or feed upon dead organic matter. Below the transitional zone, there is darkness, except for bioluminescence is profound.

(ii) *Quiescence*. There is no movement of water except the sluggish ocean currents of the greater deapth, the progress of which is very slow.

(iii) *Cold*. Below a certain deapth, the waters of all the oceans of world is nearly to the freezing point of fresh water. The main temperature of Atlantic at the surfaces 68°F, at 00 fathoms 37°F, at 1000 fathoms 35.6°F.

(iv) *Pressure*. The pressure of abyssal water is enormously increasing directly with the depth. The ratio of increase being about one ton to a square inch of surface for every 1000 fathoms of depth.

Thus the abyssal realm constitutes a simpie biotic environment of vast extent but of comparatively uniform and changeless character and hence not conducive to rapid evolution and change. None of the deep sea creature is old geologically speaking for while from 25 to 35 paleozoic genera are known in the purest shallow seas none of the animals which people the deep sea is older than the Mesozoic. They seem to be all migrants from shallow waters which have become adapted to the deep sea conditions but there is in no instance the evolution of a new race of animals exclusively restricted to the abyssal realm.

Vertical Distribution

Life exists at all the depths. *Murray* estimated the number of forms as follow:

Down to	200 meters	about	4200	species
"	200	"	600	"
"	4000	"	400	"
"	5000	"	150	"

Among the invertebrates, the sponges form an important element in the deep sea fauna, and coelentrates such as the coals, hydroids and their allies, while not so numerous as the sponges, are also well represented. Echinoderms, among which are the brittle stars and stalked crinoids are present; and holothurians or sea cucumbers are also found. The modern stalked crinoids are rarely found beyond a depth of 2000 fathoms, although free-swimming species, *Bathymetra abyssicola* has been obtained at 2900 fathoms.

Bryozoa range to 3000 fathoms. Brachiopods have been found from 2900 and probably so much deeper. Of the molluses, the Pelecypoda have a very great veritcal distribution, *Mytillus* sp. being from the shore to 3000 fathoms. In the greater depth the shells of bivalves are exceedingly delicate, being sometimes quite transparent. The majority of gastropods are shallow-water forms, although a number of them are found at depth from 1000 to 2000 fathoms. The shells and their ornamentation are more delicate than in their shoal water relatives. Cephalopods inhabit waters of moderate depths only.

Of the Anthropods, we have branacles from 3000 fathoms, ostracods from 2000, decapods such as crabs, shrimps, lobsters ranging down to 2,500, though true crabs are largely shallow-water forms.

Of the *Vertebrates* elasmobranches, the true shark have been taken at depths of from 345 to 400 fathoms and the rays down to 608 fathoms. None show deep sea characteristics, except perhaps the luminous shark, *Spinax niger,* which ranges from 500 to more than 1500 fathoms. The chimaeroids or silver sharks, on the other hand, often with huge eyes and long body and tail, are distinctly abyssal. *Chimaera affinis* ranges from 200 to 1300 fathoms, *Harriotta* from 700 to 1000 and *Chimaera monstrosa* down to 1000.

The Escociformes group of teleost fishes have a few deep sea members such as *Cetomimus,* with a huge mouth and very small eyes.

Chapter—12

Evolution by Shull

When the structure of the earth and the building and rebuilding of in crust are understood, it is easy to show that the organisms of one period of time were very different from those of earlier or later periods. That is demonstrated by the fossils, which are representative of the animals and plants of those times. If it could be assumed that early beings represented by fossils were ancestral to those which produced the later fossils, the fact of past evolution would need no other proof. The assumption of such genetic connection is more reasonable than any other hypothesis that might supplant it, so one would suppose that the facts of paleontology must early have suggested to naturalists that evolution on a large scale had occurred. Actually, however, knowledge of fossils was not one of the prime stimulants of evolutionary ideas. Lamarck, the earliest avowed evolutionist of any great significance, arrived at his concept of evolution largely on other grounds. Though he studied fossils and used them to support his evolutionary views, that development came late in his life and was a consequence rather than a cause of his advocacy of evolution. Darwin, a later most important contributor, leaned heavily on other considerations. While in his "Origin of Species" fossils are employed, his discussion of them is introduced by a chapter entitled "On the Imperfection of the Geolozical Record", which seems to indicate that he felt the contradictions offered by fossils to his theory more keenly than he felt their support. This discussion is preceded by chapters on artificial selection, the effects of domestication, classification population increase, instinct, the physiology of variation and hybridization, indicating presumably a greater reliance placed on these as evidence or as material influences. Finally, Cuvier, who was among the great students of fossils at the time when the evolution doctrine was being seriously advanced, saw fit to oppose that explanation vigorously.

There were reasons for this backwardness of paleontology as a stimulant to early evolutionary thought. Knowledge of fossils in the early nineteeth century was very meager compared with that of the present time. Moreover, geology was itself then only struggling toward a rational interpretation of earth features.The uniformitarian doctrine, according to which geological processes have been of the same general sort through long periods of the earth's development, and which has been one of the major unifying factors in geology, was first proposed and generally rejected in the eighteenth century and required the great weight of Sir Charles Lyell' authority to establish it well along in the nineteenth century. Lyell at first adhered to the view that species are permanent, and later when he adopted the evolution concept it was under the influence of Darwin's studies, with which he was acquainted long before their publication in the "Origin of Species."

At the present time fossils constitute one of the most convincing indications of the origin of species at different periods and of the general course which evolution has taken if the assumption be made that all forms, or at least large groups of them, are genetically connected with one another. Contrary assumptions have sometimes been made but appear to be untenable, for reasons to be stated later. It is the purpose of this chapter to show the unlikeness of organisms of different periods and to develop the reason for regarding them as genetically connected.

Geological Periods

In order to refer intelligibly and briefly to events in geological time, it is most convenient to employ the names of periods which are naturally marked off from one another by events that plainly occurred in the development of the earth. When stratified rocks, which must have been deposited under water, are sheared of obliquely and are covered by totally different stratified material set at a different angle, it is apparent that the construction of this portion of the earth took place during two periods of submergence separated by a period of elevation above water accompanied by erosion. When the masses of dissimilar stratified rock contain strikingly different fossil forms, it appears certain that a considerable time elapsed between the sub-mergences. On the assumption that extinct forms lived in situations similar to those in which the modern organisms most like them live, it can often be concluded that periods were characterized by mild, or cold, or wet climate. On the basis of such differ-ences geological time has been divided into eras, periods, and epo-chs, as shown in Fig. 12.1. The most

CENOZOIC	*TERTIARY*	*RECENT* *PLEISTOCENE* *PLIOCENE* *MIOCENE* *OLIGOCENE* *EOCENE* *PALEOCENE*
MESOZOIC		*CRETACEOUS* *JURASSIC* *TRIASSIC*
PALEOZOIC		*PERMIAN* *CARBONIFEROUS* *DEVONIAN* *SILURIAN* *ORDOVICIAN* *CAMBRIAN*
PROTEROZOIC		*KEWEENAWAN* *HURONIAN*
PROTEROZOIC		*ALGOMAN* *TIMISKAMING* *LAURENTIAN* *KEEWATIN*

Fig. 12.1: Geological time scale.

commonly used names of the segments of geological time are shown, with little attempt to bring those of equal rank into the same column. For the most part they are so arranged, but the important feature of such a chart is relative age. To take advantage of the similarity of the table to the column of rocks and other deposits, the odlest periods are placed at the bottom, the most recent at the top.

Earliest Life

Formations regarded as Archeozoic are known chiefly in northern North America and Europe, in China, and in Australlia. Fossils are unknown in these formations. The occurrence of graphite in them is, however, taken to mean that simple plants capable of carrying on the carbon cycle then existed. Enormous masses of limestone of the same age also probably owe their origin to secretion by plants; for, though limestone may be deposited by chemical action, deposits known to be formed in this way are of local occurrence and are derived from earlier limestone. In the *Proterozoic* there are unquestionable fossil animals and plants. Almost all the plants are algae, some of the colonies of these found in limestone being a foot in diameter. Among animals, siliceous sponges of that age are exposed in the Grand Canyon, and burrows believed to be those of worms have been found in Proterozoic sandstone in Montana.Glacial till in Canada in the Huronian formation, and in the Australian region in somewhat later deposits, indicate cold climate at least twice during the Proterozoic.

The Cambrian Outburst

Compared with the paucity of fossils in the Proterozoic deposits, the Cambrian has the appearance of pouring out a deluge of living things in great variety. As if suddenly, all the principal phyla of animals are represented in deposits of this period. Geological periods are all inconceivably long, and Cambrian was probably one of the longer ones; but even after making allowance for this great span of time the occurrence in it, for the first time, of may hundreds of species so diverse as to belong collectively to most of the phyla that have ever existed is an apparently abrupt beginning. The abundance and variety of life are rendered even more impressive by the consideration that almost certainly only a small proportion of the existent types were fossilized. That Cambrian life was not the sudden eruption which it appears to be is, of course, the view of biologists in general, for it is regarded as the result of a long period of unrecorded evolution. Since in the main only organisms with hard parts are abundantly preserved, and since by an evolutionary origin the earliest living things can hardly have possessed such hard parts, the assumption of a long period during which living things existed without fossilization is a reasonable one. No emphasis is here placed, however on the probability of an abundance of organisms in Proterozoic and earlier periods. For the purpose of this chapter is not to explain the discrepancies of geological history, by making evolutionary assumptions; it is rather to discover to what evolutionary conclusions one is driven by consideration of the patent facts regarding fossils. Viewed in light, Cambrian does present the outburst of life that is so apparent. It is from that period that the paleontological story of evolution must start. Let us see what the beginning was.

The most abundant of all fossil forms in the Cambrian were the *trilobites,* which belong to the Crustacea, and were more nearly like modern fairy shrimps than any other living group. They have long been extinct. The body was flattened and marked by longitudinal grooves into three lobes (Fig. 12.2), hence the name. The trilobites increased throughout Cambrian time, both in numbers of individuals and numbers of species; over a thousand species have been described from that period. Next in abundance were the *brachiopods.* These animals superficially resemble clams, because they possess a shell of two hinged valves. They were long classified with the mollusks, until it was appreciated that their internal structure was not a all that of mollusks, and that the two valves, instead of being right and left as in clams, were dorsal and ventral. The Cambrian brachiopods possessed shells mainly of a horny substance. Hundreds of species are known from Cambrian marine deposits.

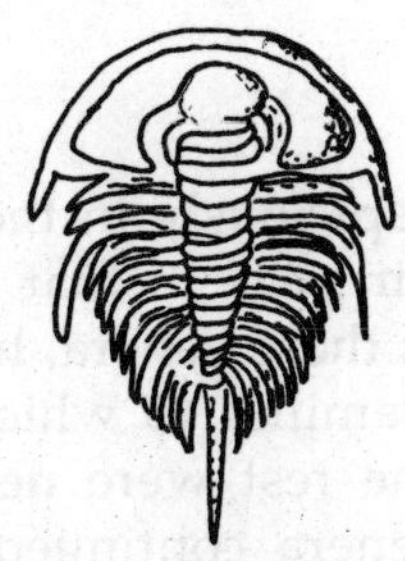

Fig. 12.2: A trilobite.

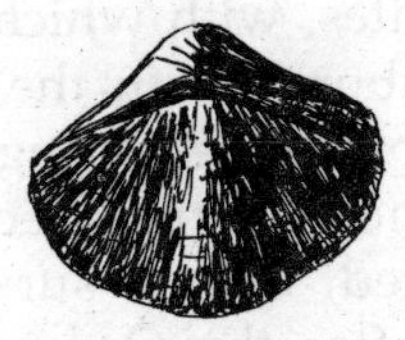

Fig. 12.3: A brachiopod.

Mollusks were less abundant than either of the foregoing groups but were represented by each of the principal classes—the clams, the snails (mostly not spiral), the pteropods, and the cephalopods. Of the echinoderms, there were a few *starfishes* and a few sea cucumbers, and a somewhat larger number of the primitive and extinct group known as cystoids. The worms, soft-bodied as they are, were not often preserved as fossils. A notable exception is the Burgess shale, an exceedingly fine-grained deposit in British Columbia, in which a great variety of impressions of worm bodies are preserved. Elsewhere their presence in the Cambrian is abundantly indicated by their tracks and burrows in the mud. Jellyfishes are even less likely to be fossilized than are the worms; yet in many places in both America and Europe are found casts of the interior cavities of their bodies. Other coelenterates, including corals and the extinct hydroidlike graptolites, the former abundant enough to constitue reefs, have been assigned to the Cambrian, though there is some doubt regarding both the affinity of the organisms and the identification of the containing rocks. The sponges, as might be surmised, are identifiable only by the needlelike speicules of their skeletons, but these are fairly common. Protozoa, the one-celled animals, were represented by both Foraminifera and Radiolaria.

The animals mentioned above belong to at least eight of the main groups or phyla. The worms probably all belonged to the Annelida, though it is possible that one of the other wormlike phyla was included. Assuming however, that the other worms are not included, there remain only two other phyla of minor importance, and one major phylum, chiefly the vertebrate animals unaccounted for. Since all these omitted groups with the exception of the vertebrates were soft-bodied, they may easily have existed in Cambrian time. Even the vertebrates may not have been absent, since fishes appear in the very next period, and an object that looks like one of their scales has been found in Cambrian rocks. It is thus clear that in Cambrian time most of the great groups of animals were already in existence.

Of plants, only fossil seaweeds have been found. Abundant plant life in the sea may be inferred on the ground that the animals must have been ultimately dependent, as they are now, upon plants for food.

The further history of life on the earth can best be told in relation to the several groups of organisms, rather than to the geological periods.

The Arthropods

The trilobites, with which the arthropods began so auspiciously in the Cambrian, waxed more abundant in the Ordovician, but reached their peak in that period with 1200 known species belonging to 125 genera. Only a few of these genera, however, are the same as those of the preceding Cambrian. Of the 13 families to which these 125 genera belonged, only 3 survived from the Cambrian; the rest were new. No new family arose after the Ordovician period, though new genera continued to appear. Only about half as many species of trilobites have been found in Silurian deposits as in Ordovician, and none of these is a survivor from the Cambrian. The group was obviously declining; only about 200 species occurred in the Devonian. New species and new genera appeared during this decline, but larger numbers of both species and genera became extinct. The decline was accompanied by curious ornamentation of the head and the development of remarkable spines on the head and tail in the new species that arose. Trilobites were rare in the Carboniferous period and none whatever later than Permian is known.

The eurypterids, resembling scorpions but living in the sea instead of on land, appeared first in Upper Cambrian, were still unimportant in Ordovician, increased in numbers and size in Silurian, and reached their maximum in both size (10-foot specimens being found) and variety in Devonian. Thereafter they declined, and the last ones known came from Permian deposits.

The horseshoe crabs, which first occur in the Silurian on Europe, never attained great variety or abundance, but they have survived to the present time. During the earlier geological periods, different genera arose successively but by Jurassic time only the genus Limulus existed, and this genus survives today on the east coast of the United States and in the Moluccas Islands. Probably only so old a genus could now occupy two such widely separated areas.

While a primitve wingless type of insect is known from Devonian, that group does not appear again until the Pennsylvanian division of the Carboniferous. The abundance and large size of insects then-2 foot dragonflies, 4-inch cockroaches-suggests they had been developing a long time. Never since Pennsylvanian and early Permian have insects been so large, and its is suggested that the decline in size was correlated with the onset of colder climates. In Permian time the dragonflies and may flies were the most conspicuous. Late in that period appeared the first beetles, in Australia. In Triassic, for some unknown reason insects were scarce, only about thirty species of all kinds being known; but they blossomed forth again in the Jurassic with multitudes of individuals and such new types as grass-jurassic with multitudes of individuals and such new types as grass-hoppers, bees, wasps, ants, flies, and butterflies. Almost all insects of Jurassic time were smaller than in the preceding ages and were of quite modern type. No very important change in them is revealed during the Cenozoic era, though fossil insects of that time are not very numerous. Insects are today more abundant in individuals than most others.

While arthropods of other kinds have been found, they were not very favorable for fossilization, and connected accounts of their changes are not so readily available.

The Brachiopods

At their beginning in the Lower Cambrian, the branchiopods were mostly of a group having horny and phosphate shells, but later in that same period they were chiefly of calcareous type. The latter became increasingly prevalent in later periods. Brachiopods increased greatly in Ordovician over 3000 species being found in this period and the Silurian together. Great beds of limestone were produced by them. Two of the families gained at the expense of all others as the group went over into the Devonian, and in this period the group attained its greatest abundance and diversity. During the Carboniferous period brachiopods maintained their individual abundance, but a decline set in the number of species and genera. New kinds were coming into existence, but they were more than offset by the loss of older forms. This decline continued, though slowly, so that throughout Paleozoic time brachiopods were so abundat that they are extensively used by paleontologists as index fossils. Being of different species and genera in successive ages, and being of well-nigh world-wide distribution, they have been used to correlate deposits in widely separated areas. Their value for this purpose was greatly diminished in Mesozoic time because, while they were fairly abundant in Europe, only a few genera were left by Jurassic time, and North America, for some unknown reason, was poor in them. Through the Cenozoic era they were inconspicuous, but about 225 species have survived to the present.

The decline of the brachiopods is notable for the tenacity with which certain genera maintained themselves. The very considerable diminution in number of species and genera which began in the Carboniferous involved the destruction of most of the genera of Devonian time; yet among the survivors were several genera that first appeared as early as Silurian. Even more striking examples are the genera Lingula nd Crania, which were among the earliest of the Cambrian forms and were of the horny-shelled type that was largely displaceld by the calcareous type before the end of Cambrian, but which are still in existence, living, at the present time. These brachiopods thus exhibit even greater permanence than the Limulus mentioned earlier among the arthropods. No satisfactory explanation of the persistence of a few genera, when nearly all genera are much more short-lived, has ever been offered.

Bivalves and Snails

As we have seen, the mollusks were already divided into their main classes when they were first preserved in the Cambrian. All these classes increased greatly in the Ordovician. For the clams, this increase was largely one of numbers; for the snails it was a change in numbers and kinds, and in size, particularly among the spiral-shelled forms. In both of these classes there was moderate further increase in variety through the remainder of the Paleozoic, and their numbers were well maintained. Clams and snails are not, however, particularly good index fossils for the Paleozoic era, because their distribution was at times limited. In some situations they were abundant, elsewhere lacking. This may be in part due to the rather ready solubility of the mother-of-pearl which lines their shells. This layer was relatively thick in the early mollusks. The outer material, which is less soluble, was relatively thicker in the later forms, and this feature may have helped to preserve them. At any rate, in the Mesozoic both clams and snails were much better represented, and were far more varied, than in the preceding era. Oysters (bivalves) first became abundant enough to form banks in the sea in Jurassic, such banks being duplicated in the Cretaceous. By the end of the later period, bivalves were largely of modern type, many of the genera then existing being still living at present. Snails reached their general modern composition by the end of Cretaceous, owing to the appearance then of genera which have persisted throughout Tertiary time to the present. Although oysters attained in the Miocene a size which they possessed neither before nor after that time (fossils from California measuring 13

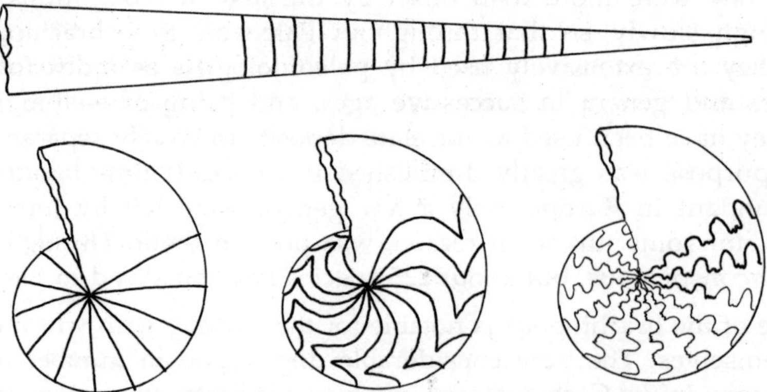

Fig. 12.4: Fossil cephalopods. Above, orthocone; below left to right, nautiloid, goniatite, and ceratite.

by 8 by 6 inches), these two groups of mollusks cannot be said to have culminated in any past period; for, with their more than 40,000 species, they are as abundant and as diverse now as they ever have been.

The Cephalopods, an Evolutionary Tree

Because of their different history and their special significance in evolution discussions, the cephalopods are separated here from the other mollusks. Reference is made only to the four-gilled types, of which the modern Nautilus is one, not to the squids and cuttlefish, whose history is not so instructive.

The early forerunners of our nautiloids had straight conical shells. At any moment of its life, the animal occupied chiefly a short segment at the wide end of its shell. Behind it was a partition which it had recently secreted. Behind that partition, toward the tip of the cone, were other partitions, each representing a former posterior limit of the animal when it was younger. As the body grew, additions to the shell were made at the wide (open) end, and periodically the animal slipped forward and secreted a new partition behind it. The partitions cannot be seen if a whole shell is present, but in fossils the outer part is frequently removed so that the sutures (edges of the partitions) are conspicuous. These orthocones, as they are called, may have begun in Cambrian and were common in Ordovician. One genus ranged in size up to 10 or 15 feet. This large orthocone persisted into the Mesozoic, though the group as a whole declined.

Fig. 12.5: Fossil ammonite.

In the Silurian the characteristic cephalopods had coiled shells. In their internal structure they were not very different from the orthocones, though the sutures were moderately curved. This is the type that resembles our modern living Nautilus, and its members are accordingly called nautiloids. Then in the following period (Devonian) the sutures were characterized by a few rather sharp angles, with saddlelike curves between. The partitions were no longer the simple saucers they had been. Cephalopods of this form are known as goniatites, from a common genus name.

Further complications of the sutures appeared in the Carboniferous, when the sutures were thrown into numerous U-shaped curves which showed, at characteristic points, a zigzag or saw-toothed form. These cephalopods are called ceratites, again from the name of a common genus. Finally beginning in the Permian but increasing in the Triassic, the sutures were transformed into the remarkable lobed and foliaceous patterns. These forms are the ammonites. They continued through the remainder of the Mesozoic but died out in the Cretaceous.

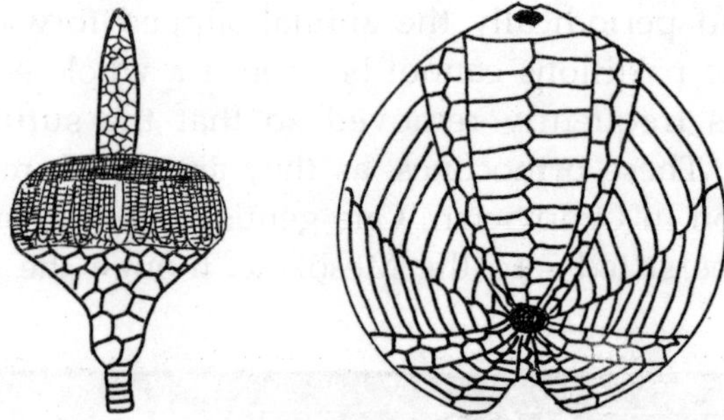

Fig. 12.6: Fossil echinoderms. Left, crinoid of early Carboniferous; right, sea urchin with 20 rows of plates and bilateral symmetry, from Cretaceous.

The chronology of the four-gilled cephalopods is not quite so simple as the foregoing account might seem to imply. The orthocones lasted into the Triassic, long after nautiloids and goniatites arose. The nautiloids persist alive today—the only type that has done so—but are limited to not more than four species at the bottom of the eastern Pacific and the Indian Ocean. Both goniatites and ceratites overlapped, in time, the ammonite group. Yet there was a succession most readily noted in the first occurrence of the several types, and in general also in their periods of greatest development. Such a series of forms, differing by only moderate steps from those which immediately precede or follow, is what is called a line, or better, a tree, of evolution, even through no genetic connection can be definitely proved. The significance of such trees is pointed out towards the end of this chapter.

The Echinoderms

Most of the Cambrian members of this phylum were of the now extinct class of cystoids, though there were some starfishes and sea cucumbers. Of this whole phylum, only the crinoids, and sea urchins appear abundantly enough to furnish a real history of their changes. The crinoids were present but not common in Ordovician, increased in numbers and variety through the succeeding periods, and reached their maximum (about 600 known species in North America) in the early Carboniferous. They were reduced in numbers but began to be radically changed in structure to conform to the modern type in Jurassic, after which the new type became in turn abundant. Free-swimming species first appeared in Jurassic and were locally abundant. Since that time they have been relatively unimportant. The sea urchins which appeared rarely in Ordovician, did not become abundant until Jurassic and Cretaceous. They are today the commonest of echinoderms. Two striking changes in their structure are demonstrated by the fossils. First, modern sea urchins have 20 rows of plates in their tests, or shells, while all Paleozoic members of the group have either more or fewer rows. Second, forms showing right-and left-sidedness appeared first in Jurassic; all earlier species had been radial. Figure 12.6 right, shows both of these features in a Gretaceous fossil.

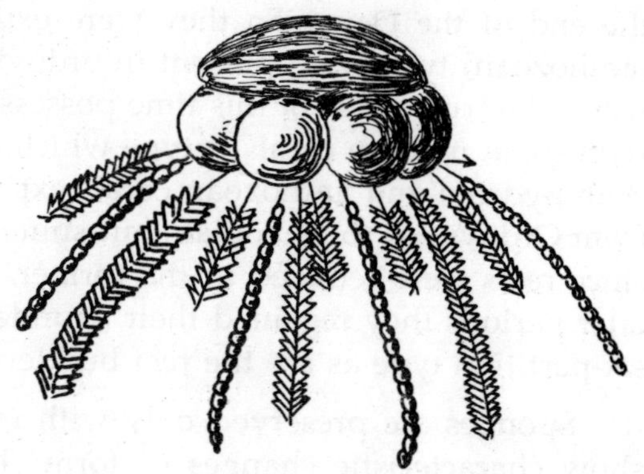

Fig. 12.7: Fossil graptolite from Ordovician.

Other Invertebrates

Of the remaining invertebrate phyla, the ones most commonly met with in marine deposits are the coelenterates, sponges, and protozoa. In the Burgess shale, where some of the most remarkable preservations of ancient life occur, there are unmistakable impressions of jellyfishes. The graptolites reached their maximum in the Ordovician, in forms arranged in rows on both sides of the stem. By Silurian time they were much reduced and were mostly those forms with only one row of individuals; and before

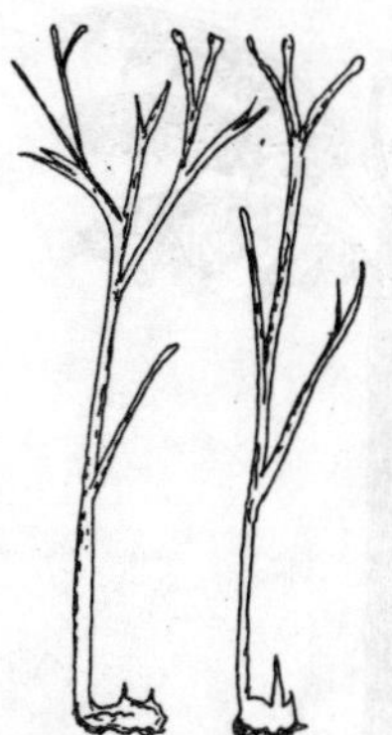

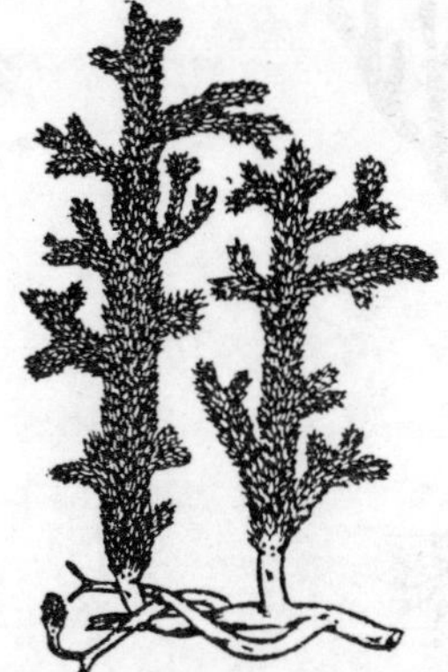

Fig. 12.8: Devonian land plants.

the end of the Devonian they were extinct. Corals of both the hydroid and the true (anthozoan) type were present in ordovician, the former abundant enough to produce reefs. The true corals of this time possessed partitions in multiples of four, as contrasted with six in modern corals, a type which did not make its appearance until the Permian. Both hydroid and anthozoan corals expanded rapidly in species rather than in genera, from Ordovician to Devonian, but suffered reverses in the Carboniferous and Permian, since reefs were reduced in the former, and wanting in the latter, of these periods. In later periods they regained their abundance, but the true corals were then of the basic six-partition type as are the reef builders of the present.

Sponges are preserved only with respect to their skeletons of spicules, and these show characteristic changes of form. The protozoa are represented only by shelled forms mostly Foraminifera and Radiolaria, which constituted so small a part of the probable protozoan life of any period as to make the history of their changes unimportant for our present purpose.

Land Plants

It is not to be expected that plants would be well preserved until they possessed woody structures. The early Devonian land plants were mostly devoid of leaves, or their leaves were mere scales. One important group of that period, the pteridosperms, was long regarded as belonging to the ferns, but it is now known that they bore seeds, as modern ferns do not. Stems of some of these were 3 feet in diameter. The lycopods, or club mosses, whose modern examples are such small plants as the ground "pine" and the creeping Selaginella, were then trees, up to 30 feet in height. Trees resembling the pines, firs, and spruces also existed then, but they did not possess needlelike leaves.

The giant treelike club mosses were very abundant in the Carboniferous and grew in all the northern continents; but by Permian time they were rare and were missing from the Triassic. The horsetails of the upper Carboniferous were trees with transverse joint, some with branches, some without. They were still abundant and large, though

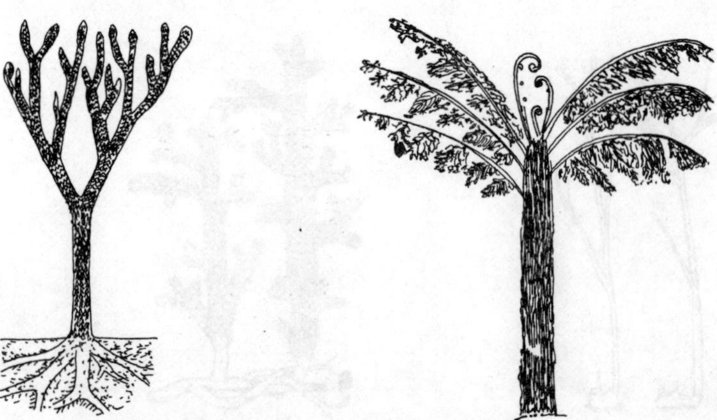

Fig. 12.9: Giant club moss (left) and tree fern from upper Carboniferous.

largely of different genera in Triassic and Jurassic. Since then they have gradually taken on the characteristics of the modern members of the group, the scouring rushes, which are reduced to one genus and some twenty-five species, very few of which grow over several feet in height.

The pteridosperms, referred to above in the Devonian as having been regarded as ferns until they were found to be seed bearers, reached their maximum development in the late Carboniferous. They were still abundant in Permian, much less so in Triassic, and after that disappeared. The ferns were common, however, all through late Paleozoic, uncommon in Triassic, but abundant again form Jurassic to Eocene. After that they were mostly replaced by modern plants.

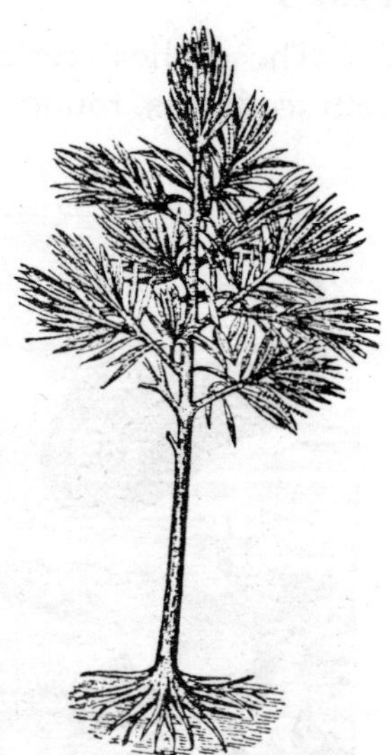

Fig. 12.10: Large-leaved late Carboniferous evergreen, Cordaites.

The Devonian "conifers" had straplike leaves, rather than needle-shaped. The same kinds of leaves are found on those of the Carboniferous. They are classed with the conifers, though their seeds, instead of being borne in cones, were situated on small budlike branches. Cones first appeared in late Carboniferous and early permian, on a tree that had narrow leaves. The splendid petrified trees of the far southwestern states were conifers of Triassic time; some of them show annual growth rings, indicating seasonal differences of climate, though perhaps not temperature extremes.

A few true cycads occur in early Triassic, many more in late Triassic, mainly represented by their leaves. In Jurassic they became so abundant that this period is often known as the "age of cycads." In late Cretaceous, however, they were largely over-whelmed by the flowrring plants, and today they constitute an unimportant group of which the sago "palm" is the most familiear example. Somewhat similar to the cycads are the ginkgoes. Appearing first in permian time, they became abundant in Triassic and especially Jurassic. With the coming of the flowering plants in Upper Cretaceous, the ginkgoes too receded in importance, and the group is today represented by only one species, widely cultivated as an ornamental plant and probably nowhere wild.

The true flowering plants appeared suddenly in such abundance and variety in late Cretaceous that it is generally assumed they originated much earlier, though little fossil evidence of them in earlier periods has been obtained. They must have migrated into the places where they are found, from other regions in which they either were not preserved or which have not been explored, possibly from arctic areas. It is the woody or otherwise hard plants that are preserved, and the late Cretaceous forests include oaks, beeches, maples, elms, poplars, magnolias, tulip trees, and palms, and grasses are also abundant in the open places. The forests were thus much like those of the present.

Fishes

The earliest certainly vertebrates are Ordovician fishes of the group known as Ostracoderms, found in Colorado, Wyoming, and South Dakota. These armored fishes, mostly of small size, reached their maximum abundance in Devonian and perished in that period. The lampreys appeared in Silurian time, as did also the sharks, and lungfishes followed in Devonian. So abundant and varied were these groups that Devonian is called the "age of fishes," There were, however, no true bony fishes (teleosts) in Devonian, the first of these occuring in Jurassic. By Jurassic time the sharks had exchanged sharp-pointed teeth for flat-topped ones, presumably used for crushing shells. Bony fishes gained the ascendancy in Gretaceous, and by early Tertiary they were of the families living today.

Fig. 12.11: Small armored fishes, Ostracoderms, from Lower Devonian.

Amphibia

The history of the amphibia is so incomplete that the chief reason for including them here is to indicate the extreme difference between the only fossil group that is known to have been abundant and the modern types. By late Carboniferous there were 88 species, belonging to 46 genera. These animals were mostly of the group called Stegocephalia, or similar to them. They possessed a tail, an armored head and belly, and short legs or none at all, and varied in size from a few inches to 10 feet in length. They became extinct in Triassic. Real salamanders, unarmored, date from the Cretaceous. The oldest known frogs lived in the Jurassic of Spain and Wyoming.

Reptiles

The reptiles first became prominent in permian. Most of them looked like large lizards, some of them with a large finlike projection upon the back. Even more extraordinary was the array of reptiles in the Mesozoic, which has been termed the "age of reptiles." It is the dinosaurs which are the conspicuous members of the group, owing to the huge size which some of them attained. Not all were large, however; they ranged from the size of a hen to 150 feet in length. The huge bodies bore ridiculously small heads. Some dinosaurs were vegetable-feeders, and these were often horned; they are found only in the Cretaceous. Others were carnivorous, and these were usually hornless. The dinosaurs as a group appeared first in middle Triassic, a little earlier in America than in Europe. By Jurassic they had become very numerous and varied. Armor plates curved over the head and neck; spines, plates set on edge along the back or along the sides, were common extravagances of structure. Some walked only on their hind legs, the forelimbs being greatly reduced. Mummified skins and dried carcasses show them to have been covered with scales as are modern reptiles. That they laid eggs is shown by fossil *eggs* discovered in Mongolia. Before Cretaceous many of the large dinosaurs had died out, and very few of the group—none in North America-survived to the Eocene. New groups kept appearing while the order lasted, for the spoon-billed dinosaurs began and ended in the Cretaceous.

Other curious reptiles of the Mesozoic were the ichthyosaurs, or fish-lizards, of the Jurassic and Cretaceous; plesiosaurs, marine predaceous swimmers of about the same time; and pterosaurs, the flying reptiles, whose wings consisted of skin stretched between the body of the hand and a greatly elongated fifth finger.

The earliest known turtles and tortoises lived in Triassic time, true lizards first appeared in jurassic, and the first snakes are found in the Cretaceous. These are the groups that have survived most plentifully to the present time. Tortoises were abundant in Eocene and huge in Miocene, and lizards increased considerably in Oligocene. Venomous snakes did not arise until mid-Tertiary.

Fig. 12.12: Archaeopteryx, fossil bird of Upper Jurassic.

Birds

The earliest fossil birds are of Upper Jurassic. They had teeth, and their beaks were not covered with horny material. They were like reptiles in these respects, and since some small reptiles were fliers, these Jurassic skeletons are assigned to the birds only because they were accompanied by impressions of feathers. The famous Archaeopteryx, about the size of a crow, with only a small breastbone (hence not a strong flier), typical perching feet, and no air-filled bones, was the principal representative from that period. Archaeornis, and a feather different from those of both these genera, are the rest of the Jurassic discoveries.

In Cretaceous, Ichthyornis was a strong flier, and Hesperornis a swimmer and diver. These forms still had teeth, though some anterior ones of the upper jaw were missing. A horny bill was also partially developed. A completely toothless lower jaw is known from the Cretaceous of Alberta, Canada.

In Eocene, most of the birds were of modern type. Not many have been preserved from later periods, so that the fossil history of birds is meager.

Mammals

Among the reptiles in the Triassic were some whose teeth were no longer of the unspecialized type characteristic of that group. These teeth showed differentiation approaching that of canines, incisors, and molars, and there were other differences in the skull which bore resemblance to features of mammals. Such differentiation increased in the Jurassic, and some undoubted mammals, of the size of large rats, existed in that period. Late in the Cretaceous were two types whose skulls resemble somewhat those of opossums and shrews. Through the whole Mesozoic, however, mammals were subordinate to the reptiles.

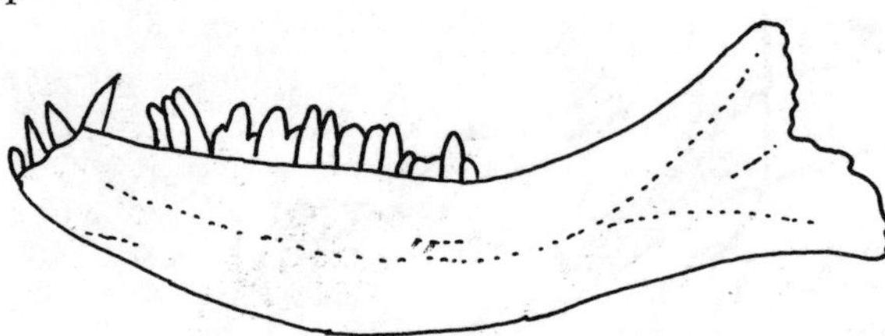

Fig. 12.13: Lower jaw of mammallike Triassic reptile, showing some differentiation of teeth.

Then followed the great development which made the Cenozoic the age of mammals. It brought a succession of forms which for size and sheer oddity almost equaled that of the Mesozoic reptiles. Seven orders of mammals are known from the Paleocene, and a few of these groups survive; the American opossum of today has changed little from its primitive appearance. Rapid changes took place, however, in Eocene time. The huge uintatheres with their curious head protuberances and curved tusks were very common. Other mammals were the horse and camel ancestors, the rodents, rhinoceroses, primitive elephants, and primitive monkeys.

Only a few of the later developments can be given in this brief account. The saber-toothed cats, with their 6-inch flattened upper canine teeth, must have been a terror around the water holes which other animals had to visit. They lived in temperate North and South America in the Pleistocene period. The ground sloths were giant immigrants from South to North America in Pliocene and Pleistocene. They must have walked clumsily on the outer edges of their feet, with their long claws bent inward. Some huge armadillos, also northward migrants from South American in Pliocene and

Pleistocene, reached a length of 14 feet. Rhinoceroses that could run with agility existed in Eocene and Oligocene, while another rhinoceros of the Oligocene and Miocene of Asia grew to the largest size of any known mammal—25 feet long, 13 feet high at the shoulders. A deerlike animals with protuberances on the front part of its nose as well as horns on top of its head is found in the early Miocene of North America.

Fig. 12.14: Eocene uintathere from Wyoming.

These curious examples are but a few of the strange creatures which made up the group of mammals during Tertiary time. In late Pleistocene they experienced a drastic and mostly unexplained destruction. Perhaps the glaciers had something to do with their disappearance. Seldom in geological history have there been comparable exterminations of great groups. Only the tropical regions, notably Africa, escaped this great destruction, and Africa's Pleistocene mammals were essentially like those of today.

The Tree of Horses

Omitted from the foregoing story are several series of changes which are especially convincing as evidences of evolution because the steps involved are small. A classical example is that of the horse.

The first mammal now recognized as horselike was Eohippus, or its near-equivalent Orohippus, in the Eocene. These animals were scarcely a foot high, and an arched back (not the sagging one of the modern horse), and had four digits on the forefeet and three on the hind feet (each digit ending in a hooflike nail). Their molar teeth had hills and valleys on their functional surfaces, which prevented them from slipping sidewise on one another. Such animals could not therefore eat grass, which can be sufficiently broken up only by a grinding movement; they must have eaten twigs or

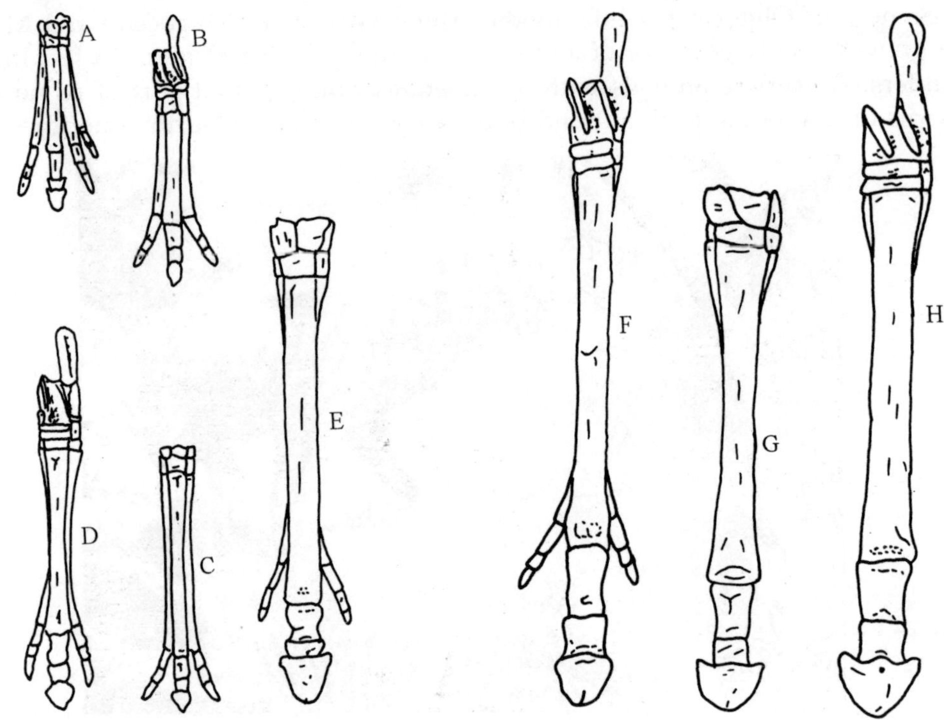

Fig. 12.15: Forefeet and hind feet of fossil horses. A, B, Orohippus; C, D, Mesohippus; E, F, Merychippus; G,H, Equus.

other vegetable matter thick enough to be crushed by mere pressure. The molar teeth also had roots that were longer than the height of their crowns.

Successive horselike animals differed from these Eocene forms in size, in the number of toes, and in the characters of the molar teeth. Four types, belonging to different geological periods, are ample to illustrate the difference.

The increase in size, to that of the 5 foot modern horse, is reflected in the sizes of the feet, where all types are drawn to the same scale. The toes are reduced in number, first to three on each foot in Mesohippus (Oligocene) with the lateral toes still partly functional, then to three with only the middle one functional in Merychippus (late Miocene and early Pliocene). The genus Equus, both in Pleistocene fossils and modern horses, has the lateral toes still further reduced to splint bones.

In the form of the molar teeth, Merychippus is transitional. Before it, the surfaces of the molars were rough; in Merychippus they were flattened enough for grinding movements; and in Equus the flattening is accentuated. In Merychippus the length of the roots and the height of the crowns are abut equal; before it the roots were the longer, after it the crowns were the longer.

The members of the horse series have been found mostly in North America, but the line became extinct there long before the coming of white men to this continent. Horses had migrated to Europe, however, and their descendants were brought back to the Americas as domesticated animals. Wild horses are the descendants of some which escaped from domestication or were abandoned.

Other Mammals

Some of the changes characteristic of the horse have parallels in successive fossils of other mammalian types. Increase in size has been fairly common. Such increase occurred in the camel and elephant series, the only others to be considered here. Molar teeth with merely crushing surfaces have often been succeeded by grinding (flat-topped) teeth; this is true of camels and elephants. The number of toes decreased in the camel line, but only from four to two. A striking change in the teeth of elephants is the conversion of the outer upper incisors to tusks weighing several hundred pounds in some forms. The great weight of these tusks requires greater leverage for the muscles of the neck, and this is provided by greater height of the base of the skull, attained by added amounts of spongy bone.

The history of the "camels" refers to the camel family, which embraces not only the Arabian one-humped dromedary and the Asiatic two-humped camel, but also the llamas and alpacas and their wild ancestors of South America. The camel family began in late Eocene in North America, mostly the United States, which has none of them

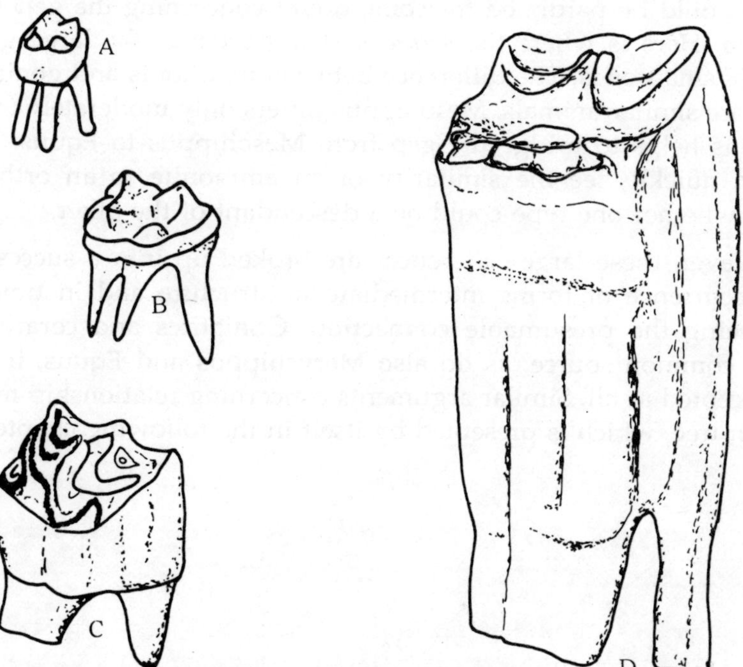

Fig. 12.16: Molar teeth of fossil horses. A Eohippus; B, Mesohippus; C, Merychippus; D, Equus.

now except in circuses and zoological gardents. No other continent received any of them until Pliocene time—at least none has been preserved and discovered.

The story of the elephants is strictly African through Eocene and Oligocene. Then these animals spread to Europe and Asia, even to North American, in early Miocene. In this continent they were the mastodons, whose skeletons are still being discovered.

For the details of these and other mammalian lines—or those of other animals—reference should be made to works emphasizing the course, rather than the causes, of evolution.

Significance of Fossil Series

The advantage of such series as those of cephalopods, horses, camels and elephats, as indications of evolution, is that the successive changes are small. It is a weakness of paleontological evidence that a genetic relationship can never be directly proved. However reasonable it may be to assume that one fossil type descends from another or that two contemporary types are cousins that is still an assumption. Biologists assume these relationships when they can see the similarities which seem to justify such a conclusion. They are doubtless joined by nonbiologists when the similarities are as close as those between living Nautilus and the fossil coiled nautiloids, or between living brachiopods and fossil Lingula. But when the layman is presented with the little elephantlike Moeritherium and a modern elephant, or with Eohippuus and a living horse, he could be pardoned for some doubt concerning the genetic connection between the two. Here is where the *series* is of importance; the large gap is divided into a number of small ones. The difference between nautiloids and goniatites is small; they obviously are similar animals. Mesohippus differs only moderately from Eohippus, and Merychippus helps to bridge the gap from Mesohippus to Equus. No one but a biologist would quickly see the similarity of an ammonite to an orthocone; others might question whether one type could be a descendant of the other.

When, however, these large differences are broken up into a succession of small ones, by the occurrence of forms intermediate in structure and in time, there is no difficulty in seeing the presumable connection. Goniatites and ceratites give plain indication of a common source, as do also Merychippus and Equus, if the theory of homology is accepted at all. Similar arguments concerning relationship may be derived from the human tree, which is presented by itself in the following chapter.

Chapter—13

Vertebrates of Paleozoic Era

Geologic Periods

The conventional subdivisions of geologic time—eras and periods—have, in general, been established on the basis of events in the marine waters of Europe and North America; and boundaries between periods were set up at points where there appeared to be major breaks in the record of deposition or marked changes in the invertebrate life. But it is obvious, a priori, that a break in the record of sedimentation in one region by no means proves that a similar cessation will appear in others, and it becomes increasingly apparent that sharp lines of demarcation do not in reality exist. Furthermore, a comparatively sudden change in marine invertebrates by no means implies a marked simultaneous change on the continents, and many of our most interesting vertebrate faunas lie on the border line between periods or even eras. The earliest Devonian vertebrates, for example, are more like those of the late Silurian than like later Devonian fishes; the late Pennsylvanian tetrapods are almost identical with those of the Lower Permian. We cannot, however, readjust the period boundaries to suit the vertebrates alone and we may retain the current nomenclature while recognizing its artificiality.

Correlation of Deposits

In assigning an age to vertebrate-bearing beds, the evidence obtained from associated invertebrates is exceedingly useful. Many of the more important deposits, however, are of continental origin and contain almost no invertebrate remains. Here recourse in often had to a direct comparison with vertebrates of other regions. If two series of deposits contain a considerable number of identical or closely related forms, it is highly probable that they are contemporaneous. But if the forms compared are few in number or but distantly related, the argument for contemporaneousness is greatly weakend.

Negative Evidence

Any contrast between the life of different periods or different geographical areas is based upon the presence at one time or place of animals apparently absent at an earlier or later date or from another region. But it must be pointed out that, while the presence of an animal may be definitely shown, absence is incapable of absolute proof; all that we can really say is that remains have not been discovered. Until recent decades no placental mammal had been identified in collections from the Cretaceous.

Table 13.1
Table of Geologic Periods

Era	Period	*Approximate Time (In Millions of Years)*	
		Duration	*Since Beginning*
Cenozoic (age of mammals and man)	Quaternary	1	1
	Tertiary	69	70
Mesozoic (age of reptiles)	Cretaceous	50	120
	Jurassic	35	155
	Triassic	35	190
Paleozoic (age of invertebrates and primitive vertebrates)	Permian	25	215
	Pennsylvanian } Carboniferous	35	250
	Mississippian } Carboniferous	50	300
	Devonian	50	350
	Silurian	40	390
	Ordovician	90	480
	Cambrian	70	550

But our knowledge of later mammalian evolution led us strongly to believe, despite this negative evidence, that there must have been Mesozoic placentals. Today, principally owing to work in Mongolia, placentals are known in the Cretaceous. Negative evidence of absence has been replaced by positive knowledge of the presence of these forms. Again, no trace of coelacanth crossopterygians has ever been discovered in Cenozoic strata; and it was long believed, with confidence, that the coelacanths had become extinct at the end of the Cretaceous. The recent discovery of a living specimen has rudely shattered this belief.

It is probable that the range in time or space of many forms with which we are familiar was greater than is now known. It is further probable that there may have existed many interesting vertebrate types of which no remains have as yet been discovered. But such probabilities are decreasing as our knowledge expands. During the present century a very great amount of paleontological work has been done, and many strange forms have been brought to light. These, however, have been almost always members of groups already or forms tending to connect such groups.

Environmental Differences

In a study of past faunas the great variety of possible habitats must be taken into consideration. Often the structure of an animal gives much evidence concerning its environment. The teeth may afford a basis for distinguishing between forest-living browsing types and plains-dwelling grazers; the limbs may show us whether we are dealing with an aquatic, terrestrial arboreal or aerial form. The nature of the sediments and the associated plants and invertebrates may give us important additional information as to its surroundings in life or death.

Among fishes the question of a fresh-or salt-water habitat is often of great interest. If a fossil types is found in strata in which marine invertebrates do not occur, it is not unreasonable to assume that it inhabited fresh waters. If it is accompanied by such invertebrates, it may well be a marine form; but this is none too certain if the association is a rare one or if the vertebrate remains are badly broken, for individuals of a river-dwelling type may drift out and become buried in the sea.

Among land animals, distinctions may be made between lowland and upland types, forest and plains dwellers. Here an unfortunate difference in the probability of preservation appears. Wooded areas are obviously regions in which but little deposition of sediments takes place and in which skeletons are usually rapidly destroyed by insects or by acids in the soil. Hence our knowledge of the evolution of arboreal and forest-dwelling animals is relatively meager; in contrast, deposits are common in plains areas and skeletons often preserved. Of the evolution of primates and racoons we know little; horses and dogs, mainly dwellers in open country, are abundant as fossils. Deposition has been common in coastal lowlands, but highland regions are areas of erosion; in consequence, our knowledge of the evolution of upland forms is slight.

Climates

At the present time differences in temperature are a major factor in the distribution of vertebrate life. Many groups are confined to the tropics, others to arctic or temperate regions. This zonal distribution forms an effective check to migration. Penguins presumably would flourish in the aretic regions, polar bears in the south; but they cannot cross the tropics. The animal life of Old and New World tropics differes greatly, for intermigration in the more recent geological epoch necessitated a passage through the artic zone. Similar situations may have strongly influenced the migration and distribution of vertebrates in the past.

Ancient climatic conditions however, appear to have varied greatly. There is considerable evidence suggesting warmer climates at times in the past in regions now in the temperate or arctic zones, while ancient glacial deposits are known in the present tropical zone. Presumably, there have been, throughout geologic history, climatic rhythms—times in which conditions were mild and equable have alternated with periods of sharply zoned climates. There are however other possible explanations. Some have suggested a wandering of the poles and consequent readjustments of climatic bands, others an actual movement of continental areas.

The vertebrates themselves offer evidence as to the climates in which they lived. Living "cold-blooded" reptiles and amphibians cannot exist in arctic or subarctic regions, and most of the extinct ones may have had similar limitations. But this type of reasoning must not be carried too far. Elephants and rhinoceroses today are tropical animals; but we know that, in the Pleistocene, hairy types closely related to living forms in both groups existed near the glacial front. The evidence of plant and invertebrate life is important in determining climatic conditions. Plant types now characteristic of the tropics once grew in Greenland. It is probable that this indicates a former warm climate there; but it is possible that plant habits have changed. The sediments themselves commonly suggest the nature of the climate under which they were formed. For example, striated rock surfaces and deposits similar to those made by recent glaciers are indicative of ancient ice ages. But the evidence of sediments is not always beyond doubt. Red beds, for example, were long thought to indicate aridity; but red deposits are now being laid down at the mouth of the Amazon and in other warm regions of abundant rainfall.

Migrations

No matter how potentially successful a newly evolved forms may be, obstacles are often present which prevent its invasion of regions in which it might flourish. The greater part of the world has always been covered by seas, and migrate is comparatively easy for marine forms. Even here, however, barriers may exist, due for example, to differences in water temperature or salinity. For terrestrial forms there are numerous types of barriers. Forests, deserts, and mountains. The well as temperature differences, are often effective bars to migration. The greatest obstacle, however, is the lack of a land connection, the interposition of a broad body of water between continental areas. Many oceanic islands have no native land vertebrates of any sort; no terrestrial mammals has ever reached New Zealand.

Viewed on the ordinary Mercator map, intercontinental migration appears a complicated process. On a north polar projection however, the situation is much simplified. Europe and Asia form a unit, and North America is separated from them only by a narrow strait across which dry land is belived to have stretched at many times. From this northern land mass depend three great southern peninsulas, South America, Africa, and Australia. The first two are connected with the north by narrow isthmuses believed to have been under water in earlier times, while Australia is separated by an island-studded region which may once have been dry land. Many of the facts of vertebrate distribution in past and present times are readily explained by a make or break of these four connections.

Land bridges have often been built in the minds of scientists to account for certain facts of distribution. Some of these structures are quite fanciful, made in an attempt to explain the real or supposed resemblances between one or two animal types in two disconnected regions but disregarding the divergences between the two regions in other animal groups. Others, however, deserve more serious consideration. This is especially true of bridges advocated between the three southern continents. The presence of carnivorous marsupials in Australia and in the South American Tertiary has suggested the possibility of a former land connection between these two areas. South America and Africa show resemblances not only in some mammalian types but also in some groups of fresh-water teleostean fishes and in the presence of related lungfish genera, and a connection between them in the early Tertiary has been strongly advocated. Elsewhere it has been pointed out that many of these facts may be accounted for by the former presence of the common types or their ancestors in northern areas from which these continents might have been reached without any major change in land areas. One must not "erect" a bridge for the convenience of one group of animals without taking into account its effect upon other elements of the fauna. For example, hystricoid rodents with relatives in Africa appear suddenly in South America in the Oligocene. A transatlantic land bridge would afford easy access for the rodents and an easy solution for the problem concerning them. But in both continents there were numerous other animals that one would think equally capable of crossing such a bridge. There is however no evidence that they did cross. Land bridges that transmit but one type of animal, and in but one direction, may have existed; but their existence is highly improbable.

Such bridges, however, may have been somwehat selective in their nature in many cases. The present "bridge" between the two Americas has allowed the passage of many animal types. Among the ungulates, deer, peccaries, tapir, and camelids have made successful southern migrations, but the plains-loving bison never penetrated beyond Honduras into this region of tropical jungles and mountains. The Behring Straits connection between Asia and Alaska was a well-traveled highway of migration in the Pleistocene. But warmth-loving animals, such as most of the antelopes, the giraffes, and hippotami, were unable to cross.

"Island hopping" is a partial solution for some problems of migration. Asia and Australia have not been in direct contact, it seems certain, at any time during the Cenozoic. Between them, however, stretch the East India, separated by short gaps, certain of which may have been, at times, dry land. Scattered down the island chain one finds a decreasing percentage of placental mammals of Asiatic origin. Certain rat types have made the whole series of "hops" to Australia; a few marsupials have worked part of the reverse journey.

It seems certain that land animals do at times cross considerable bodies of water where land connections are utterly lacking. The West Indies and Madagascar, for example, have sparse mammalian faunas which are composed of a few odds-and-ends of the numerous mammal groups found on the neighboring continental areas. Obviously, these forms have reached their island homes by some chance type of oversea transport. Floating masses of vegetation, such as are sometimes found off the mouths of the Amazon, may be one means of effecting this type of migration. Even the case of the entry of the hystricoids into South America may be a case of this sort. The South Atlantic is broad, and the chances of a crossing very poor; but Tertiary time was long, and one successful crossing might populate a continent.

A majority of geologists believe that the existing continents have preserved their identities and locations through a vast period of time, and this has been tacitly assumed above. Many workers, however, incline to a belief in floating continents; it is suggested that the land surface of the earth once constituted a single mass, from which America has broken away westward and Australia spread to the east. Most of the continental connections suggested by this theory are quite unnecessary for the explanation of vertebrate distribution, and some of the Cenozoic continental relations assumed are at variance with the facts of vertebrate history. But, in general, the hypothesis is neither strongly supported nor disproved by vertebrate evidence, and its truth or falsity must be decided from other lines of evidence.

CAMBRIAN

During the Cambrian, earliest of the Paleozoic periods, almost every major type of invertebrate life makes its appearance, and many groups have already become so highly organized and diversified that we may be sure that they have already been in existence for a long period of time. But vertebrates appear to be absent from the record. This might perhaps be accounted for by the fact that the probable ancestors were soft-bodied forms which we should not expect to find fossilized under ordinary conditions. But even in the case of a remarkable Middle Cambrian deposit in British Columbia in which the shales have preserved very delicate impressions of soft-bodied invertebrate types such as jellyfishes, no chordates of any sort have been recognized. Either vertebrate ancestors were extremely rare or else were absent from the salt waters in which all well-known Cambrian deposits were laid down.

ORDOVICIAN

Ordovician sediments are likewise predominantly marine in origin. Invertebrates are abundant and diversified, but vertebrates have been reported in but one formation, and these are fragmentary remains not giving us any idea of the appearance or anatomy of these forms. In the early Upper Ordovician Harding Sandstone of Colorado and other western states are scraps of material which in microscopic structure are seen to be pieces of the dermal armor of vertebrates, presumably ostracoderms. The sediments are considered estuarine in nature. A reasonable interpretation is that the vertebrate scraps are those of stream-living types whose remains were carried down to river mouths. This is but one of many lines of evidence strongly suggesting that the Lower Paleozoic vertebrates developed in inland waters. Their almost utter absence from the fossil record of the first half of the Paleozoic may be correlated with the fact that almost no continental sediments of such high antiquity are known today.

SILURIAN

For most of the Silurian the sedimentary record is again one of marine areas and abundant marine life, with little or no trace of vertebrates. In the late Silurian, however, began the Caledonian Revolution, a time of earth disturbances and of the building in

Europe of great mountains, of which the Scottish Highlands and the Scandinavian ranges are remnants. The erosion of materials from these mountains resulted in greatly increased continental deposits in northern Europe from Great Britain to Spitzbergen and Russia. In these beds are found the oldest adequately known vertebrate types.

The earlier Upper Silurian deposits of Europe, grouped as the near the Ludlow, tended to change in type from marine to estuarine or even continental sediments. Typical marine invertebrates becme rarer near the top of the series, and there remain only such types as the eurypterids, which may have been fresh-water forms. With this progressive change, vertebrates make their appearance and become increasingly numerous. In western England the Ludlow bone bed, a layer scarcely a foot thick but of wide extent, derives its name from the fact that it is full of tiny vertebrate plates and spines. The latest Silurian of Europe is termed the Downtonian. It consists in great measure of red sediments, "Passage Beds," difficult to distinguish from the overlying Old red Sandstone and so similar to that series that some writers have advocated placing these beds in the Devonian. Numerous vertebrate finds have been made in these "Passage Beds" in western England, southern Scotland, southern Norway, Spitzbergen, and Podolia. In general, the conditions are of a nature suggestive of fresh-water origins. An exception however, is the important locality of Oesel in the Baltic, where vertebrate-bearing layers occur in a limestone series, interdigitated with others carrying a marine fauna; condition here have been interpreted as estuarine. In the America Silurian, vertebrate localities are few in number, and finds are rare; they are for the most part in sediments comparable to those of the Downtonian. There are no records of Silurian vertebrates in any other region.

The only well-preserved vertebrates are, as might be expected, agnathous types. All major groups of ostracoderms are highly developed and diversifed. They appear to have been then at the peak of their career and to have had behind them a long history, of which, because of lack of earlier continental sediments, we know almost, nothing. There are abundant cephalaspid types, such as *Tremataspis* and *Hemicyclaspis,* and numerous primitive pteraspids (*Poraspis, Anglaspis, Cyathaspis,* etc.). Nearly all the known anaspids, such as *Birkenia* and *Pharyngolepis* are from the Downtonian, as are most of the remains of the problematical coelolepids, as *Thelodus* and *Lanarkia.*

No skeletons of jawed vertebrates have as yet been discovered in the Silurian. But such was the variety of these forms in the overlying Lower Devonian that it is certain that they were undergoing development in the Silurian. Proof of this is afforded by the fact that the late Silurian bone beds are filled with spines and plates which probably belong in some measure to forerunners of the acanthodians and arthrodires of the early Devonian. The fragmentary nature of these ramains suggests that they had been washed down from areas farther inland, in which the evolution of jawed fishes was in progresss—the higher reaches of streams in the uplands formed as a result of the Caledonian muntain-building. But it is almost beyond hope that these ancestral forms will ever be found as fossils, since such highland areas are regions of erosion rather than of deposition.

DEVONLAN

Although amphibians had evolved by the end of the Devonian, this period is often, and reasonably, called the "Age of Fishes," The short list of Silurian vertebrate localities was confined essentially to fresh-water deposits of Europe and eastern North America. In the Devonian our perspective widens. As will be seen, new continental areas are added to the list before the close of the period; further, some of the lower fish groups rapidly spread into salt-water environments, and hence marine formations of the later Devonian frequently include vertebrate remains.

Devonian Localities

The classic localities for Devonian fresh-water fishes are those of the Old Red Sandstone of Great Britian, particularly Scotland. These continental deposits embrace nearly the whole extent of the Devonian and are divided into lower, middle, and upper portions. The Lower Old Red is, we have noted, essentially continuous with the underlying Downtonian; it is well represented in southern Scotland and western England. The Middle Old Red, richly fossiliferous, is Scotland and western England. The Middle Old Red, richly fossiliferous, is found in large areas of northern Scotland—the Moray Firth region, Caithness, the Orkneys, and even the Shetland Islands. The Upper Old Red is, in contrast more limited in extent and confined to a number of small areas in England and southern and central Scotland.

Sediments comparable in nature and in faunas are found in many other regions. Lower Old Red equivalents are present in Podolia, in Spitzbergen, and in America at Campbellton, New Brunswick, and Beartooth Butte in Wyoming. Sediments of the type of the Middle and Upper Old Red are still more widespread—Spitzbergen, eastern Greenland, Ellesmere Land, the Baltic States, western and northern Russia, and Australia. Deposits equivalent to upper Old Red are common in continental North America. Scaumenac Bay in eastern Canada is a famous locality of this age, and in New York State there is a great series of Upper Devonian red beds to which the term "Catskill" is often (but loosely) applied. Fish faunas of this age and type are further found in northern and central Asia and even Antarctica and hence extend nearly from pole to pole. It may be noted that these "Old Red" faunas appear to be remarkably uniform in nature; at any given horizon the same genera may be found in areas far removed from one another.

During the course of the Devonian, marine localities become increasingly important for the vertebrate record. The mountains of the middle Rhine in Germany show a series of sediments of this type. In the late Lower Devonian there are here several beds with vertebrate remains, interlarded between purely marine layers and probably brackish in nature. In the Middle and Upper Devonian there are numerous vertebrate localities in western Germany which appear to be typically marine; Wildungen, with an abundant Upper Devonian fish fauna, is the most famous.

Table 13.2
Paleozoic Vertebrate Localities

	European Stage	*Europe*	*North America*	*Other Continents*
Permian	Thuringian (Tartarian)	Sarma fm., N. Dvina R. "Zone IV" (Russia); Cutties Hillock (Scotland); Zechstein of Germany		*Cisticephalus* and *Endothiodon* zones of Beaufort series (S. Af.); Ruhuhu, Tanga (E.Af.); Chiweta beds (Nyassaland); Mangwa (Rhodesia); Bijori beds (India); Upper New-castle Coal Measures (Aus.)
	Saxonian (Kazanian)	Kupferschiefer (Ger.); Magnesian ls. and Marl slate of Durham (Eng.); "Zones I-III" of E. and N.Russia	Coconino, Kaibab (Ariz).	*Tapinocephalus* zone of Beaufort series (S. Af.)
	Autunian (Artinskian, Kungurian)	Rotliegende at Niederhässlich, Lebach (Ger.); Braunau (Bohemia); Autun (Fr.), etc.; War-wickshire (Eng.); southern Norway.	Wichita and Clear Fork beds of Texas and other red beds in N.M. (Abo), Ariz., Okla, Kans, W. Va., etc.; Phosphoria fm. (Wyo.) *M*;* E. Greenland *M*.	Ecca (S. Af.) (barren); Dwyka (S.Af.); Itarare (Brazil); Mhum Mu (Kashmir)
	Stephanian (Uralian)	Nyrany, Kounová (Bohemia); Commentry (Fr.)	Conemaugh, Monongahela, L. Dunkard (Pa., W. Va., Ohio); Danville (Ill).	
Pennsylvanian	Westphalian (U. Muscovian)	English coal fields of Newsham, Yorkshire, Staffordshire, etc.; Kilkenny (Ire.)	Alleghany of Linton (Ohio); Mazon Cr. (Ill.); Joggins (N.S.)	
	Namurian (Lanarkian, L. Muscovian)	Scottish coal fields		
Mississippian	Dinantian-Visean, Tournaisian	Edinburgh coal field-Loanhead, Gilmerton, Airdrie, etc. (Scot.); Calciferous ss. and Cementastones of S. Scotland—Burdiehouse, Wardie, Eskdale, etc.; Limestones near Bristol (Eng.), Armagh (Ire.) *M*.	Mauch Chunk at Hinton (W. Va.); Albert Mines (N.B.) St. Louis, Keokuk, Burlington Is. (Ia., Mo., Ill.) *M*.	

Table 13.2 (Contd.)

	European Stage	*Europe*	*North America*	*Other Continents*
	Famennian Frasnian (Chautauquan, Senecan of N.A.)	U. "Old Red" of Scotland, Shetland, W. Eng.; Baltic States; Russia; Rhineland (Wildungen, etc.) *M;* Boulogne region (Fr.) *M*	"Catskill" of N.Y. and Pa.: Scaumenac Bay (Que.); Colo., Ariz.; Ellesmere Land; E. Greenland ; Cleveland Shale (Ohio) *M;* "Chemung" of N.Y. and Pa. *M;* Cedar Valley (Ia.) *M;* Milwaukee (Wis.) *M;* Hamilton of N.Y., Ont. *M.*	Antarctica; Victoria, N.S. Wales (Aus.); central Asia.
Devonian	Givetian, Eifelian (Erian, Ulsterian an of N.A.)	M. "Old Red" of N. Scotland, Orkneys, Shetlands; Baltic States; Wijhe Bay series (Spitzbergen); Rhineland *M.*	Ellesmere Land; E. Greenland; Onondagan Is. of N.Y., Ont., Ind., *M;* Columbus and Delaware Iss. of Ohio, *M*	N.S. Wales (Aus.)
	Emsian, Siegenian or Gedinnian, Dittonian (Oriskanian, Helderbergian of N.A.)	Podolia (Ukraine); L. "Old Red" of S. Scotland and W. Eng.; Wood Bay and Grey Hoek series of Spitzbergen; Rhineland *M* (in part)	Campbellton (N.B.); Beartooth Butte (Wyo.)	
	Downtonian (Cayugan of N.A. in part)	Red Downtonian "Passage Beds" (Scotland); Temeside shales, Downton ss. (W. Eng.); Podolia (Ukraine); Oesel (Esthonia); Gottland (Sweden); S. Norway; Red Bay (Spitzbergen)		
Silurian	Ludlovian (Cayugan of N.A. in part).	Lanarkshire (Scot.); Whitcliffe Flags, Ludlow bone bed (W. Eng.)	Salina (Perry Co., Pa.)	
	Salopian (Wenlockian, Niagaran of N.A.)		Rose Hill fm. (Pa.); Neperis Hills (N.B.)	
	Valentian (Medinan of N.A.)		Shawangunk (N.Y.)	

* M = Marine

Middle and Upper Devonian deposits similarly marine in nature contain vertebrates in a number of other regions, particularly North America. In the Middle Devonian there are plentiful fish remains in the Columbus and Delaware Limestones of Ohio and their equivalents in other states. In the Upper Devonian, the "Chemung" deposits in western New York and Pennsylvania contain marine vertebrates. The black Cleveland shales of Ohio, which lie close to the Devonian-Mississippian boundary, are notable for the variety of their vertebrate remains and for the delicate preservation of some of the sharks found in them.

Lower Devonian Fishes

The fish fauna of the Lower Devonian differs markedly from that of later portions of the period and is in great measure simply a continuation of that seen in the late Silurian. Ostracoderms are still fairly abundant, although not so numerous or so varied as earlier. There are a number of genera of the *Cephalaspis* and *Pteraspis* groups; anaspids and coelolepids, however are relatively rare in the Devonian.

Jawed fishes are now added in abundance to the faunal assemblage. There are, however, no Chondrichthyes identified and, except for one rare forerunner of the crossopterygians *(porolepis)*, no higher bony fishes. The early Devonian gnathostomes are predominantly placoderms. They are already diversified, and two groups are plentifully present. Acanthodian "spiny sharks" are now, in their first definite appearance, at their height in variety and abundance in such genera as *Parexus*, *Climatius*, etc. A second common placoderm group is that of *Arctolepis* and other primitive arthrodires with enormous pectoral spines.

In general, there is little reason to believe this fauna to have been other than a fresh-water one. In the Rhineland Lower Devonian, however, we find as assemblage of fishes suggestive of a change to, or at least toward, a marine life. Some of the normal "Old Red" elements are present, but, in addition, there are the curious armored but sharklike stegoselachians, *Gemuendina* and *Stensioella;* an odd, arthrodire-like type, *Lunaspis;* and an abnormal heterostracan, *Drepanaspis*. These forms seem to be representative, in part, of the ancestors of the marine forms of the later Devonian.

Later Devonian Fresh-water Fishes

Although deposits of "Old Red" type continued without interruption into middle and late Devonian times, their fauna changed greatly. Except for a few rare cephalaspids and anaspids, the once abundant ostracoderms have disappeared. The placoderm assemblage has changed in character. Acanthodians persisted through the Devonian—indeed, into the Permian—but are then and thereafter relatively rare types. Arthrodires are present in these continental sediments but are not abundant. *Coccosteus* is an "orthodox" arthrodire of advanced type; *Homostius, Heterostius*, and *Phyllolepis* are depressed forms, perhaps indicative of brackish-water conditions; ptyctodonts, such as *Rhamphodopsis*, are rare in continental deposits. *Palaeospondylus*, a tiny form perhaps belonging to the placoderms, is known from but a single quarry in Scotland but is very abundant there.

New additions to the placoderm group are the antiarch genera, such as *Asterolepis, Pterichthyodes*, and *Bothriolepis*, which were among the most common fishes of middle and late Devonian times.

The sharp decline among these lower fish groups (antiarchs excepted) is apparently due to the sudden appearance, in great abundance and variety, of the Osteichthyes, the higher bony fishes. In the Middle and Upper "Old Red" and its equivalents, there are numerous rhipidistian crossopterygians (such as *Osteolepis* and *Eusthenopteron*) which were the dominant predaceous forms of their times, and rare primitive coelacanths. Dipnoans flourished in such genera as *Dipterus* and *Scaumenacia,* similar in externals to their crosopterygian cousins but already specialized in skull and dental structure. Present, too, were primitive actinopterygians—the palaeoniscoids *Cheirolepis, Stegotrachelus,* and *Rhadinichthys*—but, in contrast to their later abundance, they were rare until late in the period.

This marked change in the vartebrate assemblage—this sudden appearance of the Osteichthyes—is an important milestone in vertebrate history. Only a few fragments of a single rare form were present to represent this entire class in the Lower Devonian, and yet, at the time of their Middle Devonian incursion, the various types were already quite distinct, with implications of a long antecedent history. Where their evolution took place—in upper river reaches or in some unknown continental area—is a major question to which we have no ready answer.

Marine Life

The seas of the later Devonian witnessed an evolutionary development among fishes quite different in many respects from that seen in inland waters. The Agnatha are rare, although the heterostracan *Psammosteus* and its relatives persisted in presumed estuarine conditions, and although we must, further assume the survival of degenerate agnathous forms ancestral to the modern cyclostomes. Placoderm history in the seas was a different story from that of streams. The presence of marine acanthodians in the late Devonian is suggested by numerous spines of appropriate character, and the stegoselachians persisted in later marine deposits. *Lunaspis* of the Lower Devonian was antecedent to *Macropetalichthys* and related forms, widespread in mid-Devonian limestones. The antiarchs appear to have, in general, kept to fresh waters. Arthrodires, however, flourished greatly in the Devonian seas. In the Upper Devonian, particularly in the Rhineland and the Cleveland shales, are found a great number of advanced arthrodires, varied in structure and varied in size from forms of modest proportions to giants like *Dinichthys.*

Parallel to the deployment of the Osteichthyes in inland waters was the rise of the Chondrichthyes in the seas. The sharklike fishes are unknown in the Lower Devonian, although the stegoselachians suggest the nature of their evolution, and even in the Middle Devonian we can identify them only from a few teeth of the primitive *Cladodus* type. By the end of the period, however, primitive sharks—*Cladoselache* and *Ctenacanthus*—were highly developed; their structure is well shown by specimens in the Cleveland shale, which have preserved in remarkable fashion even such soft and microscopic structures as muscle-fiber striations. In late Devonian deposits we further find teeth showing the appearance of the bradyodont mollusk-eaters and possibly the pleuracanths.

Fish Evolution in the Devonian

The close of the Devonian witnessed the decline or disappearance of many important fish groups. The ostracoderms were already reduced in numbers early in the period and did not survive its close. Antiarchs and arthrodires, although numerous in the Upper Devonian, disappeared before the opening of the Carboniferous. In the space of a single period, almost every fish type present in the Lower Devonian was wiped out except for a few acanthodians.

The more important events in Devonian fish history include (1) the decline and fall of the ostracoderms; (2) the rise and decline of the acanthodians; (3) the rise and disappearance of the arthrodires; (4) the appearance in the middle Old Red of the flourishing antiarchs and their extinction at the close of the period; (5) the rise of sharklike forms in the late Devonian; and (6) the sudden appearance in the middle Old Red Sandstone of the bony fishes.

For completeness, this body of knowledge leaves much to be desired. A chronological treatment tends only to throw into stronger relief the fact that we know almost nothing of the origin or real relationships of most of the early fish groups.

Amphibians

It was noted in earlier chapters that the Devonian was probably a time of marked seasonal droughts; much of the success of the bony fishes (and perhaps the antiarchs) may have been due to their possession of lungs and other drough-resisting adaptations. The amphibians appear to have been, in their initial period, merely another fish group which had developed terrestrial locomotion as a further device helpful under Devonian conditions of life, and their early development from the crossopterygians occurred in that period. The first stages in their history are, however, poorly known; we possess only a skull roof *(Elpistostege)* from the upper Devonian of Scaumenac Bay, which is that of a very fishlike amphibian or very amphibianlike fish, and, from a Greenland formation close to the Devonian-Carboniferous border several skulls of the primitive ichthyostegid amphibians.

CARBONIFEROUS

The time covered by the deposits frequently grouped as the Carboniferous Period is very great. In North America a division into two periods—Mississippian and Pennsylvanian—is now customary, for there is generally a decided contrast between a lower division in which marine limestones predominate and an upper one characterized by coal seams and shallow water deposits. Since the fish faunas which still constitue the bulk of vertebrate remains are not sharply contrasted, we shall here consider the Carboniferous faunas as a unit.

Vertebrate Localities

The Lower Carboniferous, or Mississippian, is precominantly marine and is represented, for the most part, by thick series of limestones. These frequently contain rich invertebrate faunas; less frequently they contain the remains of marine fishes, usually in the form of isolated spines or teeth. Formations and localities of this sort include the St. Louis, Burlington, Keokuk, and other limestones of Mississippi Valley; a rich horizon in the Carboniferous limestone series near Bristol, England; a similar formation a Armagh in northan Ireland; and numerous other limestone localities in western and central Europe and Russia.

Continental deposits are, on the other hand, rare. In North America there are few localities of this sort; exceptional are those at Albert Mines, New Brunswick, where large numbers of a few species of fresh-water fishes are present in oil shales and at Hinton, West Virginia, where a few fishes and amphibians have been recovered from a late Mississippian shale. In other continents there is likewise an almost complete lack of Mississippian fossils of continental origin. Scotland forms the sole exception. In the southern part of that country the Lower Carboniferous is highly developed. Some limestones are present, but materials of a more continental type are prominent calciferous sandstones, oil shales, and even coal seams. At various localities, often seeminghly estuarine in nature, there are rich fish faunas and, in addition, near the Firth of forth, occasional finds of amphibians.

Pennsylvanian fossil deposits are, in contrast to those of the Mississippian, predominantly continental in type. This was the time at which the most important coal seams of Europe and North America were formed; and numerous inhabitants of the coal-swamp pools have been preserved in the shales, ironstones, and impure cannels associated with the coals. In Europe the typical Coal Measures are included in the Westphalian floral zone. Although coals of this age are abundant on the Continent, Westphalian fossil finds are almost entirely confined to the British Isles, where Newsham near New castel and kilkenny in Ireland are important localities. In the late Pennsylvanian appears the Stephanian flora, transitional to that of the Permian. The coals of this horizon are well represented on the Continent, and Stephanian vertebrates are richly represented at Nyrany (Nurschan) and Kounova in Bohemia.

In America the earliest Pennsylvanian formations—those of the Pottsville—are, so far as in known, barren. In the somewhat later formations, equivalent to the Westphalian of Europe, there are several rich deposits, including a pocket of fossiliferous cannel coal at Linton, Ohio; nodules containing vertebrates, as well as plants, in the coal shales at Mazon Creek, Illinois; and material filling the hollow stumps of coal-swamp trees at the Joggins, Nova Scotia. Other scattered North American localities are of late Pennsylvanian age (Stephanian). In some areas and horizons typical Coal Measures conditions persist; in others there is a transition to red sediments, which continue without break into the Permian.

Despite the predominance of discoveries of continental forms of life, marine vertebrates are not uncommon finds in the Pennsylvanian. Some are present in typical marine beds. More important, however, is the fact that coal deposition was not a continuous process but a cyclical one, and layers of marine origin containing teeth and spines of sea dwellers are frequently present between successive coal horizons.

Fresh-water Fishes

The character of Carboniferous fish life in inland waters was markedly different from that of the Devonian. Of lower fish groups, the ostracoderms were extinct; gone, too, were the antiarchs so numerous in the late Devonian. Acanthodians persisted in small numbers, and, of the sharklike fishes, the pleuracanths were common in many localities. All other fresh-water forms were members of the Osteichthyes. But, although the groups represented were the same as in the later Devonian—crossopterygians, lungfish, and palaeoniscoids—their relative abundance had changed greatly in character. In the Devonian the two choanate groups were dominant, the ray-fins rare. In the Carboniferous, perhaps in correlation with changed environmental conditions, the proportions were reversed. *Megalichthys* and a few other crossopterygians, including fresh-water coelacanths, survived in reduced numbers; the dipnoans, as *Sagenodus* and *Ctenodus,* were a modest group. They were vastly outnumbered by a host of palaeoniscoids, which appear to have swarmed in every Carboniferous pool and river. About two-score genera have been described; some were still rather primitive in nature, others specialized or advanced in various degree.

Marine Fishes

In the sea as on land, the archaic fish groups were reduced in numbers. There are no ostracoderms; of the placoderms, fin spines suggest the survival of large acanthodians, but the arthrodires are absent. Nor was there any notable tendency, as yet, for a marine invasion by the higher bony fishes.

The Chondrichthyes, however developed greatly in the seas of this age. Seldom are entire specimens preserved, but abundant teeth and spines testify to the number and variety of sharks and sharklike fishes in these oceans. Most abundant were the bradyodonts, mollusk-eating forms which appear to have been depressed bottom dwellers analogous to the later skates and rays; there are four families and a variety of genera of bradyodonts, which had already reached the peak of their development in the Mississippian. These fishes were presumably a major prey of the predaceous shark types which were also common inhabitants of the Carboniferous seas. Of these sharks, some continue to represent the primitive *Cladoselache-Ctenacanthus* group, which had appeared in the later Devonian; others, however, appear to have become more "modernized" in fin structure and belong to the hybodont family, destined to continue into the Mesozoic.

Amphibians

The Carboniferous is the time of greatest development of the amphibians. Once evolved, with the possibilities of amphibious, if not terrestrial, life before them, they had spread into a host of types varying greatly both in structural features and in adaptations. Of the two major groups present, the labyrinthodonts are the better known, partly because of their tendency to grow to large size. The Devonian ichthyostegids persisted but are relatively rare finds. Embolomeres were common inhabitants of the coal-swamp lagoons; many, such as *Eogyrinus,* were elongate and slender fish-eaters and persistent water dwellers. Toward the close of the Carboniferous appear *Diplovertebron* and other forms approaching the reptiles in many structural features. The *Loxomma* group, apparently very common for much of this time, may be an offshoot of the Rhachitomi; more representative, although primitive, members of this group, such as *Dendrerpeton,* appear in the Pennsylvanian. Presumably of labyrinthodont derivation were *Amphibamus* and *Miobatrachus,* which appear to show an initial stage in the evolution of the frogs.

Less conspicuous because of smaller size but more abundant and varied were the lepospondyls. They appear in early Mississippian horizons and had already reached and passed their time of major development by the close of the Carboniferous. There were snakelike aistopods, *Dolichosoma* and *Ophiderpeton;* numerous nectridians, including "horned" forms such as *Keraterpeton,* and the elongate *Urocordylus;* and a host of varied microsaurs, such as *Microbrachis, Coccytinus,* and *Hylonomus.*

Reptiles

The development of reptiles, primarily characterized by their improved reproductive processes, took place during the Carboniferous. Unfortunately, our record of their early history is poor, owing, one may believe, to the fact that most Coal Measures localities are deposits of a nature which tends to include pool dwellers rather than inhabitants of the dry land. Late Pennsylvanian records show that the reptiles had a that time become as diversified as in the Permian; for we find not only specialized cotylosaurs of the diadectid type but pelycosaurs—even the peculiar, long-spined *Edaphosaurus.* In the earlier, Westphalian, Coal Measures there are a few obscure specimens which may be those of early cotylosaurs, but their nature is none too clear.

PERMIAN

Sediments

Toward the close of the Carboniferous a gradual change in sediments occurred in many portions of western Europe and north America. Coal seams and intervening marine beds gradually gave place to terrestrial deposits prevailingly red in color, their materials derived from the great mountain masses then being formed, such as the ancestral Alps in Europe and the ancestral Rockies and the earliest Appalachians in North America. This change in sediments was gradual, and only arbitrary lines of division can be made between the Upper Carboniferous and the Permian red beds; correspondingly, we find that the vertebrate life of the late Pennsylvanian is almost indistinguishable from that of the early Permian. In North America and abundant

Permo-Carboniferous land fauna appeared in late Pennsylvanian times and persisted well into the Lower Permian. Red beds containing these forms are found in regions a far apart as New Mexico and Prince Edward Island but are best developed as the Wichita and Clear Fork groups in northwestern Texas. The many specimens collected from the latter region give us our best picture of the world's early Permian life.

In Europe the lower Permian—the Rotliegende—is a series comparable in nature to the American red beds; the known fauna, however, is more restricted, presumably because of fewer exposures. A few specimens have been recovered from beds of this age in England, France, and Czechoslovakia; the most famous locality is that at the Plauen's schen Grunde (Niederhasslich) near Dresden, where slabs containing remains of numerous small amphibians and reptiles have been obtained.

These fossiliferous red-bed deposits cover but the earliest portion of the Permian. In America further red-bed deposition continued through much of the period, but these levels have not yielded the slightest trace of bone. In western Europe the Rotliegende is followed by the Kupferschiefer and the marine phases of the Zechstein, with a sparse fish fauna and almost no tetrapods. For land vertebrates of the Middle and Upper Permian we must turn to other areas—South Africa and Russia.

In Africa pre-Permian vertebrates are unknown; there are no fossiliferous early Paleozoic sediments of any sort, and those of the Devonian and Carboniferous are almost exclusive marine. At the close of the Carboniferous, however, there began the formation of the continental deposits of the Karroo system, some thirty thousand feet thick, which cover much of the Union of South Africa and have outliers in central and eastern Africa and Madagascar. At the base is the Dwyka series, late Pennsylvanian or earliest Permian in age, containing glacial beds and with a sparse vertebrate fauna. Following this are some six thousand feet of the Ecca series, a barren equivalent of the lower Permian red beds of Europe and North America. The Beaufort beds begin with the Middle Permian and continue without marked break, lithologically or faunally, into the Triassic. Their exposures cover a great area centering in the Karroo desert of northern Cape Colony and have a rich tetrapod fauna. Several zones, named after index reptiles *(Tapinocephalus, Endothiodon, Cisticephalus)* are distinguishable in the Permian portion of the series.

Few fossiliferous continental formation of Permian age are as yet reported from Asia or Australia; and, although Brazil has sediments of Karroo type, the permian beds are unfossiliferous except for a Dwyka equivalent. In eastern and northern Russia, however, is found a series of continental deposits which corresponds closely to the Permian portion of the Karroo system. Although known for a century, serious exploration of these beds has only recently begun; they promise to yield a fauna as interesting as that of South Africa. A single "pocket" near Elgin, Scotland, appears to be an outlying equivalent of the latest Russian Permian.

Marine Permian formations are abundant, but most are barren of vertebrates. A few fishes only have been found in early Permian limestones in scattered areas of North America (Texas, Nebraska, Wyoming, East Greenland), Russia, southern Asia, and western Australia. In the later Permian the Kupferschiefer deposits of Germany appear to have been laid down in brackish lagoons.

Fishes

As our discussion of localities suggests, marine fish life is almost unknown in the Permian. In the earliest horizons there are a few representatives of common Carboniferous chondrichthyan groups—the last cladodont sharks, bradyodonts, and hybodonts; there is little evidence of any saltwater bony fish in the early Permian. *Menaspis* and *Janassa* in the Kupferschiefer are the last of the once abundant bradyodonts, and there were a few surviving hybodonts. The old vertebrate fauna of the oceans appears to have been almost entirely wiped out at the close of the Paleozoic, although it is certain that some hybodonts and bradyodont descendants must have survived to give rise to the sharks and chimaeras of later times. The first marine radiation of the vertebrates was a an end.

In fresh waters, *Acanthodes* survived into the Lower Permian as the last of the once important placoderns. *Pleuracanthus* was a common early Permian fresh-water shark but is unknown in later times except in Australia, where the group appears to have persisted into the Triassic. The Lower Permian also sees the end of rhipidistian crossopterygians; *Ectosteorhachis* in America is the last survivor of this phylogenetically important group. Coelacanths, however, persisted in fresh waters. Lungfishes, particularly *Sagenodus*, were common at the beginning of the period but are unrecorded later.

The dominant types are palaeoniscoids. Relatively few genera are known, but they appear to have been varied in nature and included such unusual and progressive types as *Dorypterus*. A single holostean in the late Permian marks the beginning of the change in actinopterygian faunas to be witnessed in the Triassic. The Kupferschiefer deposits contain actinopterygians which may have been able to withstand brackish, if not strongly saline, waters and may represent the beginning of the first major invasion of the seas by the ray-finned fishes.

Amphibians

There is no marked break between the late Pennsylvanian and early Permian in amphibian faunas; the red beds exhibit a final phase in Carboniferous amphibian history. There are surviving embolomeres and seymouriamorphans, such as *Archeria* (*"Cricotus"*) and *Seymouria*. The rhachitomes were already present in the Carboniferous but reached the peak of their development in the Lower Permian. *Eryops* was the common and characteristic large form in the American red beds; *Actionodon* and related types were comparable European genera. In addition, there were a score or more of rhachitomes of varied character—the dissorophids, armored and with exaggerated otic notches; long-snouted aquatic types such as *Archegosaurus* and *Chenoprosopus;* the flat-headed *Trimerorhachis*. Lepospondyls were reduced greatly in numbers, but there were a few survivors, such as *Diplocaulus, Lysorophus,* and the gymnarthirds.

In the later Permian, amphibian life was much reduced. The degenerate genus *Kotlassia* represented the seymouriamorphs in the late Permian, but the embolomeres have disappeared. Gone, too, are the lepospondyls (with one possible exception). The Rhachitomi alone persisted in moderate numbers, in such forms as *Rhinesuchus,* which, with flatter heads and broadly opened palatal vacuities, were entering on a "neorhachitomous" condition foreshadowing the stereospondyls of the Triassic.

Cotylosaurs

Reptile origins took place in the Carboniferous, but it is only with the development of more terrestrial conditions in the red beds that we can gain a comprehensive picture of early reptilian life. The "stem reptiles" were an important factor in the land fauna of the early Permian. Some types, as *Limnoscelis,* appea to be very primitive; most cotylosaurs, however, are advanced or specialized in nature, showing that their point of common divergence must have been far back in the Pennsylvanian. *Captorhinus* and *Labidosaurus* are examples of the advanced American captorhinomorphs of the early Permian. The commonest of early cotylosaurs, however, are those of the *Diadectes* type, presumed herbivores with highly specialized cheek teeth.

In the later Permian, captorhinomorphs are extinct, or nearly so, and the remaining forms are herbivorous diadectomorphs. In the Russian Middle Permian are two types of procolophonids—small cotylosaurs which were destined to survive and flourish in the Triassic. More abundant in the later Permian, however, were the pareiasaurs. These ungainly-looking creatures are the commonest of finds in the Middle Permian of South Africa. Their remains are frequently found as if in the position in which they had become bogged down in the swamps where they found their food and met their death. In the Upper Permian of South Africa, pareiasaurs are relatively rare; they were, however, widespread at this time, for there are Upper Permian pareiasaurs in Russia and even in Scotland.

Pelycosaurs

It is of interest that the most prominent of early reptilian orders is not a primitive group or one which led to later typical reptiles but one destined, after many changes and vicissitudes, to give rise to be mammals. Pelycosaurs had arisen in the later Carboniferous as an early offshoot from the stem reptiles and still retained many primitive features. They were highly diversified in the red beds. Some like *Ophiacodon,* were long-snouted, semiaquatic fish-eaters. *Edaphosaurus,* with long spines and crossbars, and *Casea* were representative of a side line of herbivores. More important phylogenetically were *Dimetrodon, Haptodus,* and their relatives, the dominant carnivores of the early Permian, from which the later therapsids appear to have been derived.

Therapsids

These more advanced, mammal-like reptiles replace their ancestors, the pelycosaurs, as the common reptiles of the middle and late Permian; they are abundantly represented in the Beaufort beds of South Africa and are present in Russia as well. The great dinocephalians, including the carnivorous titanosuchids and the herbivorous tapinocephalids, are confind to, and common in, the Middle Permian; some of the Russian types appear to be slightly older in time of appearance and somewhat more primitive than those of South Africa. Dicynodonts are present in the Middle Permian of South Africa but were relatively small and rare; in the Upper Permian these nearly toothless herbivores are the commonest of vertebrates. There is, in addition, a variety of more progressive carnivores, gorgonopsians, and therocephalians, in the Middle and Upper Permian of South Africa and the Upper Permian of Russia.

Other Reptiles

It is curious that in the Permian there are few evidences of the ancestry of the reptilian groups which were to become prominent in the Mesozoic. *Eunotosaurus* of the Middle Permian of South Africa is a possible forerunner of the turtles; there are no clues as to the ancestry of the ichthyosaurs. Of the synaptosaurs, we find no forms resembling the aquatic sauropterygians of the Mesozoic, but there are a number of relatively rare protorosaurs, such as *Araeoscelis* of the American red beds and *Protorosaurus* of the European Kupferschiefer. Diapsid reptiles are notably absent in the Lower and Middle Permian; in the Upper Permian of South Africa, however, there appear a number of small primitive eosuchians, such as *Youngina*.

The late Pennsylvanian or early Permian witnesses the only appearance of that problematical order of small aquatic reptiles, the Mesosauria. *Mesosaurus* is known from the Dwyka of South Africa and an equivalent formation across the South Atlantic in Brazil; no trace of this group has ever been found in northern continental deposits.

Gondwanaland

In all three southern continents and in peninsular India as well sediments of similar types are found in the late Paleozoic and early Mesozoic; in all three continents there is evidence of pronounced glaciation in late Carboniferous or early Permian time; in all these areas there is found a common flora characterized by the genera *Glossopteris* and *gangamopteris*. This has led to a belief that these regions were then parts of a common land mass named Gondwanaland, separated from the Eurasian land areas by an east-west Tethys Sea. The distribution of the vertebrate life of the Permian, especially the Karroo therapsid fauna, has been held to support this theory. We have noted, however, that in the Middle and Upper Permian, in which therapsids are numerous in South Africa, there are few known vertebrates of any sort in the deposits of North America and in most of Europe. And we have further noted that the one good northern fauna of the later Permian, that of Russia, is highly comparable to that of South Africa. The therapsids thus furnish no proof of the union and joint isolation of the southern continents. The only favorable evidence for this theory furnished by the vertebrates is the presence of *Mesosaurus* on both shores of the South Atlantic.

Permian: Summary

The chief characteristics and events of Permian vertebrate history are: (1) the disappearance during the period of many types of fishes, including cladoselachians, acanthodians, bradyodonts, and primitive crossopterygians; (2) continued dominance of palaeoniscoids and first appearance of the Holostei; (3) disappearance of most of the amphibian types and survival (as far as known) only of rhachitomous labyrinthodonts; (4) abundance of cotylosaurs; (5) dominance throughout of synapsid reptiles—pelycosaurs in early times, therapsids in the later Permian; and (6) presence of rare representatives of other reptilian groups.

The Paleozoic as a Whole

In the Paleozoic is included by far the greater portion of the vast period of time which has elapsed since life first appeared in abundance. From the point of view of the student of invertebrates, the Paleozoic may form a not unnatural unit, since most of these various lower-animal groups appeared full fledged at its beginning and at its close underwent marked change. But with the vertebrates it is otherwise; these forms appeared at a relatively late date and evolved rapidly through a series of progressive stages; and at the end of the era our story is broken, so to speak, in the middle of a sentence. Looked at in its broadest outlines, the history of vertebrates in the Paleozoic is not that of a single chapter in their development but of two full cycles and the beginning of a third.

A first major phase in vertebrate history seemingly occurred in the earlier half of the Paleozoic. The earliest vertebrates are imperfectly known except at the very end of the cycle—at the Silurian-Devonian boundary. The vertebrates were then a rather obscure group of small, bottom-living river dwellers, playing but a minor part in the drama of life-development. There lack of biting mouth parts rendered impossible the trituration of food of any degree of size or toughness; this appears to have been a major factor militating against their development. Mud-grubbers they were, and and mud-grubbers, seemingly, they were destined to remain.

But apparently toward the end of the Silurian there took place a great advance in vertebrate structure which initiated a new cycle in the history of the group. This advance was the development of jaws; from this resulted a true conquest of the waters by vertebrates. With the development of jaws and teeth came the possibility of a predaceous existence, and active swimming succeeded the sedentary bottom-dwelling habits of the ancestral types. Jawed fishes were exceedingly abundant in fresh waters before the Devonian was far advanced and were widespread in the oceans in the Carboniferous, although a decisive and permanent conquest of the salt waters was not made until Mesozoic times. To this great cycle of aquatic conquest belongs the development of the amphibians as well, for lungs and tetrapod limbs seem originally to have been merely adaptations for a more successful life in inland waters.

In the Carboniferous there took place a second great advance among the more progressive vertebrates, resulting in the initiation of a third stage in their history. This advance was the release of tetrapods from the water through changes which eliminated the aquatic stage in individual development. The resulting new cycle of development was the conquest of the land by the reptiles.

Primitive reptiles, little known in the Carboniferous, were abundant by the beginning of the Permian. Most of them were members of two ancient groups—the cotylosaurs, stem stock of all true land vertebrates, and the synapsids, destined to give rise to mammals. The close of the Paleozoic still saw these two groups dominant. Only in the Triassic did this archaic land radiation come to an end and these forms give way to newer reptilian types which held sway during the rest of the Mesozoic.

Chapter—14

Vertebrates of Mesozoic Era

The Mesozoic era is usually known as the Age of Reptiles, for its was during this time that there flourished the great reptilian groups now extinct: the varied dinosaurs on land, the flying reptlies in the air, and ichthyosaurs, plesiosaurs, and other marine types in the seas. The mesozoic witnessed, as well, the more modest beginning of existing reptile groups turtles, lizards, snakes, crocodiles. Further, it was during this era that birds and mammals, both derived from reptilian ancestors, made their appearance. Among amphibians the Mesozoic sees the change from old to new extinction of the labyrinthodonts and appearance of typical frogs and urodeless. And although our attention naturally centers on higher groups, the fishes of the Mesozoic had an interesting development. During this time there developed modern groups of sharks and rays and the chimaeras; the ray-finned bony fishes progressed from the chondrostean to the holostean grade of evolution and finally to the modern teleost type.

TRIASSIC

The Triassic forms an introduction to the Mesozic vertebrate story. At its beginning we find faunas both on land and in the water comparable to those of the late Paleozoic. During the period, however, there were marked changes among both fishes and reptiles; by the close of the Triassic there had appeared almost every one of the striking vertebrate groups which were to play major roles in the Mesozoic story.

Vertebrate Beds

The Triassic owes its name to the sequence of fossiliferous deposits characteristic of the period in central and western Europe. It opens with the Bunter—variegated continental beds with sparse vertebrate remains, deposited on the northern shores of the seas covering the present Alpine region. These seas advanced northward; their sediments form the Middle Triassic Musehelkalk, a limestone group with a rich fauna. In the Upper Triassic, with retreating seas, there is a return to dominantly continental beds in the Keuper, a series mainly red in color, supposedly deposited under arid conditions and with a good assemblage of vertebrate remains. The Upper Triassic closes with the Rhaetic, forming a transition to the Jurassic and including a number of bone beds. The Triassic areas described are best developed in southern Germany and adjacent regions; Middle and early Upper Triassic aquatic vertebrates are abundant at Perledo and other localities in the southern foothills of the Alps, and the Upper Triassic is represented in the Bristol district of Great Britian.

A second important Triassic sequence is that of South Africa. The earlier Triassic, poor in fossils in most of Europe, is here represented by the richly fossiliferous upper zones—*Lystrosaurus, Procolophon, Cynognathus*—of the Beaufort series. The succeeding Stormberg Series covers the remainder of the Triassic; the Middle Triassic beds (Molteno) are barren (except for plants), but the Red Beds and Cave Sandstones of of the Upper Triassic have yielded vertebrates similar to those of the Keuper and the Rhaetic.

Productive Triassic areas of lesser vertical range and lesser importance are scattered through every continent. In eastern Australia deposits with rich fish faunas—Narrabeen, Hawkesbury, Wianamatta—are found at several Triassic horizons. Triassic beds important for fishes and amphibians are found in Greenland, Spitzbergen, eastern Russia, Madagascar, and India. More continental types of deposits comparable to those of the South African Lower Triassic are found in several localities in eastern Asia, and phases (Middle Triassic?) of the same fauna are seen in East Africa and the Rio do Rasto beds of southern Brazil.

Marine faunas of Middle and Upper Triassic age, more or less comparable to those of the Muschelkalk, are to be found in Spitzbergen and in Nevada and California. Upper Triassic continental deposits similar in faunal content, and even in their red color, to the Keuper and Stormberg, are plentiful in North America. They include the Newark series of the Atlantic Coast (famous for footprinst), the Chinle, dockum and Chugwater of the West. Similar red beds are present in eastern and southern Asia as well.

Fishes

We have noted a marked decrease in the numbers and variety of fishes present in the last phases of the Paleozoic: placoderms had become extinct, the Chondrichthyes and Choanichthyes were reduced to a few types, and only the actinopterygians had remained in a flourishing condition. This situation still prevailed in the Triassic. Of the cartilaginous fishes, primitive chimaeras descended from the bradyodonts were presumably present but are known only from a single poor and questionable specimen; the last lingering pleuracanths are found in the fresh-water Triassic of Australia; *Hybodus* and a few related shark types are the only known chondrichthyan inhabitants of the Triassic seas. Of the Choanichthyes, the typical crossopterygians were extinct, but the coelacanths had survived and were apparently beginning to change their habital from fresh waters to salt; *Ceratodus* represents the lungfishes and is abundant in some formations presumably deposited under arid conditions.

Table 14.1
Mesozoic Vertebrate Localities

	European Stage	*Europe*	*North America*	*South America*	*Asia*	*Africa*	*Australia*
	Upper Cretaceous inc. Danian, Macstrichtian, Senonian. Turonian, Cenomanian	Transylvania dinosaur beds; Chalk of S. Eng., Belg., Maestricht (Neth.), N. France, Westphalia *M**	Lance, Edmonton, Belly R. (Alberta-Mont.-Wyo.); Fruitland, Kirtland, Ojo Alamo (N.M.); Niobrara (Kan.) *M*; Pierre (S.D.) *M*; Chico (Calif.) *M*; Austin chalk (Tex.) *M*; Monmouth, Matawan (N.J.) *M*; Selma (Miss.) *M*	Red Beds of Patagonia; Uruguay; Ceara (Brazil) *M*	Diadochta (Mongolia); Lameta series, Trichinopoly and Ariyalur stages (Ind.); Mt. Lebanon (Syria) M	Baharie stage (Egypt); Madagascar	Opal beds (N.S.W.); New Zealand *M*
Cretaceous	Lower Cretaceous, inc. Albian, Aptian, Neocomian	Wealden (Eng., Belg., Hanover); Neocomian of Voirons (Switz.), Istria, etc., *M*; L. Greenland, Gault of Channel region (Eng.) *M*	Cloverly (Wyo.); Arundel (Md.)	Neuquem fm. (Arg.); Bahia (Brazil) *M*	Iren Dabasu, Oshih, Ondar Sair, (Mongolia); Shantung (China)	Uitenhage (S.A.f.)	Tambo ser., Rolling Downs (Queens.)
	Upper Jurassic= Malm or U. Oölites inc. Purbeckian, Portlandian, Kimmeridgian, Oxfordian, Callovian.	Purbeck beds (Eng.); Lithographite Is. of Solenhofen, Kelheim (Ger.), Cirin (France) *M*; Oxford clays and Kimmeridge beds of S. Eng. and N. France *M*	Morrison fm. of western states (Como, Lost Cabin, Vernal, etc.); Sundance (Wyo.); *M*; western Cuba *M*			Tendaguru (E.A.f)	
Jurassic	Middle Jurassic= Dogger or L. Oölites inc. Bathonian, Bajocian	Stonesfield slate, Forest Marble (Eng.)	Navajo ss. (Ariz.)			Madagascar	

* *M* = marine

Table 14.1 (Contd.)

	European Stage	*Europe*	*North America*	*South America*	*Asia*	*Africa*	*Australia*
	Lower Jurassic= Lias, with var, subdivisions	Lias of Yorkshire, Dorset (Eng.), Caen region (France), Holzmaden, etc. (Bavaria) *M*	Wingate (Ariz.)				Talbraggar (N.S.W.); Lower Walloon, Durham Downs (Queens.)
	Rhaetian, Norian	Rhaetic bone beds (S.W. Eng., S. Ger., Scania); Keuper and Lettenkohle (Ger.); Alpine region (Perledo, etc.) *M*	Chinle (Ariz.); Chugwater (Wyo.); Dockum (Tex.); Newark ser. (Mass.-N.C.); Hosselkus Is. (Calif.–Nev.) *M*		Maleri (Ind.); red beds of Yünnan	Cave Ss. and beds of Stormberg series (S.A.f.)	Wianamatta inc. St. Peter's beds (N.S.W.)
Triassic	Carnian, Ladinian	Muschelkalk of Germany and equivalents in Alpine region (Besano, etc.) *M*; Spitzbergen *M*	Shinarump (Ariz.); W. Humboldt Range (Nev.) *M*	Rio do Rasto group (Brazil)	Shansi (China)	Upper Ruhuhu bone bed. Manda beds (E.A.f.)	Hawkesbury beds of Brookvale, etc. (N.S.W.)
	Anisian, Seythian	Bunter of Central Europe; "Zone V" of N.E. Russia (Vetlugian); Mt. Bogdo (Russia); Spitzbergen *M* (in part)	Moenkopi (Ariz.); E. Greenland *M* (in part).		Sinkiang; Luang Probang (Indo-China); Mangli, Panchet (Ind.)	*Lystrosaurus*, *Procolophon*, and *Cynognathus* zones of U. Beaufort (S.A.f.); Madagascar *M*	Narrabeen, inc. Gosford (N.S.W.)

The history of actinopterygian dominance and progress seen in the later Paleozoic is continued in the Triassic. The primitive palaeoniscoids are still prominent. The characteristic ray-fins of the times, however, are those to which the term "subholosteans" is applied—forms progressive in character, structurally intermediate between palaeoniscoids and holosteans, and probably giving rise, in polyphyletic fashion, to many families of the latter group. There are perhaps half-a-hundred genera and half-a-dozen families of subholosteans in the Triassic —*Dictyopyge, Redfieldia* [*Catopterus*], *Perleidus, Saurichthys,* and *Cleithrolepis* are representative. By the end of the period, however, the subholosteans were becoming much reduced in numbers and were giving way to typical holosteans—such as *Semionotus* and *Eugnathus*—which were to dominate in the Jurassic.

With few exceptions the older actinopterygians were fresh-water forms. We have noted, however, that in the late Permian many palaeniscoids appear to have existed in brackish water. In the Triassic a strong trend is apparent among ray-fins toward a marine existence. From this time onward the sea is the major center of actinopterygian life and evolution.

Amphibians

The Rhachitomi had been the dominant Permian amphibian types; almost all the amphibians in the earliest Triassic beds are members of the same group but are "neorhachitomous" forms approaching the Stereospondyli in structure. The stereospondyls, although destined to become extinct at the close of the Triassic, flourished greatly for the time. They included *Trematosaurus* and other long-snouted fish-eaters of the early Triassic; the capitosaurs, prominent throughout the period; the metoposaurs such as *Buettneria,* with anteriorly placed orbits, abundant in the American Upper Triassic; the short-skulled brachyopids. An important document is the lone specimen of the primitive "frog," *Protobatrachus,* from the Triassic of Madagascar; urodele ancestors are known neither in this period nor in the Jurassic.

Primitive Reptile Groups

In reptillian evolution, the *Paleozoic-Mesozoic* boundary is almost without meaning. By the end of the Triassic there had appeared the typical reptilian groups that were to dominate the later Mesozoic; at the beginning of the period, however, the fauna was still essentially a Paleozoic one.

The archaic fauna is best represented in the South Africa Lower Triassic deposits and in a later phase in those of Brazil; finds from other areas are few, but enough to suggest that the South African fauna was one typical of most, if not all, of the lands of the early Trias.

Dominant in the early Triassic were the therapsids. Dicynodonts were still flourishing but restricted in variety; notable were large types such as *Kannemeyeria,* and a small amphibious genus, *Lystrosaurus,* widely distributed in the early Trias. Gorgonopsian and therocephalians were extinct and were replaced as carnivored by more progressive relatives or descendants—the cynodonts, such as *Cynognathus* and *Diademodon*, and the very advanced *Bauria* group. A few large dicynodonts survived into the late Triassic and seem to have been world wide in distribution. Otherwise, typical therapsids had disappeared by late Triassic times. Presumably descended from them were a number of obscure or aberrant late Triassic forms, the ictidosaurs, as *Tritylodon* and *Dromatherium*, advanced in nature and close to the mammalian evolutionary level.

Cotylosaurs are present in the form of *Procolophon* and its relatives, the last survivors of the order. The protorosaurs, too are a group which had arisen in the Permian and persisted into various horizons of the Triassic, where they appear in the form of such odd types as *Trilophosaurus* and the peculiar, long-necked *Tanystropheus*.

Newer reptile Groups

Other reptile types, however, were making their appearance even in the Lower Triassic to mark the beginning of the new Mesozoic radiation of reptiles. The little eosuchians, which had been present in the late Permian, continue into the Triassic and are presumably ancestral to the rhynchocephalians. This last order is represented in the early Trias by a few rare primitive types; rhynchocephalians became more abundant later in the period and developed, in *Rhynchosaurus* and its allies, a peculiar, short-lived branch of rather large, strong-beaked reptiles. Turtles such as *Triassochelys*, characteristic, if somewhat primitive, members of their order, appear—seemingly suddenly—in the late Triassic.

Marine Reptiles

In the marine Trias (Muschelkalk) of Germany and in other formations of middle and late Triassic age in Spitzbergen and western North America, we seen an early stage in the development of marine reptiles. A number of groups are already represented, and in no case have we any adequate knowledge of their ancestry. Most striking are the early inchthyosaurs, such as *Mixosaurus* and *Shastasaurus*, already well adapted for a marine life but somewhat less advanced than their Jurassic descendants in development of paddles and tail fin. Varied sauropterygians are present. Particularly common were the nothosaurs, forms already aquatic but much less specialized than the plesiosaurs which were to replace them in the Jurassic. The Triassic was the period during which there developed the placodonts—amphibious mollusk-eating sauropterygians with curious resemblances to the turtles. *Thalattosaurus* of the American Triassic is a poorly known marine reptile of obscure relationships.

Archosaurs

The archosaurs, or ruling reptiles, were destined to be the dominant land vertebrates of the later Mesozoic; the stem order of this sub-class, the Thecodontia, made its appearance with a few genera in the early Triassic and became prominent in the later part of the period but was destined to the disappear at its close. Most representative of thecodonts were the pseudosuchians—small reptiles with bipedal tendencies, such as *Euparkeria* and *Ornithosuchus*. The phytosaurs are among the commonest of late Triassic fossils in northern continents—aberrant, long-snouted thecodonts adapted to a predaceous amphibious life. *Protosuchus* of North America and similar types from South Arfica are transitional in structure between primitive thecodonts and the crocodiles which were to replace the phytosaurs in the Jurassic.

Primitive Dinosaurs

By the latter part of the period the dinosaurs, evolved from the pseudosuchians, had made their appearance. Of ornithis-chians there is but a single rather dubious fragment from South Africa, but saurischians are abundant in upper Triassic formations, particularly in Europe and South Africa. A number were small lightly built coelurosauriand, such as *Podokesaurus* and *Saltopus*. *Teratosaurus* is typical of a series of large forms leading to the large carnivores of the Jurassic. A third type, of which *Yaleosaurus ("Anchisaurus")* and *Plateosaurus* are representative, consists of heavily built dinosarus which are the ancestors of the huge Jurassic sauropods.

Triassic: Summary

The vertebrate life of the Triassic is characterized by (1) rarity of cartilaginous fishes; (2) replacement of chondrosteans by holosteans as the dominant fish group; (3) a trend toward marine life on the part of bony fishes; (4) last appearance of the labyrinthodonts; (5) appearance of rhynchocephalians and turtles; (6) development of primitive reptilian marine life including ichthyosaurs and varied sauropterygians; (7) the abundance and radiation of the thecodonts, followed by their extinction; (8) appearance and rapid development of the theropod dinosaurs; (9) extinction of typical mammal-like reptiles, but (10) persistence of forms perhaps transitional to mammals (ictidosaurs).

JURASSIC

The Jurassic was a time of widespread seas and reduced land areas. For most of the duration of the period the principal deposits are of marine origin, and it is only toward the close of the Jurassic that continental sediments become prominent. In consequence, our knowledge of aquatic vertebrates is much more complete than is that of terrestrial forms.

Vertebrate Localities

The typical Europen Jurassic section is almost exclusively marine in nature. The Lower Jurassic—the Lias—contains numerous localities rich in marine vertebrate remains, particularly on the Yorkshire coast; the region of Lyme Regis on the English Channel; the opposite coastal region of Normandy; and in southern Germany, particularly the famous Holzmaden quarries in Bavaria. The later European Jurassic—frequently termed the Dogger and Malm or (combined) the Oolites—in general contains far fewer ramains of vertebrates. There are, however, important exceptions. The Oxford and Kimmeridge clays are rich in fish and reptilian remains. Prominent late Jurassic localities are the lithographic limestone deposits of Solenhofen and Kelheim in Germany and Cirin in France. Here fine-grained sediments, presumably laid down in quiet, coralreef lagoons, have preserved numerous specimens of fishes and occasional tetrapod intruders. Jurassic marine deposits are present in other parts of the world and may occasionally produce vertebrates.

Land vertebrates are, in general, notable for their absence. A few are present in the marine European records, as at Solenhofen, and in Middle and Upper Jurassic estuarine localities in England, at Stonesfield, and in the Purbeck beds. An abundant record of terrestrial forms is found only at the very end of the period and then only in two regions. The Morrison beds, exposed over large areas of western North America from Montana through Wyoming to Utah, Colorado, and even Oklahoma, appear to have been laid down under conditions similar to those of the modern Mississippi delta and contain a particularly rich dinosaur fauna. Very similar in nature and in fauna are the Tendaguru beds in Tanganyika Territory, East Africa.

Fishes

The Jurassic, with seas full of invertebrate food, saw the beginning of a great recrudescence of cartilaginous fish life. There appear, in the Lias, *Squaloraja, Myriacanthus,* and other primitive members of the mollusk-eating chimaeras, destined to play a modest part in marine life from this time forth. Much more important were developments among the selachians. From the surviving hybodonts there began to branch off various types of modern sharks. *Heterodontus* and *Hexanchus,* relatively primitive living forms, had appeared before the end of the Jurassic, as had half-a-dozen genera representing more modern shark families. Somewhat slower to develop were the skates and rays, but the guitarfish, *Rhinobatis,* a primitive skate type, was already evolved by the late Jurassic.

Of fresh-water bony fishes we know almost nothing. The sea, however, was swarming with members of this class. Apart from *Undina* and a few other coelacanths, all these forms were actinopterygians. There were surviving palaeoniscoids; the earliest sturgeon, *Chondrosteus;* and a few subholosteans. Almost all, however, were holosteans. The semionotids *Dapedius* and *Lepidotus;* pycnodonts such as *Gyrodus* and *Microdon; Aspidorhynchus;* amioids such as *Caturus, Eugnathus, Pachycormus,* and *Macrosemius,* advanced types such as *Pholidophorus.*—these are but a few of the many common forms And in addition, we find, particularly toward the end of the period, swarms of small primitive teleosts, such as *Leptolepis,* which herald the approaching rise of this final major division of the ray-finned fishes.

Amphibians

The Jurassic marks the low point in our knowledge of amphibian life. Unless a doubtful fragment from Australia belongs to a labyrinthodont, we have only a few bits of frog remains to represent the entire class.

Marine Reptiles

The nothosaurs and placodonts and the obscure thalattosaurs of the Triassic were absent in Jurassic waters, but the seas contained an abundance of marine reptiles. The ichthyosaurs of the period were far advanced in their readoption of a fusiform fish body and very fishlike fins. In the Lias they appear already to have reached the peak of their development. Most of the ichthyosaurs of the period have been included in the single genus *Ichthyosaurus,* but they appear to have actually included a wide range of genera varying in paddle and jaw development; *Eurthinosaurus,* for example is a form closely analogous to the modern swordfish. The sauropterygians are represented by a variety of plesiosaurs, including shortnecked types, such as *Pliosaurus* and *Peloneustes,* and more abundant longer-necked genera, such as *Plesiosaurus* and *Cryptocleidus*. The only instance of invasion of the seas by the archosaurs is the development in the later Jurassic of the short-lived group of marine crocodiles, such as *Metriorhynchus* and *Geosaurus*.

Dinosaurs

But few remains of land vertebrates are known from the earlier marine deposits of the period; most of our knowledge of dinosaurs in gained from the faunas of the Morrison of North America and the Tendaguru beds, both of late Jurassic age.By this time the dinosaurian groups had developed in spectacular fashion. Among the saurischians, the coelurosaurs were present in forms of modest size, as *Compsognathus* and *Ornitholestes*. Other bipeds had, however, increased in size to produce such large types as *Allosaurus, Ceratosaurus,* and *Megalosaurus*. Parallel to this development had been the evolution of their gigantic cousins, the amphibious sauropods. These appear to have been at the peak of their development in the late Jurassic—*Diplodocus, Apatosaurus* [*Brontosaurus*], *Camarasaurus, Cetiosaurus,* and *Brachiosaurus* are among the familiar names.

The development of the ornithischians appears to have been a slower process than that of the saurischians. But by the time of deposition of the Morrison and its equivalents a variety of these forms was in existence. Most were bipedal ornithopods of relatively small size, of which *camptosaurus* is representative. A striking Morrison form is the quadruped *Stegosaurus,* with a double row of plates and spines on its back; *Kenturosaurus* is a closely related East African genus.

Other Reptiles

Our knowledge of the smaller reptiles of the Jurassic is very limited. We find a variety of turtles, mainly of the primitive amphichelydian group but including, in *Thalassemys* and its allies, primitive fore-runners of the cryptodires. Most were presumably amphibious in nature; it is possible that some were well along in adaptation to marine existence.

Homoeosaurus is a small Jurassic reptile exceedingly close to the surviving rhynchocephalian *Sphenodon* of New Zealand; *Pleurosaurus* and *Sapheosaurus* are small types of doubtful position but often included in the Rhynchocephalia. Lizards, ralated to this last order and like them, presumably derived from an eosuchian stock first appear in the form of two rare Upper Jurassic genera.

Crocodiles are abundant as fossils. We have already noted the development of a marine group; there were four other families showing more normal amphibious structures. All were in the mesosuchian stage of evolution, with the palate less developed than in modern forms.

The Jurassic was the time of the appearance and greatest development of the pterosaurs. In the Lias several genera of long-tailed forms, such as *Dimorphodon,* are present and are already completely adapted for flight. In the late Jurassic long-tailed pterosaurs, such as *Rhamphorhynchus,* are still abundant, but we find, in addition numerous speciment of *Pterodactylus,* with a short tail and more elongated wings.

Birds

Our first glimpse of bird life comes from two specimens, *Archaeopteryx* and *Archaeornis,* from the Solenhofen Upper Jurassic lithographic stone. It is certain that birds are of thecodont derivation, but earlier stages in their development are unknown and would be difficult to recognize if found in the absence of feathers. There typical avian structures are here present to prove the nature of the specimens despite their retention of teeth, long bony tails, and typically dinosaurian "arms."

Mammals

We have noted in the late Triassic the presence of rare and obscure forms which are seemingly transitional to mammals, and some of which may perhaps belong to this group. In the Jurassic, mammals of primitive but varied types are definitely present; they are, however, small rare, and fragmentary. Almost all known specimens come from three sites—Stonesfield in the middle Jurassic of England, the late Jurassic Purbeck beds of that country, and a single small pocket in the Morrison at Como Bluff, Wyoming. In the Stonesfield locality there is a seeming survivor of the tritylodont group; its definitely mammalian fauna is extremely limited and includes but two triconodonts and a single pantothere, *Amphitherium.* The two Upper Jurassic sites have a much more extensive fauna; those of the two sides of the Atlantic show relatively small differences from one another. There are further triconodonts and numerous pantotheres; the symmetro—donts make their only known appearance at this time. *Plagiaulax* and several relatives are the earliest representatives of the curious aberrant multi-tuberculates destined to persist to Eocene times.

Jurassic: Summary

Prominent features of Jurassic vertebrates were: (1) among fishes, the height of the holostean supremacy the appearance of the first teleosts and the beginning of the radiation of modern elasmobranch types; (2) appearance of frog, first of the modern amphibians; (3) among dinosaurs, the continued expansion of bipedal types in both orders and the evolution of armored forms and of sauropods; (4) a great development of marine reptiles, especially ichthyosaurs and plesiosaurs; (5) appearance of lizards, typical crocodiles, and pterosaurs; (6) the first birds; (7) development of small primitive mammals of several types.

CRETACEOUS

The Cretaceous is among the longer periods of the earth's history; by some authors it is divided into two periods, the Comanchean and the true (Upper) Cretaceous. This scheme will not be followed here, but there is considerable contrast between the vertebrates of the earlier and the later parts of the period.

Vertebrate Localities

Lower Cretaceous marine life is best represented in the Neocomian beds of Europe (particularly Dalmatia), and there are equivalent deposits containing fish faunas and occasional aquatic reptiles in North America, Bahia (Brazil), and other regions. Much more prolific are Upper Cretaceous marine beds: the Chalk and related deposits of southern England and Westphalia; the similar Niobrara Chalk of North America; Australian marine deposits; Ceara (Brazil); etc. Notable are the limestones of Mount Lebanon, Syria, with a wealth of varied fishes.

In the Lower Cretaceous, continental deposits are none too abundant. The Wealden of southern England, northern France, and Belgium has a terrestrial fauna of very early Cretaceous age, and there are other, if less important, areas of this age in North America and in China and Mongolia. In the Upper Cretaceous the most famous of continental formations are the great dinosaur beds of western North America—Belly River, Edmonton, Lance, and equivalent—which extend over great areas from Alberta down the flanks of the Rockies to New Mexico. In Europe. Upper Cretaceous continental sediments are rare, except for an area in Transylvania. Dinosaur faunas are, however, present in India, China, and Mongolia and in a variety of localities in the southern continents—Patagonia, South Africa, eastern Australia.

Fishes

The selachians have expanded greatly. The stem hybodonts make their final appearances in the Cretaceous. Their descendants are numerous. Of some sixteen or so families of living sharks, all but four had representives in the Cretaceous seas. Many of the genera living today, such as *Carcharias, Isurus, Carcharodon, Galeocerdo,* etc. were already identifiable; since the opening of the Jurassic the major part of modern shark evolution had been accomplished. The skates and rays had also reached essentially modern conditions before the end of the Cretaceous, and there were already present such familiar living types as *Pristis,* the sawfish; *Raja,* the common skate; *Trygon,* the sting ray; and *Myliobatis,* the great eagle ray.

In the Cretaceous, *Mawsonia* and a few other rare types are the last of fossil coelacanths and were, indeed, thought to represent the termination of crossopterygian existence before the recent discovery of a living coelacanth. This period sees the replacement of the holosteans by the teleosts as the dominant group of fishes. Of lower actinopterygians, the seas contained the last rare palaeoniscoid and a few holosteans. Most of these last disappeared before Upper Cretaceous times, and only the pycnodonts remained in a flourishing condition. The teleosts were diversifying rapidly and replacing them. Most typical of Cretaceous teleosts were primitive isospondyls, clupeoids related to the tarpons and herrings. Early Cretaceous genera were few, but in the late Cretaceous there were swarms of such fishes, some (as *Portheus)* of very considerable size. More advanced groups were quick to make their appearance. In the Upper Cretaceous more specialized isospondyls of various types were present. A dearth of fresh-water deposits is perhaps responsible for our lack of knowledge of the important Ostariophysi—the carp-catfish group. There are, in the Upper Cretaceous, primitive eels, representatives of the deep-sea groups of Heteromi and Iniomi, and even of the great group of the Acanthopterygii. These spiny-finned fishes are, however, rare and include but a few genera of berycoids, perches, mackerels, pompanos, and a member of the specialized plectognaths.

Amphibians

To the meager record of the history of modern amphibian groups, we may add in the Cretaceous the first appearance of urodeles.

Marine Reptiles

Aquatic reptiles continue in abundance in the Cretaceous, although with some change in the membership of the groups concerned. The marine crocodiles disappeared early in the period. Ichthyosaurs, although seemingly highly adapted to a marine life, were on the downgrade. They appear to have become quite rare by Upper Cretaceous times; not a one, for example has ever been found in the richly fossiliferous Niobrara Chalk of North America. Plesiosaurs, on the other hand flourished vigorously, and the Upper Cretaceous chalk deposits contain abundant remains of a score of genera, among which *Polycotylus, Trinacromerum,* and *Elasmosaurus* are familiar forms; *Kronosaurus* of Australia, with a skull nearly 10 feet long, was the giant of the order.

The turtles, amphibious by nature, took to the sea in Upper Cretaceous times. Several families of cryptodires, including such genera as *Archelon, Protostega,* and *Toxochelys,* adopted this habitat and tended to modify themselves to it by paddle-like limb developments and lightening of armor.

Another interesting Cretaceous development was the appearance of marine lizards. In the Lower Cretaceous of Europe, *Dolichosaurus* and *Aigialosaurus* are representatives of monitor-like lizard families which were becoming adapted to an aquatic habitat. From such forms sprang the mosasaurs, such as *Mosasaurus* of Europe and *Tylosaurus* and *Clidastes* of North America, large and common predaceous reptiles of the Upper Cretaceous.

Dinosaurs

Of Lower Cretaceous dinosaurs we know but little, apart from those of the Wealden of western Europe, where *Iguanodon,* a large ornithopod, was common and of eastern Asia, where carnivores, sauropods, and ornithopods were present. Apparently, the fauna was, in general, similar in content to that of the Morrison.

Late Cretaceous dinosaur beds are highly developed in western North America, and more scattered beds and finds are present in every other continent. In many respects this Upper Cretaceous fauna is a very different assemblage from that seen a period earlier. Coelurosaurs still survive, but even these slenderly built saurischians have tended to grow in size and specialize into such forms as the toothless ostrich dinosaur, *Struthiomimus.* The larger carnivores are now represented by ponderous giants, such as *Tyrannosaurus* and *Gorgosaurus.* The sauropods had regressed in numbers and variety during the Cretaceous. Among all the numerous dinosaur finds from North America there are but two fragmentary specimens of these amphibious forms. It appears, however, that this paucity of sauropods is less pronounced in other regions, for there are a number of records of *Titanosaurus* and related sauropods from the Upper Cretaceous of the Southern Hemisphere and even in India and China.

Ornithischians abound in the Upper cretaceous. Little *Thescelosaurus* is a survivor of a very primitive ornithopod group. Of these bipeds, however, the dominant forms of the times were the amphibious duckbills, the hadrosaurs or trachodonts, abundantly represented by a score of genera which show numerous variations in the curious crests which frequently crown their heads. The stegosaurs had disappeared but have been replaced by another group of armored dinosaurs—*Ankylosaurus* and related types, low-bodies, large-skulled genera covered with a solid dorsal armor of bony plates. A final group of ornithischians is that of the horned dinosaurs, the ceratopsians, with a bony neck frill in the primitive genus *Protoceratops* and with large horns developed in addition in more progressive genera, such as *Triceratops* and *Monoclonius.*

Other Reptiles

Turtles are numerous in Upper Cretaceous deposits. Many are still amphichelydians of Mesozoic type, but representatives of the two existing suborders are present as well. There are a few pleurodires, in cluding even representatives of the living South American genus *Podocnemis.* More numerous are cryptodires. In addition to the marine forms already mentioned and survivors of the Jurassic *Thalassemys* group, there are relatives of the living marsh turtles and the soft-shelled trionychids. There is almost no evidence of rhynchocephalians. Sometimes placed in this order but of doubtful position is *Champsosaurus,* a small, long-snouted amphibious diapsid of the late Cretaceous and earliest Tertiary. in the Upper Cretaceous, lizards are, for the first time, abundant as fossils. Apart from the mosasaurs, half-a-dozen families are represented, including forms typical of most of the major subdivision of the group. In this period, too, there arises the last of the reptile groups to make its appearance—the snakes, represented by the rare and primitive types *Pachyophis* and *Simoliophis.*

As in the Jurassic, the crocodiles are well represented. Most are still in the mesosuchian atage of development but there are a number of more progressive genera leading to the modern crocodiles and alligators.

Pterosaur remains are rare except in the Niobrara Chalk of the Upper Cretaceous. They rhamphorhyncoid type is extinct, and by the Upper Cretaceous we find surviving only short-tailed pterodactyloids, such as *Nyctosaurus* and the giant, toothless *Pteranodon*.

Reptillian Extinction

By the end of the Cretaceous the greater part of the reptilian life of the Mesozoic had become extinct—all ichthyosaurs, all sauropterygians, all the dinosaurs and pterosaurs. Surviving into the typical Cenozoic we find only the turtles, lizards, snakes crocodilians, and a lone rhynchocephalian. We must not overemphasize the rapidity of this extinction; for some of the groups—such as the ichthyosaurs and perhaps the pterosaurs—had become reduced in numbers well before the end of the period. Many dinosaur groups, however, appear to have flourished greatly in very late Cretaceous deposits and their extinction at the Mesozoic-Tertiary boundary is one of the most dramatic events in vertebrate history.

The reasons for this reptilian catastrophe have been much debated, and the causes may have been complex in nature. It has however been reasonably argued that geological processes may be in great measure fundamentally responsible. In the late Cretaceous there came about the Laramide revolution a time of mountain-building during which there began the elevation of the Rockies and other mountain chains. This condition of rising lands might well have affected markedly the more amphibious dinosaur groups, such as the sauropods and duckbills by limiting the areas of the swamps and lagoons in which they made their livelihood. Food materials, too would be affected, for these geological disturbances brought about climatic shifts, which are seen reflected in marked changes in vegetation in the late Cretaceous. Many herbivores are narrowly restricted in their diets, and floral changes may be responsible for much extinction. Finally the disappearance of carnivores would inevitably follow the disappearance of the herbivores upon which they fed.

The reasons for the elimination of marine forms are more difficult to deduce. It may be however, that the replacement of holostean fishes by progressively higher and presumably more efficient teleost types was a factor, in addition to changes in invertebrate life.

Birds

There is a gap in our history of birds between the late Jurassic and the late Cretaceous. Even in deposits of the latter age our knowledge of birds is almost entirely confined to oceanic or water-dwelling forms, such as the ternlike *Ichthyornis* and the swimmer *Hesperornis* of the Niobrara Chalk. These were typically birdlike in most skeletal features, but they still retained teeth. Possibly, however, other diversified and progressive forms may have been included in the unknown inland avifauna.

Mammals

Of Lower Cretaceous mammals we know exactly three teeth (of a multituberculate). In the late Cretaceous mammals are again found but are restricted almost entirely to a few localities in the Lance formation of Wyoming and to one in Mongolia. The fauna is markedly different from that seen in the Jurassic. The multituberculates are still present and are plentiful, but the other Jurassic orders—triconodonts, symmetrodonts, and pantotheres—have disappeared. The common types are marsupials, some strikingly similar to the living opossums. And, in addition, there are a few primitive placentals prophetic of the emergence of this group in Tertiary days.

Cretaceous: Summary

Among the principal features of Cretaceous vertebrate history may be cited the following: (1) among fishes further deployment of the selachians and a dwindling in importance of the holosteans, which were being replaced by primitive teleosts; (2) appearance of the first urodeles; (3) among marine reptiles, continued importance of plesiosaurs, appearance of marine lazards and turtles, reduction of ichthyosaurs; (4) progressive modernization of the crocodiles and turtles and development of primitive snakes; (5) climax in size of pterosaurs; (6) among dinosaurs, development of the largest carnivores, dwindling importance of sauropods, appearance of duck-billed ornithopods, heavily armored ankylosaurs, and horned dinosaurs; (7) birds reaching modern development except for the retention of teeth; (8) replacement of most Jurassic mammal groups by small primitive marsupials and placentals; (9) and finally the complete extinction by the end of the period of most of the characteristic reptilian types of the Mesozoic, including all the dinosaurs, the ichthyosaurs, plesiosaurs, marine lizards and pterosaurs.

Mesozoic History

The reptiles are the most interesting and most important vertebrates of the era. The Mesozoic includes the full cycle of existence of many major reptilian types—the dinosaurs, pterosaurs, and the varied marine reptiles—as well as the age of origin and expansion of still surviving groups, such as the turtles, lizards, and crocodiles. But it includes as well, in the early Triassic, the close of an earlier cycle in reptilian history which had begun near the end of the Carboniferous, with cotylosaurs and mammal-like reptiles as its chief characters. The earliest Mesozoic is an age of transition from the old to the new. On the other hand, the end of the era is a seemingly natural point of cleavage as regards vertebrate life.

With other groups, too, it seems that the Mesozoic includes not merely one typical cycle but, at its beginning the final chapter in the old order. Among the bony fish the palaeoniscoids were still dominant at the beginning of the Triassic, just as they had been since the beginning of the Carboniferous. Before the close of the Mesozoic not only had the holosteans replaced them, but they in turn had been driven into insignificance by the progress of the teleosts. Again, the only known Triassic amphibians were the last survivors of the old labyrinthodonts of the Paleozoic and only in the Jurassic and Cretaceous do we find the first characteristic members of the modern groups. Among cartilaginous fishes, also the Triassic was a transitional period, while the Jurassic and Cretaceous witnessed a great revival in shark development. Primitive birds and primitive mammals were characteristic of the Mesozoic; but this, as far as our present knowledge goes, means merely the Jurassic and Cretaceous, for these groups are unknown in the Triassic.

The Triassic was thus a boundary period—a time of transition. In the Jurassic and Cretaceous we witness a full and characteristic Mesozoic life. And with the coming of the Cenozoic we enter a new age, the Age of Mammals.

Chapter—15

Vertebrates of Mammalian Era

With the extinction of the great reptiles there began a new chapter in the history of land vertebrates. The mammals had until then remained small, rare, and inconspicuous but had seemingly progressed far in structural organization. once the way was cleared, they emerged from obscurity and began a spectacular radiation into a host of types, many of which survive to the present day. The account of the Cenozoic given in this chapter will be based mainly on the history of the Mammalis; other groups are treated but briefly.

This last phase of the earth's history, comprising the Cenozoic era, is usually divided into a Tertiary period of considerable magnitude and a second period, the Quaternary, barely begun. The Tertiary may be divided into some five epochs—Paleocene, Ecocene, Oligocene, Micocene, and Pliocene; the Quaternary includes only the short Pleistocene epoch and Recent times. These divisions are based in most cases upon the degree of modernity reached by invertebrates in European marine deposits. In Europe these epochs have been subdivided into a series of ages listed in Table 4. In many instances the terrestrial deposits of Europe can be correlated with marine beds and hence placed in the appropriate stage. But in North America this is not so readily dine, for most American mammalian deposits are in the interior western states, far from marine fossil-bearing sediments. A series of ages based on the mammals contained in the various North American beds has, however, been established, and approximate correlations can be made between these ages and the European ones. A similar series has been erected for South American deposits. These three series of ages are given in Table 4, together with data on the best-known localities, areas, or formations in the various continental areas.

Cenozoic Deposits

The lost best and most complete series of mammal-bearing beds is that found in western North America, where sediments, spreading out from the Rocky Mountains and Sierras, have preserved a good record of numerous faunas ranging from the beginning of the Cenozoic to the Pliocene. Europe in Tertiary times was a peninsular area frequently flooded by shallow marine waters; the deposits along the shores of these seas or derived from the rising Alps include numerous fossiliferous formations which have been studied intensively for more than a century. In Asia early Tertiary deposits are well represented in Mongolia; later epochs are best represented in the Siwalik Hills and other areas of sediments from the rising Himalayas; Burma supplies further data for the early Tertiary and Pleistocene, China for the later Cenozoic, Java for the Pleistocene. In Australia, Pleistocene fossils are rather plentiful, but only a single terrestrial mammal is known from the entire Tertiary. Africa appears to have been a stable land mass during most of the Cenozoic; apart from later Eocene and early Oligocene faunas from Egypt, Tertiary continental deposits are extremely restricted in area and faunas. In South America there are extensive sediments in Patagonia, which have preserved much of the curious fauna of mammals which evolved there during Tertiary times.

Cenozoic Conditions

We have little reason to believe that there has been much alteration in the extent position of the continental areas during the Cenozoic, and almost all known facts in the distribution of mammals may be accounted for by small changes in the present continental relations. It is probable that australia has been in its present condition of isolation since the end of the Mesozoic, for almost none of the Cenozoic types of terrestrial mammals has reached that region. The independent evolution seen in South American mammals indicates that continent became separated from North America very early in the Cenozoic and that an easily traversed connection was not rebuilt until about the end of Tertiary times. Although, as noted above, the history of Tertiary Africa is poorly known, it is very probable that this region was isolated from Eurasia during the earliest Cenozoic and that opportunity was given there for the initiation of peculiar and characteristic placental groups. Comparison of successive North American and Eurasian faunas indicates that the Bering Straits connection between the two areas was made and broken a number of times.

Various Tertiary land bridges have been proposed between continental areas to account for real or fancied similarities between the regions so connected. Most of them rest of flimsy evidence and need not be considered here. Stronger support has been given to a proposed connection between South America and Australia, either directly or via Antarctica. This is based upon the similarity in the marsupials of the two continents; but this similarity can be as readily explained as being due to parallel evolution from common ancestors, which were presumably widespread in northern continents in late Mesozoic days. The most serious problem, discussed earlier, is that regarding a Tertiary bridge between South America and Africa.

Early Tertiary times seem to have had an equable, warm climate, which gradually gave place to the extreme temperature differences found in the Ice Age, with a cold arctic zone grading rapidly southward into warm tropics, a situation still existing at the present time. This climatic change seems to have resulted in a slow southward retreat of many forms the better-known regions of Eurasia and North America and the survival in the tropics of many groups long after their extinction in the present north-temperate zone. The peccaries and tapirs of south America are relatively recent migrants from the north; the antelopes and associated mammals of tropical Africa form a fauna strikingly similar to that of Europe in the early Pliocene.

The geographic regions customarily used by zoogeographers in interpreting the distribution of life today reflect accurately the geological history. Some six regions are generally accepted: the Palaearctic, for Europe and the Mediterranean region and northern and central Asia; Nearcotic, for North America; Oriental, for Southern Asia and the East Indies; Ethiopian, including most of Africa; Neotropical, for South and Central America; and Australian, including Australia, New Guinea, and certain adjacent islands.

These six regions are not all equally distinct from one another. The Australian region stands in sharp contrast with all five other regions in its faunal content, a feature associated with the separate Cenozoic history of that continent. The Neotropical region also shows very individual characteristics, due to the development of a peculiar fauna during the Tertiary isolation of South America. The Palaearctic and Nearctic regions are identical in climate and separated only by the variable make-and-break between Alaska and Siberia. Their faunas are so similar that they are frequently considered as a single Holarctic area. The distinction of Ethiopian and Oriental regions from the Palaearctic is due primarily to the relatively recent establishment of sharp temperature gradients; during much of the Tertiary the three were essentially a single faunal unit.

The major interest in the Cenozoic lies in mammalian history; in most of the other vertebrate groups relatively little evolution took place during this era, and a treatment of them by successive epochs is unnecessary.

Fishes

Shark and ray teeth are common in many Tertiary marine beds, and chimaeroid toothplates are occasionally present; only in exceptionally favorable circumstances are skeletons of cartilaginous fishes discovered. All, or nearly all existing groups of sharks skates, and rays appear to have been established in the early Tertiary, and but little further evolution appears to have taken place. Actinopterygians below the teleost level are rare in the Cenozoic. The pycnodonts persisted into the Eocene; today there survive only the African polypterids, the sturgions paddlefishes, garpike *(Lepidosteus)*, and *Amia*, the bowfin. The last three are represented by characteristic specimens from the Eocene, the first two by fragmentary remains of equally early date. A single living coelacanth crossopterygian has been recently discovered; but no Tertiary fossils are known, presumably because this group had taken up life in the deep seas, an environment of which there are few known geological records. Of the three surviving lungfishes, there are Tertiary records of the Australian and African genera but not of the South American forms.

The teleosts are the dominant Tertiary fishes and are found in a variety of deposits both fresh and salt in origin; the shales of the Green River region of North American and of Monte Bolca in Italy are famous Eocene fish localities; diatom beds in California and marine deposits in Algeria and at Licata in Sicily are among prominent late Tertiary deposits. In fresh waters the dominance of the Ostariophysi, the carp-catfish group, is a striking feature of the Cenozoic; in marine deposits there is an increasing variety of higher; spiny-finned teleosts, although the old-fashioned isospondyls are still abundant.

Amphibians and Reptiles

Amphibian remains are relatively rare in the Tertiary and shed little light on the history of the frog and salamander groups represented. Turtles are abundant in many formations, and most of the living types are already represented by close relatives at the very beginning of the era. In the Eocene the primitive amphichelydians became extinct. The pleurodires, subsequently rare as fossils, have been confined to the southern continents since the earlier Cenozoic. Lizards are not infrequently discovered. The snakes are the one progressive group of reptiles during these latter times; poisonous snakes are apparently a Tertiary reptilian innovation. Crocodiles, mostly related to living forms, are fairly numerous throughout the period. *Champsosaurus* of the late Cretaceous persists into the early Eocene. Fossil representatives of the Typical Rhynchocephalia are unknown.

Birds

It has been said that the Cenozoic is as much the Age of Birds as it is the Age of Mammals; the birds are the dominant group in the air during the entire era. The fragile nature of their bones and their usually small size have rendered the unraveling of their history difficult. The principal points in their Tertiary history, including the sporadic development of large flightless forms, have been noted in an early chapter. Toothed forms appear to have become extinct by the end of the Cretaceous; except for some tropical types; almost every large group of modern birds, and even most of the major families, were already present in the early Tertiary.

PALEOCENE

This oldest of Tertiary epochs is relatively poorly represented in the marine sequence, but the existence of a pre-Eocene chapter is abundantly proved by numerous older continental deposits, particularly well developed in North America. In the San Juan basin of New Mexico and southern Colorado, Lower, Middle, and Upper Paleocene deposits are represented by the Puerco, Torrejon, and Tiffany beds; the Polecat Bench formation of North-western Wyoming appears to cover the entire time interval between the end of the Cretaceous and the True Eocene; "Fort Union" beds in eastern Montana and adjacent states have produced Paleocene mammals in various areas, particularly in Middle Paleocene deposits near the Crazy Mountains; the Dragon fauna of Utah is an early Middle Paleocene assemblage.

In other continents Paleocene mammals are unknown, rare, or of late date. In Australia, as we have noted, the Paleocene fauna is unknown, as is that of the later Tertiary eras; and there are no African Paleocene mammals. In Europe there are a few late Paleocene forms mainly from Cernay near Rheims; in Asia s small assemblage, apparently equally late in time, is found in the Gashato of Mongolia. In South American a handful of specimens has recently been discovered in the Rio Chico of Patagonia, some of which come from an obviously late horizon; a few scraps may be earlier. In consequence, our discussion of the Paleocene is based mainly on the American record as seen in the Puerco and Torrejon and their equivalents; the late American Paleocene and that of other continents is considered separately.

The terrestrial fauna of the Paleocene, with the dinosaurs absent, is mainly a mammalian one. But if we were able to visit the Paleocene, the aspect of the fauna would be of a strange and unfamiliar sort. Many of the mammals then present are assigned to orders still existing—Insectivora, Carnivora, Primates—but these Paleocene representatives of the modern orders were for the most part archaic or aberrant types. Not a single living family is present in the characteristic Paleocene, and much of the fauna belonged to orders now entirely extinct.

The few mammalian types known in the Cretaceous are still present. Opossum-like marsupials persist in insignificant numbers. The archaic multituberculates, on the other hand, are present in abundance; most were small members of the *Ptilodus* group; but, in addition, we find *Taeniolabis* and its allies, forms of considerable size and still higher specialization.

The insectivore types persist from the Cretaceous into the Paleocene but are not particularly prominent; once having played its part as ancestral placentals, this order is tending to sink into the relatively insignificant role it plays today. There are a few zalambdodonts and a number of genera of the leptictids, forms leading toward the dilambdodont families of later epochs. Rather more prominent are some side branches, such as *Mixodectes* and the pantolestids, which had departed far from the typical insectivores but were hardly important enough to merit ordinal separation.

Much more prominent were early carnivores, represented by a score of genera. Almost all are members of the Acreodi, without specialized carnassials and still very close to their insectivore ancestors; from them by the Middle Paleocene had developed *Dissacus* and other early members of the blunt-toothed mosonychids. The pseudocreodonts, characteristic of the Eocene, are not present in typical Paleocene horizons, but the progressive miacids, destined to give rise to modern carnivores, were represented, although sparsely.

A third placental group is that of the primates, which appear in the Middle Paleocene. The forms present are more or less transitional in their cheek teeth between the insectivores, on the one hand, and lemurs and tarsioids, on the other. A puzzling feature is that nearly all the Paleocene genera, in several different families, have developed large and often rodent-like incisors, perhaps in relation to some type of feeding habit now difficult to interpret.

Ungulates were abundant but ungulates of unfamiliar types, of primitive structure, and for the most part, of small size. Almost all may be included in the Condylarthra as a basal order of hoofed mammals. The low-crowned, "squared-up" molars indicate a shift to a herbivorous diet; hoofs are present in some cases, but no all; in many respects, however, the early condylarths were still close to the insectivore-carnivore stock. Four families are present, of which *Mioclaenus, Periptychus, Meniscotherium,* and *Tetraclaenodon* are representative. The first three are apparently rather aberrant; the last is a member of the phenacodont family which may be close to the ancestry of more advanced ungulate orders.

The orders so far mentioned are still so close to the ancestral placentals that the group boundaries are difficult to determine; were we ignorant of their later histories, we should be justified in including almost the entire assemblage in a single ordinal group.

More divergent forms were, however, already making their appearance. The distinction is often made between "archaic" and "progressive" mammal groups, the former being types which tended to a high degree of specialization at an early stage but were destined for early extinction; the latter, forms which tended to evolve more slowly but with greater adaptability and greater success. Among the earlier Paleocene mammals, archaic types are already distinguishable in the taeniodonts, such as *Conoryctes,* the giants of their times, and *Pantolambda,* and ungulate which had already grown to the size of a sheep and which in its relatively heavy build foreshadows the larger amblypods of later times. A tendency toward prematurely large size is also seen in some of the condylarths and certain of the carnivores.

Following the typical faunas of the epoch, we find in the late Paleocene of North America a fauna in which most of the groups previously present still persist but are often represented by larger types and in which the mammalian assemblage is more varied and foreshadows that of the true Eocene. The Acreodi are already reduced in numbers, and the presence of *Oxyaena* marks the appearance of the Pseudocreodi. The early ungulates were increasing in size and variety; particularly prominent were great amblypods, such as *Barylambda* and, at the end of the Paleocene,*Coryphodon*. Several new mammalian groups appear, however, whose earlier history is unknown. These include the first of the uintatheres among archaic ungulates; the first of the peculiar tillodonts; *Palaeanodon,* an ancient relative of the South American edentates; and *Paramys,* an ancestral rodent.

The few Paleocene mammals of Europe, late in late, add little to our knowledge of mammalian evolution. Such froms as have been discovered appear to be close to American mammals of similar age, although generically distinct, and indicate that there was a common fauna widespread in northern regions. The known Paleocene of Asia, equally late in date, is puzzling and disappointing. It was long believed that Asiatic beds of this age, when discovered, would reveal the ancestors of some of the groups, particularly of "modern" orders of ungulates absent in the American Paleocene.

Table 15.1
Cenozoic Vertebrate Localities

Epoch	*European Ages*	*European Localities*	*Asiastic Localities*	*African Localities*	*North American Ages*	*North American Localities*	*South American Ages*	*South American Localities*
Pleistocene		Caves, river terraces, etc., early Pleistocene inc. Villafranchian of of Val d'Arno (Italy), Perrier (France), Norwich Crag (Eng.)	Pinjor (Siwaliks); fissure deposits of Sze-Chuan and loess of China; Irrawaddy beds (Burma); volcanic ash deposits of Java	Caves and superficial deposits in Algeria, E. Af. (Oldoway, etc.), S.Af.		Caves, postglacial; La Brea, McKittrick (Calif.); Hay Springs Broadwater (Neb.)		Pampean beds (Arg., etc.) Tarija (Bolivia); Brazilian caves
	Astian	Roussillon, Montpellier (France)	Dhok Pathan (Siwaliks)		Blancan	Hagerman (Idaho); Benson, San Pedro (Ariz.); Rex Road (Kan.); Blanco (Tex.)	Chapadmalalan	Chapadmalal fau -na of Pampas
Pliocene	Plaisancian	Perpignan, Trévoux (France); Red craig (Eng.);Piemont region (Italy) *M**	Nagri (Siwaliks)		Hemphillian	Rattlesnake (Ore.); Hamphill (Tex.); Long Island (Kan.); Ash Hollow (Neb.)	Monteher-mosan, Tunuyanian	Tunuyan beds and part of Araucanian series (W. Arg.); Monie Hermoso (southern pampas)
	Pontian	Concud (Spain); Curcuron, Mt.Léberon (France); Antwerp (Belg.) *M*; Eppelsheim (Ger.); Vienna basin (Austria) *M* (in part); Pikermi, Samos (Greece)	Maragha (Persia); Chinji (Siwaliks); Honan, Shansi, red beds (China)	Wadi Natron (Egypt)	Clarendonian	Santa Fe. in part (N. M.); Clarendon	Huayquerian, Mesopotamian	Huayqueria beds and part of Araucanian series (W. Arg.); Mesopotamian beds (Entre Rios region, Arg.)
	Sarmatian	Antwerp (Belg.) *M*; Sevastopol (Russia)	Kamlial (Siwaliks); Tung Gur (Mongolia)	Moghara (Egypt) Uganda	Barstovian	Pawnee Creek (Colo.); Madison Valley (Mont.); Mascall (Ore.); Barstow, (Calif.)	Chasicoan, Friasian	Chasico beds (of southern Pamps); Rio Frias beds (of western Patagonia)
Miocene	Vindobonian	Sansan, Simorre, Grive, St. Alban, St. Gaudens (France); Eibiswald, Eggenburg, etc. (Austria); Molasse of central Europe in part; Steinheim, Oeningen (Ger.), Mte. Bamboli (Italy); Belg. *M*	Loh (Mongolia)		Hemingfordian	Sheep Creek, Marsland (Neb.); Hawthorn (Fla.); Temblor (Calif.); Calvert (Md). *M*	Santacrucian	Santa Cruz beds of Patogonia

* *M* = marine

Table 15.1 (Contd.)

Epoch	European Ages	European Localities	Asiastic Localities	African Localities	North American Ages	North American Localities	South American Ages	South American Localities
	Burdigalian, Aquitanian	Faluns Touraine, St. Gérand-le Puy, Agen region, Orléans sands (France); Molasse of central Europe in part; Mainz basin, Ulm (Ger.)	Bugti beds (Baluchistan)	Kenya Colony; Namib (S.W.Af.)	Arikareean	Gering, Monroe Creek, Lower Harrison (Neb.); Rosebud (S. D.); M.-U. John Day (Ore.)	Patagonian	Patagonian formation *M*
	Chatian	Sandstones of Beauce *M*, La Rochette, La Milloque, Agen reg ion (France); Mainz basin (Ger.); Cadibona (Italy)	Hsanda Gol (Mongolia) Turgai Turkestan)		Whitneyan	Upper Brulé (*Protoceras* beds) of White River series (S.D., N.D., Neb., Wyo., Colo.)	Colhuehuapian	Colhué Huapi beds (*Colpodon* fauna) of Patagonia
Oligocene	Stampian (Rupelian)	La Ferte—Aleps, Bournoncle–St. Pierre, Aveyron, Quercy phosphorites in part (France); Flonheim, Weinheim (Ger.)			Orellan	Lower Brulé (*Oreodon* beds) of White River series (S.D., N. D., Neb., Wyo., Colo.)		
	Sannoisian (Lattorfian)	Quercy phosphorites in part, Ronzon, Brie limestone Lobsann (France); Hampstead beds (Eng.)	Ulan Gochu Mongolia)	Upper beds of Fayûm (Egypt)	Chadronian	Chadron (Titanothere beds) of White River series (S.D., N.D., Neb., Wyo., Colo.)	Deseadan	Deseado beds (*Pyrotherium* fauna) of Patagonia
	Ludian, Bartonian	Gypsum beds of Montmartre, Quercy phosphorites in part, Débruge, Gargas, St. Hippolyte de Caton, Robiac, Castres, Calcaire de St. Ouen (France); Mormont (Switz.); Isle of Wight (Eng.)	Ardyn Obo, Shara Marum, Irdin Manha (Mongolia); Pondaung (Burma)	Qasr-el-Sagha beds of Fayûm, Upper Mokattam beds *M* (Egypt)	Duchesnean, Uintan	Duchesne River, Uinta (Utah); Washakie in part (Wyo.); Sespe in part (Calif.); Jackson of southeastern states *M*		

Table 15.1 (Contd.)

Epoch	*European Ages*	*European Localities*	*Asiastic Localities*	*African Localities*	*North American Ages*	*North American Localities*	*South American Ages*	*South American Localities*
Eocene	Auversian, Lutetian	Calcaire grossier, Issel, Argenton, Buchsweiler (France); Bracklesham (Eng.); Egerkingen (Switz.); Mte. Bolca (Italy)		Mokattam beds *M*., Birket-el-Qurûn (Egypt); Nigeria *M*	Bridgerian	Bridger, Washakie in part (Wyo.); Green River in part (Wyo., Colo., Utah); Huerfano in part (Colo.)	Mustersan	Musters beds with *Astraponotus* Fauna (Patagonia)
	Cuisian, Sparnacian	Argile plastique of Soissons region, Meudon, Epernay (France); London clay (Eng.); Erquelines, Orsmael (Belg.)			Wasatchian	Largo, Almagre (N. M., Colo); Huerfano in part (Colo.); Gray Bull of Bighorn basin Lysite and Lost Cabin of Wind River basin (Wyo.)	Casamayoran	Casa Mayor beds with *Notostylops* fauna (Patagonia)
	Thanetian	Cernay (France); Thanet sands *M* (Eng.)	Gashato (Mongolia)		Clarkforkian Tiffanian	Clark Fork, Silver Coulee, Polecat Bench in part (Wyo.); Fort Union in part (Wyo., Mont.); Plateau Valley (Colo.); Tiffany (N.M.); Paskapoo (Alberta)	Riochican	Rio Chico (Potogonia); Pernambuco *M* (Brazil)
Paleocene	Montian				Torrejonian, Dragonian	Torrejon (N.M.); Polecat Bench in part (Wyo.); Crazy Mountain Field, Fort Union (Mont.); Dragon fauna (Utah)		
					Puercan	Puerco (N.M.); Polecat Bench in part (Wyo.)		

The single fauna so far found, the Gashato, has not realized these hopes. It reveals, much as in North American beds of the same horizon, multituberculates, insectivores, carnivores, and condylarths and a uintathere. A welcome find is the oldest fossil hare, *Eurymylus*. The commonest fossil is an ungulate, *Palaeostylops;* but, startlingly, this belongs not to any normal northern group but to the characteristic South American order of Notoungulata.

In South American itself the Rio Chico beds appear to be Paleocene. A few fragments may come from fairly low levels; most, however, appear to be late in date. The material is poor but shown, nevertheless, that most of the groups which were to characterize the later Tertiary of that continent were already present. These forms include marsupials, edentates (represented by armadillo scales), litopterns, condylarths presumably related to litoptern ancestry, notoungulates, and possibly astrapotheres and pyrotheres. We are accustomed to think of the Northern Hemisphere as the area of origin of mammalian groups and hence tend to interpret the presence of these forms as due to early migration from the north. It is, however, equally possible that part or all of the distinctive South American mammals had developed locally.

It is obvious that our known Paleocene faunas, while giving us interesting chapters in the early deployment of the placentals, do not give us a complete story. As we have said, neither the Paleocene nor any other Tertiary epoch gives us any idea of marsupial evolution in Australia. There are no Paleocene deposits in Africa, where one may expect that the early development of subungulates and possibly other types was under way.

We have, further, no knowledge of stages in the evolution of rodents and lagomorphs, which appear fully developed at the close of the Paleocene, or of the perissodactyls and artiodactyls, which appear abruptly at the beginning of the Eocene. Possibly these orders may have been evolved in some area of Eurasia or North American unrepresented in our present records; possibly in familiar regions but in environments not represented by fossiliferous deposits.

EOCENE

In the Eocene, North America is, again, the continent in which the faunal sequence is most abundantly represented. Lower Eocene faunas, usually termed "Wasatch," are present in abundance in the Bighorn and Wind River basins in Wyoming; the Bridger basin of southwestern Wyoming is the classic Middle Eocene collecting ground; that for the Upper Eocene is to the south in the Uinta basin of northeastern Utah.

In Europe the Lower Eocene is known from relatively poor beds in the London basin and in related deposits in northern France and Belgium; later Eocene horizons are increasingly fossiliferous, including such well-known localities as Egerkingen in the Swiss Middle Eocene and the Upper Eocene gypsum beds of Montmartre (Paris). The Quercy phosphorites of south-central France are cave fillings in older limestones containing numerous small mammals. Part are late Eocene; the later deposits extend into the Oligocene. In Asia there are few fossils before the latter part of the epoch, when several faunas are present in Mongolia and in Burma. In South America the Casa Mayor beds are probably of early Eocene age; the Musters, also in Patagonia, rather later. African now enters the Tertiary terrestrial record. The Mokattam beds of Egypt contain a marine mammalian of Middle Eocene age, and continental deposits begin in the Fayum of western Egypt in late Eocene times.

In the Lower Eocene, it appears, the northern continental areas formed, much as today, a single zoogeographic region, for Europe and North America had many genera in common. There was no sharp break between the life of the Paleocene and the new epoch; for with the exception of the *Periptychus* type of condylarths, every major group present here in the Paleocene was still present. Most of the older groups were destined to become extinct; but their disappearance was a gradual one. The earliest Eocene contains the last survivors of the long-lived multituberculates; opossums (*Peratherium)* persist from the Eocene onward into the Miocene of both Europe and North America. There are insectivores—primitive, typical, and aberrant. Creodonts dominate the scene among carnivores. The Acreodi, however, were rapidly replaced by numerous pseudocreodonts of the oxyaenid and hyaenodont families, with well-developed carnassials. The aberrant primates of the Paleocene are still present and, in addition, more typical lemurs, such as *Notharctus*. Condylarths persisted in the Eocene, where *Phenacodus* is a familiar Lower Eocene ungulate and *Hyopsodus,* a late survivor, represented the group in America until the end of the epoch. *Paramys* and other rodents are present in both continents, but the order is relatively rare in the earlier European Eocene. The amblypod *Coryphodon* is found in the Lower Eocene in Europe and American alike. A second curious appearance of the notoungulates (in addition to *Palaeostylops)* is that of a single fragment from the Lower Eocene of Wyoming; except for these occurrences, notoungulates have never been reported beyond the confines of South America. Uintatheres were never present in Europe but developed further in North America. The tillodont *Esthonyx* was present in both hemispheres; taeniodonts were present in the American Lower Eocene but not in Europe.

The diagnostic feature marking the beginning of the Eocene is the presence of the two great modern groups of ungulates—the perissodactyls and the artiodactyls. Both appear simultaneously and seemingly abruptly in Europe and North America in Lower Eocene times. In both groups the ordinal characters are fully established, but their ancestry and areas of development are unknown. In the early Eocene the artiodactyls are rare. The perissodactyls, on the other hand, were at once abundant; *Eohippus* and primitive tapiroids are common finds in the "Wasatch,"and close relative were present in Europe; in North America the first titanotheres and chalicotheres were present before the close of the Lower Eocene.

For the later history of the Eocene in the Holarctic area, Asiatic evidence is available, as well as that from the European and American extremities of the region. In some groups the entire area shows a fairly uniform picture, in others—notably the ungulates—there are marked differences which suggest regional isolation in the Middle Eocene, partially corrected by intermigration toward the end of the epoch.

Bats appear full-fledged in both hemispheres in the Middle Eocene; various primitive and aberrant primates persisted throughout the epoch. Among the carnivores, typical acreodonts became extinct early in the Eocene, but other creodonts flourished in all three northern areas. The Pseudocreodi are the characteristic carnivores; some remained small but others, such as *Pterodon* and *Hyaenodon* of both hemispheres and the oxyaenid *Sarkastodon* of Mongolia, tended to grow to great size. The blunt-toothed mesonychids are absent in Europe after the early Eocene but remained until the end of the epoch in North America and Asia, *Andrewsarchus* of the latter continent being a giant form. Miacids are common in both land masses. In the late Eocene this family comes to an end, seemingly by the evolution from it of the first of the more typical fissipedes, with such forms as *Cynodictis* and *Procynodictis* representing the forerunners of civet, weasel, dog, and cat families. Primitive aplodontoid rodents are present in both hemispheres but are more common in North America, where rodent families of this group continued to flourish in the later Tertiary. We have noted the occurrence of a rare lagomorph in the Paleocene; the next record of hares is in the Upper Eocene of eastern Asia and North America; the former continent may well have been the place of their development.

Many features in the history of nonungulate mammals in the later Eocene are thus similar in Eurasia and North America. But there were notable differences; the former region shows, on the whole, the more progressive situation. Shrews and moles are present in Europe by the end of the Eocene but do not appear in North America until later; on the other hand, the aberrant mixodectid insectivores persisted in North America in the later Eocene, as did the taeniodont *Stylinodon,* tillodonts, and palaeanodont edentates. Among the rodents the pocket gophers appeared in the late Eocene in North America; in Europe there developed at this time anomaluroid families posibly ancestral to hystricoids and myomorphs.

It is among the ungulates the Europe and North America show the greatest divergences in the course of the Eocene. Most of the archaic ungulates of the Paleocene and early Eocene have become extinct in both areas. In North America, however, the ungainly uintatheres and the small condylarth, *Hypopsodus,* survived until the end of the epoch. Perissodactyls flourished in both regions but followed different lines of development in many respects. In North America the horses, in the form of the relatively rare genera, *Orohippus* and *Epihippus,* remained conservative in size and structure; in Europe, on the other hand, there was a rapid trend toward large size and specialization, some specimens of the three-toed *Palaeotherium* of the late Eocene being even larger than modern horses. Primitive tapiroids were present in both continents but more abundant in Europe, where *Lophiodon* was a typical large form. Primitive running rhinoceroses are absent in the European Eocene; *Hyrachyus* and its relatives are confined to North America. More progressive running rhinoceroses—the hyracodonts—and such amphibious forms as *Amynodon* inhabited both North America and Asia. In the late Eocene of all three northern continents are found the first rare specimens of the true rhinoceroses which were to become abundant in the Oligocene. The oldest recorded chalicothere is from the Middle Eocene of North America; but by the late Eocene this family is represented in the Asiatic area in which it later flourished. Titanotheres, too, are first found in North America but by the end of the epoch were developing vigorously in Asia as well, and two stragglers have been discovered in eastern Europe.

Still more divergent in development were the artiodactyls, which, we have noted, were rare in early Eocene days. Little progress appears to have taken place in the group in North America. In Europe and Asia, however, middle and late Eocene faunas show a remarkable expansion. Some of the types developed were bunodonts suggestive of later pigs and peccaries, although apparently not the actual ancestors of these forms. Most, however, are primitive selenodonts—caenotheres, anoplotheres, and xiphodonts are the dominant European types. In the late Eocene of that continent are found, further, primitive pecorans, in such forms as *Amphimeryx* and the tragulid *Gelocus*. In the Upper Eocene there appears in North America, apparently as the result of invasion from Asia, a similar series of artiodactyls, which, however, belong to different families. They include some rather piglike piglike types, primitive selenodonts ancestral to the camels, oreodonts and agriochoeres, and a considerable number of pecorans of the distintive hypertragulid group.

We have noted above some features in which the later Eocene of Asia is comparable faunally with Europe or North America or both. Imperfect, however, as our knowledge of Eocene history there is, we can see that in certain respects Asia had developed differently from the regions to west or east. Among the primitive ungulates, large condylarths and uintatheres persisted there through the Eocene. Horses were absent, but tapiroids were apparently common. Anthracotheres were very abundant in the late Eocene and may have developed in that area. *Archaeomeryx* of Asia is the most primitive known hypertragulid; presumably this group reached America from an earlier Asiatic home, and quite probably the camels and oreodonts were of Asiatic origin.

In the middle and later Eocene appear the first representatives of the two purely marine orders of mammals, the Cetacea and the Sirenia. We are ignorant of their terrestrial forebears and cannot be sure of their place of origin. But although archaeocete whales (such as *Basilosaurus)* and sirenians are found in the American Eocene, most of the earliest and most primitive types are from the Mediterranean region and especially from the Mokattam beds of Egypt. This suggests that these groups originated in Africa during the earliest Tertiary. The only known terrestrial African mammals of the Eocene are the early proboscideans *Moeritherium* and *Barytherium* from the Fayum. Not improbably, proboscideans, sirenians, and hyracoids (which appear slightly later in that area) are products of early Tertiary African evolution from a primitive subungulate stock.

In South America the Casa Mayor beds of the early Eocene give us our first adequate picture of the specialized Tertiary faunas of that continent; the Musters beds of the Middle Eocene show a slightly more advanced stage. Except for rodents and primates, all members of the Tertiary assemblage are represented. Marsupials were abundant and include not only opossums but larger and more strictly carnivorous borhyaenids (one already as large as a wolf), caenolestids, and *Polydotops* and its relatives which paralleled the multituberculates in dentition. Edentates are represented by armadillos of a primitive nature, such as *Utaetus. Didolodus* and similar types are currently regarded as condylarths related to the ancestry of the litopterns; of these latter there are genera ancestral to the two families prominent in the later Tertiary. The Notoungulate are, as later, the commonest of ungulates; the Eocene genera appear to include forms antecedent to the various subgroups of later epochs as well as numerous primitive notioprogonians such as *Notostylops;* all are small primitive little differentiated, and with low-crowned teeth. The astrapotheres and pyrotheres, as archaic ungulates, appear to have played in South America the role of amblypods and uintatheres of the Northern Hemisphere; large representatives of these two groups were the giant of the fauna.

OLIGOCENE

European Oligocene deposits are numerous. A portion of the Quercy phosphorites are of early Oligocene date. Ronzon in France and the Mainz basin in Germany are among the better-known localities, a large proportion of which are located in France. In Asia, Oligocene continental beds are almost entirely confined to Mongolia, with Ulan Gochu and Hsanda Gol as the main sites. The earliest of important Tertiary terrestrial African deposits is that of the Lower Oligocene of the Egyptian Fayum. In North America the main fossil localities are those in the White River series in South Dakota and adjacent states, with successive titanothers (Chadron) and *Oreodon* and *Protoceras* beds (Brule). In South America, Patagonia continues to be the exclusive area of interest. The Deseado beds, with the *Pyrotherium* fauna, are Lower Oligocene; the late Oligocene Colhue Huapi beds contain essentially a Miocene assemblage.

In the northern continents the Oligocene fauna has a much more modern appearance than that of the Eocene. This is not due in any measure to the appearance of new types; for hardly a third of the living families of terrestrial mammals were then present and, except in the case of the carnivores, there were few new groups introduced at this time. This more modern aspect is rather due to the extintion during or at the close of the Eocene, of a great number of families and even orders which were characteristic of Paleocene and early Eocene days. To cite this list is to cite the greater part of the roster of the typical Paleocene mammals, including multituberculates, primitive insectivore groups, arctocyonid creodonts, miacids, condylarths, and plesiadapid primates and, in addition a number of the late Paleocene and Eocene families, such as mesonychild and oxyaenid creodonts, adapid lemurs, and a few of the more primitive forms among the modern ungulate orders.

In the Holarctic land areas, there continues, in the Oligocene the faunal situation established in the later Eocene, with geographical continuity in some groups (particularly in some of the carnivores) and independent paralel evolution in others, notably the ungulates. Europe and North America offer two extremes; Asia has characteristics of both, although showing greater affinities with Europe. There are few features suggesting any great interchange of faunal elements between the two hemispheres during the Oligocene; most of the common characteristic appear to have resulted from a late Eocene interchange.

Common types include such forms as oppossums, the major families of modern insectivores, and primitive beavers, which appear in the late Oligocene. In both hemispheres there appear in the Oligocene the first representatives of the great murid group, destined to play an increasingly important role in the later epochs. Among carnivores both hemispheres show the persistence in the early Oligocene of the last surviving group of the creodonts, in *Hyaenodon* and its relatives, later confined to tropical regions. A new development of importance is the rapid deployment of progressive fissipede families developed from the miacids. Canids of various types mustelids, and felids, both saber-tooths and false saber-tooths of the *Dinictis* group, were abundant in both hemispheres.

A limited number of ungulates are common to both hemispheres. Among the perissodactyls, chalicotheres were present in Eurasia and doubtfully in North America; *Protapirus,* first of true tapirs, also appears in both areas; amynodont rhinoceroses are widespread, with *Cadurcotherium* as European representative, *Metamynodon* the American genus. A notable perissodactlyl development is the expansion of the true rhinoceroses during the Oligocene in such primitive genera as *Caenopus* in North America and *Aceratherium* in Eurasia. Among the piglike artiodactyls there are common bunodont families. *Anthrocotherium* and other members of its family are abundant in Eurasia, and *Bothriodon* penetrated to America, where the group failed to survive. Pigs and peccaries make their appearance; both are found in Europe, and the peccaries reached their later American center in the Oligocene.

Among the nonungulates Eurasian Oligocene reveals various groups absent from America; the first of viverrid carnivores a group which never reached the Americas; ochotonids, which arrived there much later; anomaluroid rodents, the first dormice and dipodoids, and forms which may be the first bathyergoids. Among the perissodactyls the only distinctive European forms are *Palaeotherium* and its relatives, which persisted until the Middle Oligocene. There were, however, numerous selenodont artiodoactyls peculiar to Eurassia. *Dichobune, Caenotherium, Anoplotherium,* and *Xiphodon* are examples of archaic forms "held over" from the Eocene and for the most part exterminated before the close of the epoch. More important however, were primitive pecorans, such as *Amphimeryx* of the early Oligocene, and tragulids, such as *Gelocus* and *Bachitherium.*

North American shows equally distinctive features. There are various rare hold-overs here of the archaic Paleocene and Eocene groups—the last anaptomorphid and apatemyid primates, leptictid insectivores, palaeanodont edentates; surviving ischyromyd rodents, and their derivatives, such as the mylagaulids and ancestral sewellels; ancestors of the pocket gophers and kangaroo rats; hares, present here and in Asia but not as yet in Europe.

Among ungulates, the perissodactyls are more prominent in North America than in Oligocene Europe. Small three-toed horses, *Mesohippus* and *Miohippus,* flourished; *Hyracodon* is a common rhinocerotoid; primitive tapiroids survived; the lowest Oligocene beds are characterized by the presence of numerous giant titanotheres, which disappeared abruptly before the Middle Oligocene. Artiodactyls are increasingly important, however. Camels, such as *Poëbrotherium,* are common; *Protoceras* represents a peculiar American family of primitive pecorans; *Leptomeryx* and other hypertragulids are common Oligocene finds; *Agriochoerus* is present. The oreodonts become in the Oligocene the commonest of ungulates; *Oreodon* and other members of the family are the most frequent fossil finds in middle and late Oligocene deposits.

Known Asiatic deposits show that many of the European Oligocene groups were represented in that continent, and accidents of preservation and collection may be responsible for the seeming absence of other European families. Distinctive Asiatic features include the absence of equoids of any sort, with a variety of surviving tapiroids in their place; the presence of a last lingering amblypod; the development of giant rhinoceroses, such as *Baluchitherium,* which persisted until the Miocene. Other features, implying a connection with North America in the late Eocene or early Oligocene, include the common presence of hares and of abundant giant titanotheres.

We have noted the almost complete lack of Paleocene or Eocene African terrestrial records. In the Lower Oligocene we at last find an African fauna; this, unfortunately, lies at the northern margin of the continent, in the Egyptian Fayûm, and shows evidences of invasion from Eurasia. In the known fauna there are no fissipedes, but hyaenodont creodonts such as *Plerodon* and *Hyaenodon;* a considerable variety of anthracotheres; *Phiomys,* representing the anomaluroid rodents destined to survive in to the African Miocene. All these forms may be recent immigrants, since they are members of contemporary Eurasian families. More doubtful as to origin are anthropoid primates, unknown earlier (except for possible Burmese forms). These include a very primitive anthropoid, *Parapithecus;* in *Apidium* a true Old World monkey; in *Propliopithecus* an early, manlike ape. Definitely autochthonous forms are represented by *Arsinoitherium;* a considerable variety of hyracoids; and proboscideans, which now include not only the archaic *Moeritherium* but the mastodonts, *Palaeomastodon* and *Phiomia*.

The Deseado beds of Patagonia show an important phase in the development of the isolated South American fauna. Except for marsupials of the *Polydolops* type and a few families of ungulates, all Eocene groups are still present. Marsupial carnivores are numerous and varied (one was larger than a cave bear). Edentates were still rare but were increasing in variety, for glyptodonts had appeared in the late Eocene and ground sloths of modest size had also made their appearance. The native ungulates had already reached a climax in size and variety. Both major families of litopterns were advanced in their development. Notoungulates were numerous and exceedingly varied; high-crowned teeth had developed in many lines. Most characteristic of the notoungulates of the age were *leontinia* and large related types. Huge astrapotheres were present, and the giant of the fauna was the elephant-like *Pyrotherium*.

A startling phenomenon is the abrupt appearance of hystricoid rodents. There is no trace of rodents in the Eocene of the continent, and it is highly improbable that these forms, so successful from this time on, could have been present earlier and escaped discovery. As we have noted in an earlier chapter, an African origin seems probable, but their means of transport offers a difficult problem.

Oligocene marine mammals are rare, perhaps because of the rarity of appropriate types of sediments. Sirenians are known from the European Oligocene. An archaeocete whale is known to have survived, and we are sure that the differentiation of more modern whale groups was well advanced, for a lone squalodont represents a primitive toothed-whale group and a cetothere announces the presence of early whalebone whales.

MIOCENE

The Miocene is an epoch of considerable length and one represented by numerous and varied faunas in a number of regions. European localities are most abundant in France and Germany. St. Gerand, Sansan, and Simorre are among famous French localities of varied ages, and the various Molasse deposits and those of the Mainz basin and Steinheim are among the familiar fossiliferous beds of the German region. The Miocene is poorly represented in Africa, although there are scattered areas of this age in Egypt and in eastern and southwestern Africa. In Asia, too, the Miocense is present in but few places. The Loh and Tung Gur beds represent the last of the sequence of Mongolian deposits; the Bugti beds of Baluchistan have yielded a few mammals. In the late Miocene, however, we see the beginning of the deposits of the Siwalik Hills in India, which continue from this date to the Pleistocene.

In North America a major portion of the Miocene beds lies in the Great Plains area; they include such well-known deposits as the Harrison, Rosebud, Sheep Creek, Marsland, and Pawnee Greek. In Oregon the fossiliferous portion of the John Day beds are early Miocene. Florida in the Miocene has yielded the only good fauna found in eastern American in any stage of the Tertiary.

In North America a major portion of the Miocene beds lies in the Great Plains area; they include such well-known deposits as the Harrison, Rosebud, Sheep Greek, Marsland, and Pawnee Greek. In Oregon the fossilliferous portion of the Johan Day beds are early Miocene. Florida in the Miocene has yielded the only good fauna found in eastern America in any stage of the Tertiary.

In South America the Miocene opened with a marine invasion represented by the Patagonian formation; following this were deposited the Santa Cruz of Patagonia, with a rich and varied fauna, and later beds of lesser importance.

The Miocene faunas are greatly "modernized" as compared with conditions at the beginning of the preceding epoch. During or at the end of the Oligocene there had become extinct a score of families of more or less archaic types, particularly among the ungulates. Among the victims were the last amblypods, the pyrotheres, older proboscidean types; titanotheres, hyracodonts, and lophidonts; primitive bunodont artiodactyls and primitive ruminants such as anoplotheres, xiphodonts, and amphimerycids; several primitive rodent families, the last of the archaic primate groups which characterized the Paleocene and Eocene. As a result of these extinctions and of new evolutionary developments we find that, in the known Miocene faunas as a whole, half of the living families of terrestrial mammals were then in existence and that these likewise constituted a good half of the faunal assemblage. If the peculiar South American types be excluded, the modern note is more marked; in the Northern Hemisphere only a quarter of the fauna is made up of families which have failed to survive.

We have noted the nearly complete isolation of the two hemispheres during the Oligocene; as a result, the Palaerctic and Nearctic faunas at the opening of the epoch differed widely in many respects. There are, however, indications of some intermigration during the early Miocene and again toward its close, which tended to reduce to some degree the faunal differences. The evidence suggests free communication between Africa and Eurasia during this epoch, but South America and Australia were still entirely separate evolutionary areas.

As before, the differences between North America and Eurasia were relatively small among the nonungulate orders, although even in cases where families were common to both areas the genera present were usually distinct. The creodont carnivores had disappeared except for lingering hyaenodonts in the Old World tropics. Varied dogs were present including large "bear-dogs" and, in the late Miocene such forms as *Hemicyon,* which appear close to the origin of the Ursidae. Mustelids, too, flourished, and felids, including true saber-tooths and pseudo saber-tooths, such as *Nimravus* of North America and *Pseudaelurus* in Europe. In the Old World the civets persisted, and from them had evolved, by the end of the Miocene, the first hyenas; in North America, in contrast, the early Miocene shows the beginning of the procyonids in *Phlaocyon* and other genera. Lagomorphs, rats, and primitive beaver-like forms (*Steneofiber, Palaeocastor)* are among common rodents.

Before the close of the Oligocene, primates had disappeared from North America. In the Eastern Hemisphere, however, the evidence, scanty though it is, tells of a considerable amount of evolution among the higher primates. *Pliopithecus,* a primitive gibbon, and *Dryopithecus,* an advanced manlike ape, appear in the European Miocene, and the discovery of several genera of these apes in the early Miocene of Africa suggests that this continent may have been an important center of primate evolution.

As in the Oligocene, there is a marked difference between the two northern continental areas in ungulate evolution, and there are marked differences between early and late faunas. Among the perissodactyls, rhinoceroses are present in both hemispheres. *Diceratherium* is the common American genus in the early Miocene; *Baluchitherium* persisted into the Miocene of Asia; more normal genera, *Aceratherium* and similar forms, are found in both areas but are more common in Eurasia. A last amynodont appears in the Miocene of Asia. *Macrotherium* and *Moropus* represent the chalicotheres in Eurasia and North America, respectively; tapirs are present in both areas.

Miocene horse history is mainly an American story. *Miohippus* persisted into the early Miocene and gave rise to *Anchitherium* and *Hypohippus* as larger forms of the same primitive type. *Parachippus,* first of horses with high-crowned teeth and reduced side toes, appeared at the beginning of the Miocene and still more advanced genera, *Merychippus* and *Pliohippus,* are characteristic of the America late Miocene. With the extinction of the palaeotheres, Eurasia had become barren of equoids, but *Anchitherium* successfully migrated to the Eastern Hemisphere and is present there in many Miocene localities.

The two areas have little in common as regards artiodactyls. Entelodonts made their last appearance in both areas, however, in the Lower Miocene; *Dinohyus* was a giant American form. The common bunodonts were swine in Eurasia, peccaries in North America; *Listriodon* and *Palaeochoerus* were representative piglike forms. Anthracotheres are absent in North America and in Europe, as well, beyond the Lower Miocene, but they survived in southern Asia and in Africa. In Europe *Caenotherium* was present in the Lower Miocene as the last of nonpecoran selenodonts in that areas; in America, in sharp contrast, there were abundant and varied camels, such as *Stenomylus* and *Oxydactylus,* and numerous oreodonts, including *Promerychochoerus, Merychyus,* etc. In both hemispheres primitive percorans of the tragulid or hypertragulid groups survive but are uncommon.

In the Miocene are found the first stages in the deployment of higher ruminants. The earliest Miocene of both hemispheres witness the appearance of those types here included in the Palaeomerycidae: in Europe, *Dremotherium* and *Amphitragulus* and, somewhat later *Palaeomeryx;* in America, *Blastomeryx,* followed by other deerlike forms. Later in the Miocene appear the first (and relatively rare) representatives of the living higher pecoran families. In Eurasia are primitive antelopes such as *Protragocerus,* primitive deer as *Dicrocerus,* and *Palaeotrayus,* the first giraffid. These forms are absent from America; instead, we find *Merycodus* as the first of the prong-buck family.

A new element in the ungulate fauna of the northern continents is the advent of the proboscideans, previously known only from Africa. *Dinotherium,* with its peculiar lower tusks, bunodont mastodonts, such as *Gomphotherium* and *Serridentinus,* and the lophodont *Mastodon* all appeared in Europe early in the Miocene and persisted throughout that epoch. In America their advent was retarded until the middle or end of the Miocene; *Dinotherium* failed to penetrate to America.

The Miocene Santa Cruz fauna is the best known of South American mammal assemblages. Marsupial carnivores and hystricoid rodents were numerous and varied. Edentates were abundant for the first time; forms, mostly of relatively small size, represent not only armadillos but anteaters, glyptodonts, and all three families of ground sloths. Litopterns were at the peak of their development, including, for example, *Diadiaphorus* and the pony-like *Thoatherium* among the proterotheres and *Theosodon* of the more heavily built macraucheniid family. Pyrotheres were extinct, but *Astrapotherium* was a giant end-form of another archaic line. Notoungulates were now much less varied in the number of families represented and were, in general, small in size; in number of individuals present however, they bulk large in the faunal picture. *Homalodotherium,* the toxodonts *Adinotherium* and *Nesodon, Pachyrukhos,* and *Protypotherium* are representative genera.

In *Homunculus* we see the first representative of the platyrrhine monkeys of South America. The early history of this group is as puzzling as that of the hystricoid rodents. Perhaps the advent of the platyrrhines took place by invasion in some such fashion as that suggested for the rodents an epoch earlier. But primate remains are rare also in later South American sediments, and it is possible that ancestral primates were present earlier in the Tertiary in more wooded tropical areas.

Marine mammal-bearing deposits are present in many regions; the Calvert cliffs of Maryland, the Patagonian beds of South America, the Aquitanian basin of France, and the Antwerp basin of Belgium are among the prominent sources of fissil marine mammals. A lingering archaeocete was present in the early Miocene, but the epoch was one of rapid expansion of higher cetacean groups; the order was seemingly never more abundant or varied than at that time. Nearly every family of toothed and shalebone whales had appeared by the end of the Miocene. The squalodonts, primitive toothed shales are common even in the earliest Miocene; and dolphins, ziphiids, and sperm whales are found in increasing numbers in later formations. The cetotheres, primitive Shalebone whales, were abundant in the late Miocene, and there were even rare representatives of the living families of this suborder. *Halianassa* and other sirenians were present. The Miocene is further notable for the appearance of the first pinnipeds, including both eared and earless seals, and the first walrus, *Prorosmarus.*

PLIOCENE

The Fliocene is a relatively short epoch. Late beds, forming a transition to the Pleistocene, will be mentioned separately; the characteristic Pliocene fauna is that of the early part of the epoch, known in Eurassia as the Pontian age. Deposits containing this fauna are found widely distributed throughout the Mediterranean region and in eastern and central Europe; Eppelsheim, various localities in the Vienna basin, and Pikermi and Samos in the Aegean are among the best-know localities. To the eastward this same fauna extends through Persia (Maragha) and the Siwaliks to rich early Pliocene deposits in Honan and Shansi in China. Somewhat later European levels are best represented in southern France, as at Perpignan; in India the Siwalik series continues throughout the Pliocene; and there are scattered later Pliocene deposits in China. The Pliocene of Africa is almost unknown.

In North America there are numerous Pliocen deposits, the greater part of them in the Great Plains region. The earlier part of the epoch is represented in the Clarendon beds of Texas, the Snake Creek of Nebraska, and the Santa Fe beds of central New Mexico; a later stage in such formations as the Hemphill of the Texas Panhandle and the Republican River beds of Kansas.

In South America the long series of Patagonian sediments causes at about the end of the Miocene, and the principal Pliocene areas are found farther to the north in Argentina. In the Andes foothills a set of beds often termed the Araucanian series covers nearly the whole length of the Pliocene; Monte Hermoso in the southern Pampas and the Entererrian region of northeastern Argentian also have characteristic deposits of this epoch.

The Pliocene fauna of the Northern Hemisphere is in great measure one of modern type. There had been a considerable elimination of old families, particularly of ungulates, during or at the close of the Miocene. The last amynodonts, entelodonts, caenotheres, and hypertragulids had been eliminated in the northern continental areas; the last archaic archaeocetes had disappeared from the oceans; and in South America astrapotheres, had disappeared from the oceans; and in South America astrapotheres, homalodotheres, notohippids, and interatheres are absent from the ungulate roster. The fauna is in the main one of modern families. Excluding South America and Australia, 80 per cent of the families of terrestrial mammals then present are types still extant; and while a number of living families are not represented by Pliocene fossils, these are mainly rare and obscure groups. The Pliocene faunas further show much similarity generically to modern times; many familiar living genera, particularly among the carnivores and rodents, had arisen before the period was over, and in other cases generic differences between modern and Pliocene types are small.

There is evidence that a limite amount of faunal interchange took place between Eurasia and North America at the beginning of the Pliocene, but for most of the duration of the epoch there is little evidence of further connections. Old and New Worlds show many faunal similarities, but these can be attributed in the main to persistence of types interchanged during the late Miocene, and the genera present are usually distinct in the two regions.

The Carnivores have run a parallel course in both continents. There are numerous canids and mustelids. Felids close to the living genera are present, addition to saber-tooths. Primitive bears have appeared. Hyenas are prominently represented in Eurasia. There are a variety of bunodont mastodonts in both continent. *Hypohippus*, descended from *Anchitherium* of the Miocene, is present in Asia as in North America; and *Hipparion*, a lightly built, there-toed horse, reached Eurasia from North America and became widespread and common there. Rhinoceroses are present in both regions but are rare in North America; chalicotheres survive in both hemispheres, as do a few palaemerycids. In Pontian times a nearly homogeneous fauna is found all the way from Spain to China and may well have been characteristic of Africa as well, for it is one quite similar in many respects to that found in Africa today. Elements in common with north America were noted above, but the presence of *Hipparion* as a characteristic Pontian from deserves emphasis. A striking feature is the abundance of bovids; several score of genera of antelopes have been described from the Pontian. Primitive genera of deer are common, as are the pigs and, in Asia, varied giraffids, including giant-horned forms. The peculiar proboscidean *Dinotherium* was still present, and in the eastern area primitive elephants such as *Stegodon,* make their appearance.

In North America we find an ungulate assemblage still differing markedly from that of Europe. Not a bovid is present, or a true cervid. Instead, there is a great radiation of antilocaprids with varied types of horns, a last protoceratid, and a few final representatives of the oreodonts. Camels still flourished in a variety of forms, and peccaries were not uncommon. Horses of varied types were especially abundant, including hold-overs from the Miocene, as well as the *Hipparion* group and types progressing toward the modern one-toed forms.

There is evidence that the South American land bridge was now in process of re-establishment after the lapse of most of the Tertiary, for a first ground sloth had appeared in North American by mid-Pliocene times and a glyptodont at the end of the epoch.

In South American the Araucanian faunas of the early and middle Pliocene show evidence of this faunal interchange in the presence of a number of raccoon types, which may have used a Central American island chain as stepping stones. The fauna was otherwise purely native in character. A number of Miocene families of ungulates had disappeared, but other types of hoofed mammals still flourished, and the edentates and the hystricoid rodents were highly varied; in many instances there was a strong tendency toward large size. The borhyaenid marsupials still functioned as the carnivores of the faunas, and the evolution of a marsupial saber-tooth is an interesting development.

In the sea the archaeocetes were extinct, and the primitive squalodont whales were reduced and had vanished by the end of the Pliocene; ziphiids were fewer, and the various dolphin types were less numerous. Sperm whales, however, continued in fair numbers. Among the whalebone whales, a few final members of the cetotheres survived; they were being replaced by more abundant balaenids and balaenopterids.

In most continents beds are present which ar transitional from Pliocene to Pleistocene; they are here included in the former epoch, but their position is much disputed. In Europe this is the Villafranchian stage, best known in the Arno Valley in Italy. Corresponding horizons are known in southern and eastern Asia; in America this terminal Pliocene is known in the Blanco beds of Texas and similar deposits in

Kansas, Idaho, and Arizona. The fauna includes a mixture of characteristic Pliocene types with genera typical of Pleistocene and modern times. The period was apparently one of rising lands, with the re-establishment of the Bering Strait bridge. This is indicated by the fact that the modern horse genus, *Equus,* appears to have developed at this time in North America and hard rapidly spread through the Old World to become a "type fossil" for this stage and the typical Pleistocene beds which follow.

In South America, too the latest Pliocene, represented by the older beds of the Pampas—the Chapadmala—is a time of transition. The earlier Pliocene "natives" are still present, except for the proterotheres. But in these deposits are also found numerous invaders from the north. Felids, normal and saber-toothed, and bears, as well as procyonids, are among the invading carnivores; horses, peccaries, and deer have entered to compete with native ungulates, and varied mice to rival the hystricoids. The end is in sight for much of the native population.

PLEISTOCENE

This shortest of Cenozoic epochs includes only the period of the Ice Age of northern regions and was, at the most, a million years or so in length. In the Pleistocene we reach a climax in the development of steep climatic gradients—a development which appears to have been in process of establishment during the later Tertiary. Arctic, temperate, and tropical zones were well differentiated. Their boundaries fluctuated considerably during the course of the Pleistocene. This evidence indicates that there were four successive southward advances of ice caps which covered much of Europe and North America. contemporaneous with these advances were the development of great areas of tundra, local glaciations in more southerly mountain masses, and a shifting of climatic zones so that areas such as the Sahara and the deserts of the southwestern United were, for the time, well watered. Between the periods of glaciation were three interglacial stages, during which zonal boundaries shifted far to the north, so that essentially tropical conditions were to be found over great parts of Europe and North America.

As a consequence of these climatic conditions, our treatment of faunal assemblages in the Pleistocene must be on a somewhat different basis from that used in the discussion of earlier Tertiary epochs. There we were able to consider the fauna of a continental area as essentially a unit; in the Pleistocene, however, divisions by climatic zones are prominent in the faunal picture. In both Eurasia and America the Arctic fauna was (and is) a very meager one; the temperate-zone fauna was much restricted in many ways, particularly as regards ungulates; various groups once widespread over much of the northern continental areas had retreated to the tropics, to reappear in the north, if at all, only during the warm interglacial stages.

Complete extinction of mammalian groups—families or higher categories—was relatively rare during or at the close of the Pliocene. Definite extinction may be noted in the case of the borhyaenid marsupials in South America, replaced by incoming fissipede placentals; of the last tropical survivor of the old creodonts; of the proterotheriid litopterns in South America; and, in North America, of two native ruminant families—the oreodonts and the protoceratids. There is, however, a very significant reduction in many groups—a reduction in numbers of genera or of areas of distribution. Rhinoceroses had made their last appearance in North America; *Dinotherium* had gone from Eurasia; in South America the macraucheniid litopterns were reduced in numbers; and still further reduction took place among the notoungulates. The variety of mastodonts had decreased before the opening of the Pleistocene. In the Old World, anthracotheres and chalicotheres were close to extinction; rhinoceroses, tragulids, and giraffids reduced. The typical Pliocene horses had become extinct, except for sparse records of *Hipparion* in the early Pleistocene or in transitional beds.

In any region where the fauna is adequately known, we find, in the Pleistocene, representatives of all the types now present. But, in addition, there are invariably numerous animals now extinct; and most of these are of large size.

The Pleistocene giants include representatives of almost every order of mammals: giant deer in both Eurasia and North America; the largest of proboscideans and of edentates. Large types were present among the South American rodents, and there were giant beavers in both Europe and North America. In Australia there were giant kangaroos and wombat-like forms and even an "outsized" *Ornithorhynchus*. Nor were the primates exempt. Madagascar lemurs included forms as large as the great apes, and there is recent evidence in southeastern Asia of giant human or subhuman types.

Apart from these large forms, since extinct, the Pleistocene assemblage—particularly that of the later Pleistocene—was one with an air of familiarity about it. The time since the close of that epoch has been to short for any major evolutionary advance to have taken place, and hence it is not surprisng that not merely most of the genera but also many of the species living today were common members of the Pleistocene faunas.

In the Pleistocene the northern continents were more united faunally than they had been since the earliest Tertiary. There had been considerabel exchange between the two hemispheres in late Pliocene days, and the Bering Strait bridge appears to have been in existence during much of the Pleistocene, permitting further intermigration. Climatic conditions, however, were a strong limiting factor here. Arctic types and temperate-zone forms accustomed to cold climates and rugged terrain could cross this region readily; plains animals would find passage difficult; tropical forms were barred.

The Palaearctic and Nearctic regions had, in the Pleistocene as today, many groups in common and with generic identity in many cases. A few, such as the woolly mammoth and the musk oxen, found from England to New Mexico, were arctic types; others were groups familiar in temperate zones or climatically ubiquitous. Among essentially temperate-zone types of Eurasian origin we may note the deer and more northerly types of bovids, such as sheep, bison, and goats; the first reached North America in late Pliocene and Pleistocene times, and the bison became the commonest of America ungulates; and, while true goats did not make the passage, the related *Oreamnos*, the mountain "goat," did.

Groups already present in both areas and, in general, more tolerant of changing climatic surroundings are numerous. Such include varied sciuromorph and myomorph rodents; lagomorphs; canids; true cats, large and small; persistent saber-tooths; many mustelids. Pliocene mastodont types persisted into the early Pleistocene in both continents, and the true *Mastodon* was a late survivor in North America. Elephants, presumably of Asiatic origin, made the crossing to North America, and several types of mammonths, in addition to the woodly variety, were present in Old and New Worlds. Tapirs were present in both Eurasia and the Americas; true horses of the genus *Equus* flourished in both areas; camels, still present in their American homeland, made the passage to Eurasia.

Despite the many resemblances due to earlier as well as to Pleistocene interchanges, Old and New World faunas differed in a number of respects. Some of the contrasing types, such as the woodly rhinoceros and the giant tundra-dwelling *Elasmotherium* of Eurasia, were cold-climate forms. Most, however, were warmth-loving animals many of them relatively recent in their development; even in the Pliocene, in would seem, the Siberian-Alaskan bridge was a difficult route for such animals to travel and in the Pleistocene was impossible.

With the coming of the Ice Age, many elements of the Pliocene Eurasian fauna were driven southward. A few of these forms, such as the hippopotami, hyenas, and southern genera of rhinoceroses ventured north again during the warm interglacial stages. Others never returned. They were, for the most part, confined to the southern regions of Asia and of Africa. Many of the types found in the Pleistocene in these areas have survived to modern times. Among these may be mentioned a varied series of primates, including great apes, cercopithecid monkeys, and lemurs; abundant antelopes; viverrids; pangolins; aardvarks. As has been said, this tropical fauna is markedly similar to the older Pontian fauna once widespread in more northerly regions. This similarity is enhanced by the persistence in the tropics during the Pleistocene of a number of archaic mammalian types which have since become extinct, such as *Dinotherium* and the last anthracotheres and chalicotheres.

In North America as in Eurasia peculiar indigenous mammalian types persisted in the Pleistocene. Such forms were, however fewer in number than in the Old World; the more prominent forms are the peccaries, native geomyoid rodents, and antilocaprids, of which several genera were present. More notable was a considerable influx of South American mammals which had traveled northward over the Central American land bridge. A variety of ground sloths—*Megatherium, Mylodon, Megalonyx, Nothrotherium*—were common mammals; and glyptodonts and armadillos, large and small, were present in the Gulf region. Of hystricoid reodents only the porcupine entered northern areas. It is not improbable that the Pleistocene and Recent opossums of North America are reimmigrants from the south.

A great host of northern invaders entered South America in the Pleistocene and are abundant fossils in such superficial deposits as those of the Argentine and Uruguayan pampas and the Tarija beds of Bolivia. Of carnivores, the mustelids, the dogs, and the saber-tooth *Smilodon* had joined the procyonids, cats, and bears already present and presumably made havoc among the native animals. Competition of newcomers with the native herbivores was increased with the incoming of tapirs, peccaries, and llamas to join the horses and deer, which had entered in the late Pliocene. Mammoths failed to gain entry, but several mastodonts were present, and among the smaller mammals there were numerous invading mice and lagomorphs.

Of the old natives of South America, the hystricoid rodents, the monkeys, and the opossums appear to have been little affected. The edentates were seemingly in vigorous condition; for in addition to the smaller types—armadillos, anteaters, tree sloths—the Pleistocene fauna included a variety of ground sloths and glyptodonts, unaware (so to speak) fo their approaching extinction. On the other hand, the native ungulates were already close to the vanishing-point. Of the great array of families and orders present in the earlier epochs, there are but half-a-dozen genera still present—these including the litoptern *Macrauchenia* and the notoungulates *Toxodon* and *Typotherium*. Pleistocene and Subrecent cave deposits in the West Indies indicate that by some route a variety of small sloths and hystricoid rodents, nearly all now extinct, had reached those islands.

The Tertiary history of Australia is practically unknown; in the Pleistocene, when we first see the Australian mammal assemblage, it was in great measure that found today: monotremes; a great variety of marsupials, both carnivorous and herbivorous; of terrestrial placentals only a restricted series of rats and—presumably fellow-entrants in the late Pleistocene—man and dog. A feature of the Pleistocene here as elsewhere, is the development of giant types, such as the marsupials *Diprotodon* and *Thylacoleo*.

Man appears to have been essentially a Pleistocene development. His homeland was surely some portion of the Eurasian-African land mass, and this area marks the limit of his spread for most of the Pleistocene. Australia, as far as we now know, was entered late in that epoch, and America not until the end of the Pleistocene, at the earliest.

RECENT

The Recent fauna of the world is marked, apart from the spread of mankind, by a relative paucity in its mammalian population. There are presumably no living types which were not also present in the Pleistocene, but many forms then present are now extinct. These were mainly mammals of large or more than average size. We may list the more conspicuous of the absent groups: the giant marsupials and monotremes of Australia; the saber-tooths; the last of the South America ungulates; the ground sloths and glyptodonts; all the proboscideans except the two surviving elephants; the chalicotheres; the sivatheres; the anthracotheres; giant cervids; giant beavers. Regionally too, there has been much extinction among groups formerly widespread. The range of the surviving proboscideans in a greatly restricted one; camels and tapirs disappeared from North America, peccaries from much of this region, horses from both Americas; rhinoceroses and hippopotami from much of Eurasia.

The reasons for this extinction of large mammals in difficult to discern. It was not altogether a matter of climatic vicissitude due to the Ice Age, for many of the extinct proboscideans survived to the end of the Pleistocene; and horses, camels, and ground sloths may have persisted in the American Southwest until very late times. Nor can we believe that disease can have been a major factor; for a disease which would attack such a variety of large animals and leave small ones untouched would be peculiar indeed. The spreading of mankind, *Homo sapiens,* is the only new feature seen in the late Pleistocene. Man may perhaps be responsible in some degree for this catastrophe; but his influence at the most must have been an indirect one.

Chapter—16

The Origin and Adaptive Radiation of Reptiles

The land dwelling vertebrates evolved from crossopterygian fishes. It was also noticed that the immediate descendants of the latter were the labyrinthodont amphibians that in turn gave rise to the cotylosaur reptiles. All this occurred before the beginning of the Mesozoic. Terrestrial vertebrates at that time were probably carnivorous and they could not respond directly to the energy supply offered by terrestrial plants, but they could and apparently did respond to the opportunities presented by the rediation of insects. Probably for the first time in evolutionary history there was an adequate food supply to support fully terrestrial predators. Two forces primarily responsible for the origin and radiation of reptiles were directly related to the exploitation of this new energy source. One was the evolution of a more effective jaw mechanism specifically adapted to feeding on insects. The evolution of more effective locomotion on land. A final change, the evolution of the amniotic egg, was the definitive step that separated reptiles from their amphibian ancestors. The cotylosauria were contemporaneans with numerous primitive amphibians during late Carboniferous and early Permian times, but before the end of the Permian the amphibians had been out-competed and nearly exterminated by the reptiles, and the latter had become dominated.

The best known and most primitive of these is *Seymouria,* a small tetrapod found in the upper portion of the lower permian sediments exposed to the north of the town of *Seymour,* Texas. It is a good example of a persistent structural ancestor. It is also a good example of a "connecting link" of which there are many in the fossil record. It

Fig. 16.1: Seymouria.

had but one condyle, five digits on the hand (manus), but had, like Amphibia, only one pair of sacral ribs, whereas nearly all reptiles have at least two. The number of joints is the fingers is also more than that of Amphibian than that of reptile that is what *Seymouria* may be regarded as incompletely reptilian.

AMBHIBIAN AFFINITIES OF *SEYMOURIA*

Where the *Seymouria* was an amphibian or a reptile, its answer depends on whether *Seymouria* laid an amniote eggs on the land, like modern reptiles, or whether like modern frogs, it returned to the water to deposit its eggs. This problem remained unsolved till *Dr. T.E. White* made a detailed study of *Seymouria*. On the basis of the occurrence of haemal arches he tried to prove the egg laying nature of the animal like large amniotes.

Seymouria also resembled early Amphibia in many features as given below:

(*i*) Flat skull with reduced ossification.

(*ii*) Presence of inter temporal bone.

(*iii*) Occurrence of primitive palate.

(*iv*) Labyrinthine teeth on vomers and palatines.

(*v*) Position of fenestra ovalis below the basal level of the brain is amphibian.

(*vi*) Traces of lateral line canals in head region.

(*vii*) Presence of one pair of sacral ribs.

(*viii*) Due to the short neck pectoral girdle lies close behind skull.

(*ix*) Skull is without temporal vacuities (an amphibians condition).

These characters of *Seymouria* rather securely on the side of the amphibians. And added to this evidence is the clear indication that some genera closely related to *Seymouria*, notably *Discousouriscus* from Europe, were characterized by gill-bearing larvae.

REPTILIAN AFFINITIES OF *SEYMOURIA*

The mixture of amphibian and reptilian characters seen in *Seymouria* is indicative of the gradual transition that took place between the two classes during the evolution of the vertebrates. Because the change was gradual rather than abrupt it is difficult to draw clear cut distinctions between amphibians and reptiles when all the fossil materials are taken into consideration. There are, however, certain character typical of reptiles found in *Seymouria* as given in the following lines:

(*i*) Presence of short, stubby, and muscular limbs.

(*ii*) Monocondylic skull.

(*iii*) Skull is without vacuity as in living chelonians.

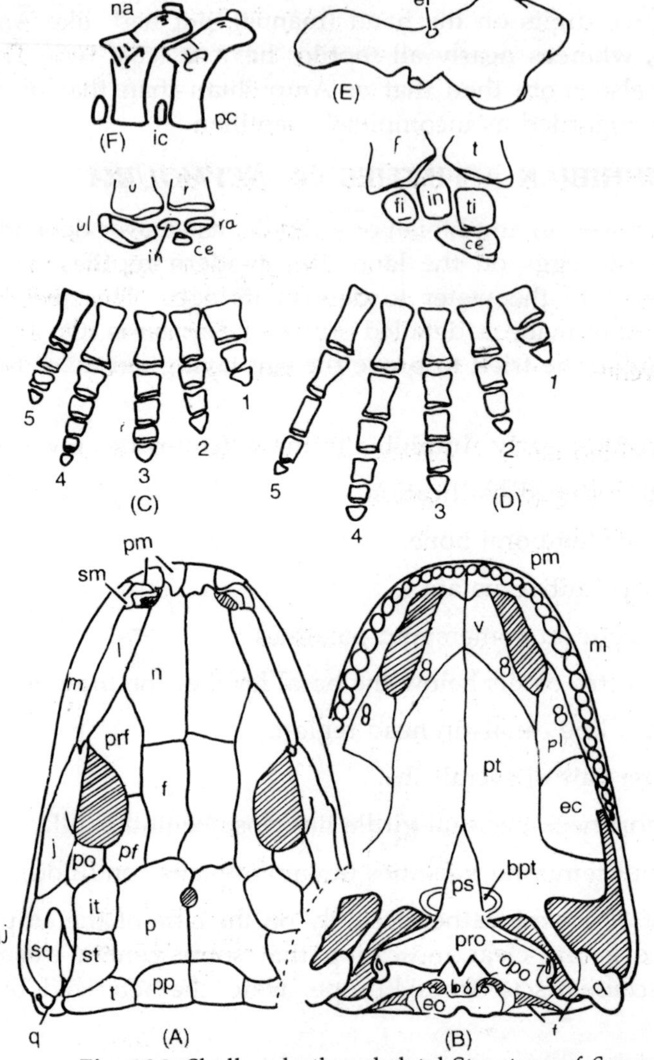

Fig. 16.2: Skull and other skeletal Structure of *Seymouria.*

(*iv*) Pelvic girdle is attached to the vertebral column by means of sacral vertebrae.

(*v*) Number of phalanges is greater and more reptilian with a basic formula of 2-3-4-5-3.

(*vi*) Presence of a large pineal eye emerging through a hole between the parietal bones.

Due to having mixed characters of amphibians and reptilians, the Symourians are supposed to be the connecting links between amphibians and reptiles. *Romer* treats it as a reptile under a separate heading the *Cotylosauria,* whereas others classify the Seymourians under the head *Seymouriomorpha.*

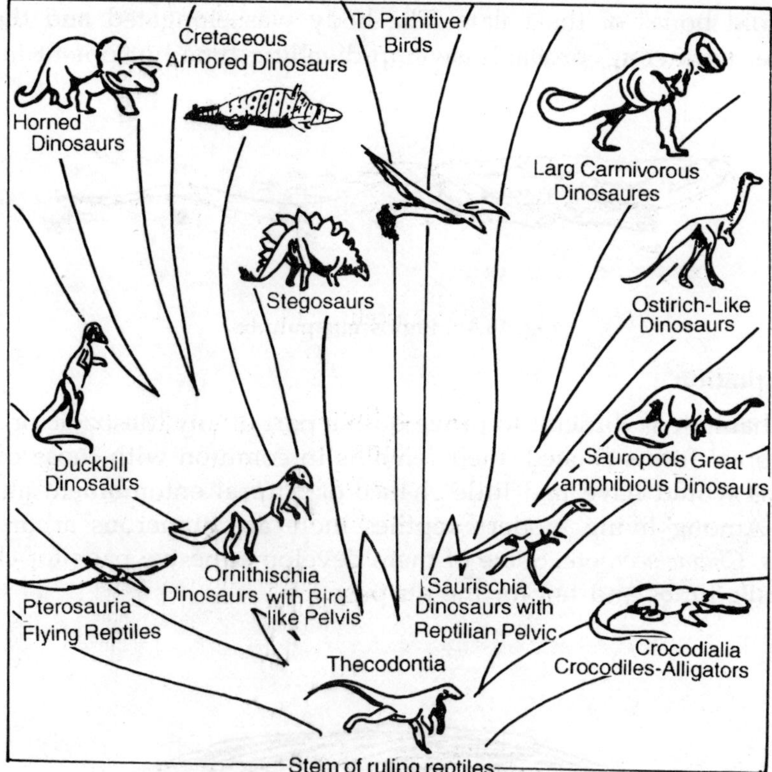

Fig. 16.3: Evolutionary tree showing adaptive radiation.

ADAPTIVE RADIATION

The Mesozoic era has been called the `Age of Reptiles', for while higher forms the birds and mammals, make their appearance during its course. During Permian Period the stem reptiles underwent a considerable radiation, some types becoming pecularly specialized. Food and safety including living space are the basic causes responsible for the entire evolutionary changes. Perhaps the reptiles have shown the greatest evolutionary diversity and adaptive radiations of all vertebrate groups. They were adapted to live in different ecological conditions. Some important reptilian adaptions are taken here under the following heads.

Central Forms

The central form was doubtless a short legged crawling cotylosaur such as *Limnoscelis* found in the lower Permian sediments of New Mexico. It was reptile some five feet or more in length. The skull was solidly roofed, somewhat elongated and rather deep. The eyes were placed laterally. On the skull roof was a well developed pinceal opening. There were sharp teeth on the margins of the jaws, and small teeth on the pterygoid bones of the palate. The body was elongated and the limbs were stout but rather sprawling, probably swamp-dwelling type but potent in evolutionary possibilities.

Fig. 16.4: Limnoscelis paludis.

Arboreal Adaptation

Arboreal habitate is difficult to prove on the part of any Mesozoic or older reptiles, for if any arboreal forms existed, their remains in common with those of other forest-dwelling forms would have had little chance of natural entombment and subsequent preservation. Among living modern reptiles there are numerous arboreal forms like geckos, *Calotes, Chamelaeon* etc. Some of them develop adhesive pads for climbing while others prehensile limbs and tail for the purpose.

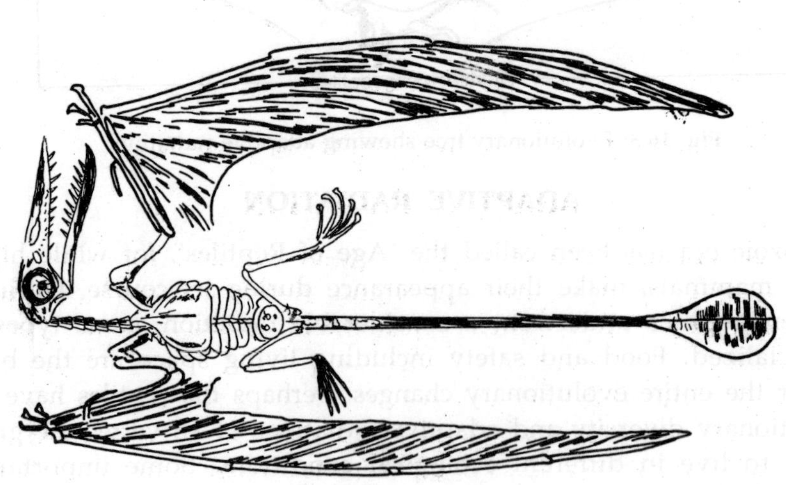

Fig. 16.5: Skeleton of Jurassic Pterosaury.

Aerial Adaptations

Aerial or volant or flying reptiles such as the pterodactyls were finally adapted to their habitat, ranging as they did from the size of a sparrow to the largest of nature's flying mechanism with ample power both of varied and sustained flight, but their origin is lost through the unperfection of the record of Triassic life.

Among living reptiles true reptilian fliers have not yet been reported. Some reptiles like *Draco* simply glides over from branch to branch with the help of its patagi.

Aquatic Adaptations

There are many aquatic forms including both living and extinct, properly adapted for aquatic habitat. Plesiosaurs, Mesosaurs, Ichthyosaurs, Mosasaurs (see lizards) of the among the extinct forms, and turtles, tortoise, terrapins, crocodiles, *Alligators* and *Gavialis* etc. Among the living forms showing aquatic adaptations. For this they have or had paddle like limbs as in *Ichthyosaur, Plesiosaur,* turtles, tortoise etc. The *Hydrophis* or sea snake has a laterally compressed tail.

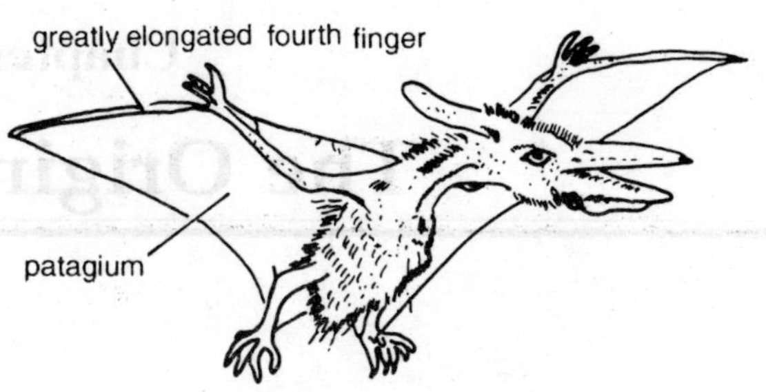

Fig. 16.6: Pteranodon.

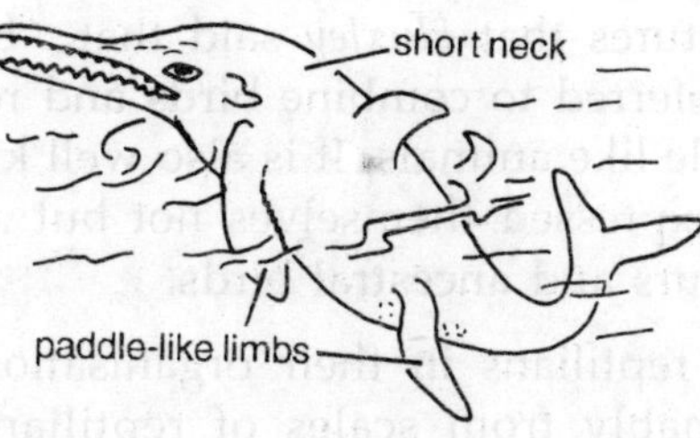

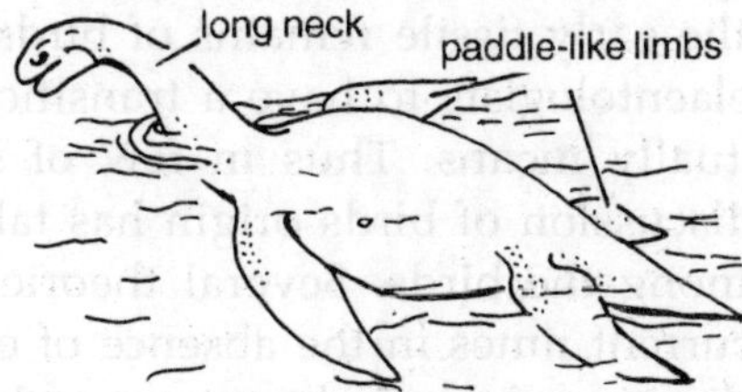

Fig. 16.7: Ichthyosaurus, Plesiosurus.

Fossorial Adaptation

Fossorial reptiles are rare as fossil for while burial is a prime requisite to fossilization, self burial rarely carries with it the necessary imperviousness to air to insure preservation. Hence we cannot point to a single ancient reptilian group as of extensive fossorial habit, although in the *Cotylosaurs* and *Pelycosaurs* there are certain forms whose powerful implied musculature of arm and leg points to digging powers of no mean degree. Among living reptiles there are many fossorial forms like *Uromastix,* snakes, limbless lizards etc.

Chapter—17

The Origin of Birds

INTRODUCTION

Of all the animals the birds are the best known and most early recognise. The study of comparative anatomy and of the fossil evidence of ancient life has convinced zoologist that birds are unquestionably descended from reptiles. The bird skeleton shows so many reptilian features that *Huxley* said that, "birds are glorified reptiles" and many zoologist have preferred to combine birds and reptiles in a single class the *Sauropsida* which means reptile like animals. It is also well known that the evolutionary potentialities of archosaurs expressed themselves not but also in two experiments in flight represented by pterosaurs and ancestral birds.

The birds are far more reptilians in their organisation than the mammals. The evolution of feather presumably from scales of reptilians type conferred obvious advantages for the avian flight. The *Archaeopteryx* and *Archaeornis* though bridge the reptiles and birds, but still the early fissile remains of birds are scare. Therefore, there is a still demand by the paelaentologists to have a transitional fossil to record exactly what evolution of birds actually means. Thus in few of sufficient transitional type between reptiles and aves, discussion of birds origin has taken the form of theories as to how flight got started among the birds. Several theories exist to account for this. Thus the origin of birds in current times in the absence of exact fossil link is generally dealt on the anatomical peculiarities of the *Archaeopteryx* and the theories of hypothetical proaves.

Affinities of Birds with Archosauria

(1) The birds are far more reptilian in their organisation than mammals. The structure of dinosaurs and of pterosaurs, and the birds shows many similarities which point the common origin of all three groups from the primitive thecodont stock.

(2) The establishment of bipedalism (walking on two legs) among early thecodont left the fore limbs free for conversion into wings, by the ancestors

Fig. 17.1: Restoration of probable fore-runners of birds.

of pterosaurs and the birds. But as the contrasting wings structure in the two groups clearly shows that they must have arisen independently and evolved along the different line from the very fast.

(3) Resemblances such as the pneumatisation of the bones, the tendency to develop toothless beak which was present among some dinosaurs, the enlargement of the sternum and the form of brain are doubtless showing parallel evolution.

(4) Birds are like the ornithichians dinosaurs in the structure of their pelvic but are more like sauroschian dinosaurs in their structure of both arms and legs and this fact misled early writers who believed that birds were descendent from the birds like dinosaurs. But on the contrary the above dinosaurs lack a clavicle and they differed widely from birds in other skeletal peculiarities.

Affinities of Birds with Living Reptiles

Birds show many resemblances to the recent existing reptiles in the anatomy, embryology, physiology, palaentology and these are enumerated below:

ANATOMY

Exoskeleton

In both the birds and reptiles there is an exoskeleton formed from the epidermis, the feathers of birds are nothing but, the modified epidermal skeleton of reptiles.

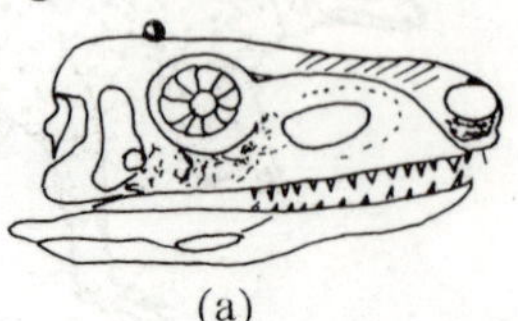

(a)

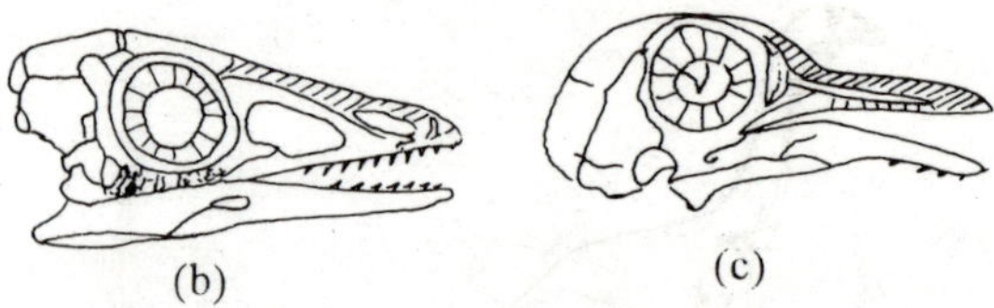

(b) (c)

Fig. 17.2: Comparison of reptile and bird skull to show evolutionary stages: (a) The pseudosuchian reptile. (b) Archaeopteryx. (c) The common pigeon.

Endoskeleton

(*i*) *Skull.* In both the cases the skull articulates with the vertebral column by means of single occipital condyle. The diapsid condition of skull is formed in both reptiles and birds.

(*ii*) The lower jaw is similar is both the groups.

(*iii*) The suspensorium is biquardrate and is autostylic.

(*iv*) The epiphysis are absent in both the cases.

(*v*) The ribs, as in *Rhyncocephalia,* have uncinate processes.

Visceral Organs

(*i*) Respiration in both is pulmonary and there is no gill in the adult. The air sacs which formed in the birds are also present in the lung of *Chameleon*.

(*ii*) The heart is similar to crocodiles.

(*iii*) The red blood corpuscles are oval and nucleated.

(*iv*) There are 12 pairs of cranial nerves in both the groups.

(*v*) The eye is surrounded by sclerotic bones as formed in some dinosaurs.

Embryology

(*i*) Birds are oviparous as reptiles are, with certain exceptions.

(*ii*) The ova are very large with much yolk usually with calcareous shells.